Early Statistics Fundamentals

Antonio Curtis

Preface

After teaching introductory Statistics for several years I found that no book on the market (that I encountered—and I reviewed many, but certainly not all) approached the topic in a manner that was well-suited to my approach. Some books went into too much detail about technical points or exceptional cases and/or included too many non-essential topics for an introductory course. That material is all very interesting and useful for practitioners of Statistics, but becomes confusing and counterproductive for students being introduced to Statistics. With any book I used I had to sift through it and tell students what not to worry about because we were not going to cover that material in our class. We would have a huge textbook but not use a significant portion of it. There is only so much material that can be sufficiently covered in an introductory Statistics course in a manner that students (with a minimal math background) can grasp and retain. Some texts had too many non-applications and/or did not emphasize the real-world meaning of statistical results. Many books presented the material in an order that did not seem optimal to me. I found myself having to provide a lot of supplemental material and explanations which motivated me to write a book. This is not a criticism of those many fine books that I used for many years in my classes, but I needed something different. As a new teacher, I let whatever textbook was popular in the department guide my delivery of the material. After some time, however, I thought it would be easier and better for students to have a book that matched my style and philosophy of the course. As an instructor, if you do not like this book, that is fine. No hard feelings. I am sure you can find one of the many other books currently on the market to suit your needs and preferences.

After many years of teaching I also found that many (from my observation, it's actually most) instructors of basic Statistics never learned any Statistics as students. They had absolutely no Statistics education at all, not even the basic introductory Statistics course they were "teaching". Their knowledge of Statistics does not extend beyond the page of the textbook they last read (the textbook they were handed before being sent into the classroom). These instructors are generally faculty members with pure math degrees working in math departments. Their department offers a basic Statistics course so these faculty members are allowed to teach the class with no background in the subject. That still horrifies me (Can you imagine someone being allowed to teach a calculus class when that person never actually passed a calculus class him/herself? I digress...) but there is nothing I can do about it. This book may help those students with the misfortune of being in a class with such an unqualified instructor.

The order of the topics you may find different than in other books. In many books, there is a chapter discussing confidence intervals for one population parameter, a chapter about basic hypothesis testing for one population parameter, and then a chapter where hypothesis testing and confidence intervals for two parameters are discussed at the same time. It makes more sense to me to discuss confidence intervals all in the same chapter, discuss all basic hypothesis tests in one chapter, and then discuss more exotic hypothesis tests (ANOVA, correlation, etc.) after that. Some books cover correlation very early in the text and then, many chapters later, formally consider hypothesis testing. Really, though, testing for a correlation is just a form of hypothesis testing so it makes more sense to me to cover that with the other hypothesis testing material.

This book focuses on applications. For example, instead of several problems where the confidence level, sample size, and sample proportion are provided as isolated pieces of information, my focus is on students being able to assess situations, extract relevant information, and perform appropriate analysis. I tried to avoid a formulaic way of presenting word problems so that students need to really understand what is going on as opposed to memorizing the way the author writes questions. As I stress to my students, a Statistics situation is all about the data and the research question. Most exercises in this text have "extra information." In reality, there is always "extra information" when statistical processes are used. Part of using Statistics correctly is being able to ferret out what is necessary to answer a particular question and what is not. Statistical problems do not present themselves in such a way that only the relevant information for each research question is present and everything else disappears. Unlike many other textbooks, I keep the exercises to a minimum in each section. It's more important for students to thoroughly understand a few exercises than to power through 20 exercises without really knowing what's happening. Besides, in each section the problems are basically the same so, after figuring out how to do the first couple exercises correctly, correct answers can be obtained by just using the same formula and strategy over and over again. At the end of major chapters, I included a section where different types of problems are mixed together so students can see if they really understand what is going on as opposed to blindly repeating a series of steps from one problem to the next in the individual sections. There is also an emphasis on real world conclusions and not just getting correct numerical answers. Students should not pass a Statistics class without really understanding how to employ statistical methods and completely understand their results. A Statistics class should be about more than simply plugging numbers into formulas.

Probability is minimally discussed in the book. Probability is different from Statistics. Many basic Statistics instructors make the mistake of spending too much time on probability which 1) confuses students because the thinking is different in probability than it is in Statistics, 2) is largely unrelated to what students have to do in inferential statistics, and 3) uses valuable lecture time that should be used discussing and practicing statistical procedures. Introductory classes that attempt to be both Statistics classes and Probability classes usually do a poor job of both. What we do in Statistics is based on probability; however, one does not have to be an expert in probability to use Statistics just like one does not have to be a master mechanic to drive a car. This book gives just enough probability background to enable students to implement inferential statistics.

This book is designed for students taking a first course in Statistics with only an algebra background (no calculus) and no science background. The answers are not expressed using significant figures because students are not expected to have learned that topic previously nor is that a topic that is learned as part of a Statistics class. Answers provided in this book will always be rounded to four decimal places after any leading zeros. Obviously, an instructor can make his/her own rules about decimal places.

Much appreciation to Minitab (version 16 was used to create technical diagrams and output) as well as the General Social Survey (which inspired some of the exercises). Examples and exercises in the text were all created for the purpose of learning procedures, not dispensing factual research results. For that reason, no effort was made to fill the examples with real data but, rather, to present a wide variety of types of problems and answers.

Answers provided in the back of the book are obtained by doing the problems by hand and relying on critical values from statistical tables, the only exceptions being problems in which Minitab output was provided as part of the given problem. If technology is used to complete problems that were intended to be done by hand, answers will usually be slightly different from what is provided.

As is true with many first editions of a book, there will likely be typographical errors and errors in solutions. While I made every effort to avoid these, some undoubtedly slipped in. This will give me something to improve on in the second edition as I also find substantive improvements to make. In the meantime, I am confident that this book will serve as a solid introduction to essential elements of Statistics and will build the foundation for many students' future Statistics studies. I am confident that, after a student passes a Statistics class taught using this book, the student will be able to assess a research situation, choose the correct statistical procedure (at the level of basic Statistics), execute that procedure, and understand the real world meaning of the results.

Dear Student,

You probably do not realize it yet, but you are about to take the best math class you have taken so far, and likely the best you will ever take in your college career. All math is important and useful, but those qualities are not always immediately obvious to students. When you take a Statistics class, you will never ask, "when will I ever use this in the real world?!"

To be successful in a Statistics class, you must put in work. There are no shortcuts. Attend each class session. Before class, use the restroom, eat, and put away your phone and any other distractions. During the lecture, pay close attention, take good notes, ask questions when you are confused, and work along with any exercises that are completed in class. Do not distract yourself with assignments for other classes or devices (phones, etc.) during your Statistics lecture. Work diligently on homework, seek help from peers, seek help from your instructor, and seek help from knowledgeable tutors. Practice at least a little Statistics material each day. Read the examples provided in each section of this book along with the detailed explanations that follow. Make an honest attempt at each exercise before looking at the answer. Try to actually learn the course material, and not just "pass" the class. Even if you think you do not like math or are not good at it, approach the course with a positive attitude and open mind. No matter how great or horrible you perceive your instructor to be, your success ultimately depends on your hard work and dedication to your success. During the semester in which you take a Statistics class, plan to spend more time on that class than any of your other classes. As is true in all math classes, in the class you are taking topics build on each other. It will be much easier for you to learn each topic when you have consistently and persistently been learning the previous topics.

You will need a solid understanding of the material in this course in order to succeed in later classes. This is true even if you do not plan to major in a math-related field. If this introductory Statistics class you are taking is a prerequisite for upper division studies in your major, then you will definitely use the material you learn in this class later (that's why this course is a prerequisite!). If you think, "This is my last math class and I will never take another math class ever again!!!!" you might be surprised later when you find yourself needing math, and Statistics in particular, in later classes.

The technology-based statistical output in this text all comes from Minitab. In the future you may use one of the many other statistical software packages (SPSS, STATA, R, S-plus, SAS, etc.), or a later version of Minitab but, once you learn to read statistical software output, you can read the output from any source.

When you perform calculations, don't round until the very end. If you have to stop and write down numbers from your calculator, and then use those written numbers in additional steps to get a final answer, retain all the numbers the calculator displayed. Sometimes answers you get may differ a little from the answers shown in the back of book due to the number of decimal places your calculator holds or if your calculator allows you to enter complicated expressions as one big entry. Also, if you try the exercises using technology, answers may be different from the answer you get by hand due to rounding and our reliance on statistical tables.

I once took an introductory Statistics course like the one you are about to take. I knew nothing about the course and was certainly not intending to major in it or any other math-related field. I took the course, fell in love with the topic, and that completely changed my goals and career path. Best case scenario—I hope the same for you. No matter what, I hope you learn a lot in this course that you use in later classes and possibly in your career.

Let the learning begin!

Contents

1 Introductory Concepts

The whole world of Statistics is based on data. **Data** is just information that has been collected. Responses to survey questions, measurements of objects, test scores, stock prices, crime rates, approval ratings, etc. are all examples of data. **Statistics** is the field that concerns itself with methods of obtaining the best data for a research situation and then correctly handling that data to get useful conclusions about the source of the data. In any situation involving Statistics, there is a group that is being studied. That complete group is the **population**. A population does not have to be people—it can be non-humans or objects (fish, houses, lawn chairs, cars, etc.). If data is collected from each member of the population, that collection of data is called a **census**. A number that describes the whole population is called a **parameter**. In research of any great interest or importance, all of the members of the population usually cannot be observed or measured due to the size of the population, lack of access, or other reasons. So, researchers can only directly observe some of the population called a **sample**. Informally, you sometimes hear the word "statistic" to mean any piece of numerical information. In the field of Statistics, though, that term has a very technical definition. A **statistic** is a number that describes a sample.

Suppose 30% of all Americans find life exciting. In that case, the population is all Americans and 30% is a parameter. There is no sample, because information was obtained from all Americans. When there is no sample, there is no statistic.

Suppose in a group of randomly selected Americans 30% of them find life exciting. The population is all Americans. This time, we don't know the value of the parameter (the percentage of all Americans who find life exciting). The sample is the group of randomly selected Americans. The statistic is 30%.

[Some books get a little more technical when it comes to identifying population and sample. For example, instead of saying that the sample in the previous paragraph is the group of randomly selected Americans, they would say the sample is the life view of the members of the group of randomly selected Americans. In this and most introductory books, we will just consider the group and not the specific measurements we are taking from each member of the group as the population or sample.]

Every situation using Statistics has a population and a parameter. Otherwise, there would be no need to use Statistics. The parameter's value may be known or unknown. There may or may not be a sample. If there is a sample, it has a statistic that measures the same thing as the parameter of interest.

Data can be divided into two types: quantitative and categorical. **Quantitative** data involves numbers that represent counts or measurements. With **categorical** data, also called qualitative data, data can be separated into categories. For our purposes, categorical data is anything that is not quantitative. Heights, weights, incomes, commute times to work, temperatures, number of children living in the household, number of cars you own, etc. are all examples of quantitative data. Religious affiliation, favorite dessert, type of occupation, typical mode of transportation, etc. are all examples of categorical data. Note that zip codes are categorical data. Sure, they are numbers, but they don't represent counts or measurements. For quantitative data you must have numbers AND those numbers need to represent quantities—how much of something.

Quantitative data can be further divided into two subtypes: continuous and discrete. With **continuous** data, possible values are from a scale without gaps, interruptions, or jumps. More simply, continuous data would allow any number of decimal places. With **discrete** data, possible values can be counted. Heights, weights, commute times to work, temperatures, etc. are examples of continuous data. In these measurements any number of decimal places are *possible*. The device you are using to get the measurements may be limited to a certain number of decimal places, but the actual measurements themselves could have any number of decimal places. If you say it took you 20 minutes to get to work today, that might be true. It's more likely that the actual time was something like 20.234576633648... minutes. If one person took 20 minutes to get to work and someone else took 21 minutes to get to work, are there any possible commute times between those two? Yes, infinitely many. The "how long did it take you to commute to work" scale has no gaps, interruptions, or jumps because, for any two commute times, there are always values between them. Incomes, number of children living in the household, number of cars you own, etc. are examples of discrete data. If one person owns 20 cars and another owns 21 cars, there is no number of cars between those two, that's why that's discrete. If one person has an income of $50987.17 and another has income $50987.18, there is no amount of income between those two (also, if you have lots of time on your hands, you could count those incomes—using pennies), so that is discrete.

We can also consider levels of measurement: nominal, ordinal, interval, and ratio. Nominal data has non-ordered categories, like: cake, candy, ice cream, pie ("What is your favorite dessert?") or walk, car, bus ("How do you typically commute to school?"). At the ordinal level of measurement there are ordered categories but there is no quantitative difference in the categories, like: first place, second place, third place (contest rankings) or loved it, liked it, did not like it, hated it (movie reviews). Interval data has an order, and there is a quantitative difference between values, but there is no natural zero, like temperatures measured in °F. There is a meaningful quantitative difference between 60°F and 80°F (the quantitative difference is 20°F), but 0°F doesn't represent the total absence of heat—

there's no natural zero. With data measured at the ratio level, the data has an order, there is a quantitative difference between values, and there is a natural zero, like commute times to work or weights of alligators.

For all practical purposes, the most important difference is categorical versus quantitative. That distinction will make a huge difference in selecting the correct statistical procedure later in the course. Knowing the difference between discrete and continuous comes into play when talking about probability distributions (more on that later). The levels of measurement aren't that critical but they are usually mentioned in introductory Statistics texts.

There are processes that we use to organize, display, and summarize a set of data. These are called **descriptive statistics**. As alluded to earlier, we usually cannot get a census for an entire population because that would require too much time, money, personnel, etc., we cannot gain access to the whole population for whatever reason, or the research we are doing requires destroying the items being tested. Most of Statistics involves getting a sample and then using that sample to make an estimate of a parameter or test a claim about a parameter. When we do that, we are working with **inferential statistics**. The inferences made are only as good as the data collected and the techniques used to acquire and analyze that data.

Exercises

For 1) – 4) indentify the population, parameter, sample, and statistic.

1) In a local market, 12% of all the perishable items have expiration dates that fall within the next two days.

2) In a local market, a cart-load of perishable items was collected and 12% of them have expiration dates that fall within the next two days.

3) While looking at the records for a few minor league pitchers, it was found that they gave up an average of 3.72 walks per game.

4) Minor league pitchers give up an average of 3.72 walks per game.

For the following exercises, identify the type of data as categorical or quantitative. For quantitative data, also classify it as continuous or discrete.

5) The weight of a person experiencing binge eating disorder.

6) The number of people in a city experiencing binge eating disorder.

7) The average number of people experiencing binge eating disorder for three cities.

8) A therapist helping people with eating disorders diagnoses each patient as experiencing anorexia, bulimia, or binge eating disorder.

9) Responses to a "how many marbles are in this jar?" contest.

10) Time it takes to feel relief after taking a pain medication.

11) Patients at a doctor's office are asked to provided the type of primary insurance they have.

2 Organizing and Displaying Data

We begin talking about descriptive statistics by considering ways we can organize and display data. We will by no means consider all the techniques for organizing and displaying data, but we will focus on the most basic and common. Once these are understood, the other techniques are easily grasped as they are simply variations on these themes. We'll start with frequency distributions.

Example: A health and wellness researcher randomly selected 62 Statsville residents who exercise. The amount of time (to the nearest minute) each resident spent during a recent exercise session was recorded. The results are shown below:

60 62 64 63 72 71 73 60 65 88 87 77 85 76 65 88 76 90 105 70 62 61 60 68
84 120 61 70 100 144 63 47 61 150 66 61 62 80 63 60 74 63 81 81 98 90 70
60 72 73 110 90 95 71 61 87 72 60 63 61 120 72

Create a frequency distribution for this data starting with lower class limit 41 using class width 17.

Answer:

We'll start by making a tally. If we go one-by-one through the data, and place a tally mark next to the class in which each value belongs, we get:

exercise time	tally
41-57	|
58-74	𝍸 𝍸 𝍸 𝍸 𝍸 𝍸 𝍸 ||
75-91	𝍸 𝍸 𝍸
92-108	||||
109-125	|||
126-142	
143-159	||

To make the frequency distribution, we replace the tallies with the whole number frequencies they represent:

exercise time	frequency
41-57	1
58-74	37
75-91	15
92-108	4
109-125	3
126-142	0
143-159	2

Now that you have seen a frequency distribution it will be easier to consider the associated vocabulary. Each of the exercise time intervals is called a *class*. This frequency distribution has seven classes (41-57, 58-74, etc.). Notice that they do not overlap. The smallest number that can belong to a class is called the *lower class limit*. The lower class limits are 41, 58, 75, 92, 109, 126, and 143. The largest number that can belong to a class is called the *upper class limit*. The upper class limits are 57, 74, 91, 108, 125, 142, and 159. The *class width* is the number of units between consecutive lower class limits. From 41 to 58 is 17 units, from 58 to 75 is 17 units, from 75 to 92 is 17 units, and so on. This pattern needs to continue through the whole frequency distribution. That's how we knew the correct way to construct the classes, based on the directions to start with lower class limit 41 using class width 17. In fact, it is usually easier to start with the lower class limit given in the instructions, use the class width to determine the rest of the lower class limits, go back and fill in the upper class limits, and stop when your last class includes the biggest number in the data (for this data, 150). The upper class limits will be the number just before the next class begins, going out to however many decimal places are used for the data (in this case, none). In this data, if values went to three decimal places, the upper class limits would be 57.999, 74.999, and so on.

There is more frequency distribution vocabulary, starting with class boundaries. *Class boundaries* provide the halfway points between the end of one class and the start of the next class. We use the established pattern of those class boundaries to get a class boundary that is just before the first class and just after the last class. Throughout the text we will find ourselves needing to get a number that is halfway between two other numbers. From your previous math classes you know you get that number by adding the two original numbers and dividing by two. The number halfway between 57 and 58 is $\frac{57+58}{2} = 57.5$, the number halfway between 74 and 75 is $\frac{74+75}{2} = 74.5$, and so on. To get our class boundaries we would continue with that process until we landed at $\frac{142+143}{2} = 142.5$. Looking at our pattern, we see that the consecutive class boundaries have the same width as the lower class limits. Using these results, we get 40.5 as our first class boundary (it is just before the first class) and 159.5 as the last (it is just after the last class). Putting it all together, our class boundaries would be 40.5, 57.5, 74.5, 91.5, 108.5, 125.5, 142.5, and 159.5. [Of course, it may be possible for you to figure out the class boundaries without any calculations, just by looking at the upper and lower class limits.]

The *class midpoints* are the midpoints of the respective lower and upper class limits for each class. The midpoint for the first class is $\frac{41+57}{2} = 49$, the midpoint for the second class is $\frac{58+74}{2} = 66$, and so on. Doing this calculation for each class gives class midpoints: 49, 66, 83, 100, 117, 134, and 151.

When creating a frequency distribution, always follow given directions! If there are no directions (besides, "make a frequency distribution") then just do what seems to make sense for the data. In cases where you are deciding what makes sense for the data, your first and last classes should not have frequency zero. The first class should include the smallest value in the data and the last class should include the largest value in the data. Also, the class width must remain constant throughout the distribution. Otherwise, the true distribution of the data is distorted and misrepresented. These are general rules of thumb that may be overridden in practice (but not in this course) by some compelling research reason.

When we look at a display of data, like a frequency distribution, we often ask the question, "Is the data approximately normal?" Normal has a very specific meaning in Statistics. For our purposes at this point, when data is approximately normal the frequencies start low, increase to some maximum frequency, then decrease to a low frequency. Also, the frequencies are approximately symmetric, which means that the maximum frequency is approximately in the center of the distribution. The starting and ending frequencies do not have to be identical, simply "low" relative to the other frequencies.

The exercise time data is not approximately normal. These frequency distributions represent data that is approximately normal:

scores	number of students
0-19	3
20-39	12
40-59	22
60-79	14
80-99	1

tensile strength	number of bolts
21300-21309.9	22
21310-21319.9	49
21320-21329.9	116
21330-21339.9	78
21340-21349.9	36
21350-21359.9	11

The first frequency distribution represents scores of students on an assessment and how many students earned scores in each interval (class). The second frequency distribution represents the tensile strength (in foot pounds) of bolts and shows how many bolts had tensile strengths belonging to each class. As you can see, the frequency column does not have to be labeled "frequency". How many values belong to each class constitute the frequencies. The tensile strengths are measured to one decimal place so that is reflected in the upper class limits. In the tensile strength distribution, there is an even number of classes so the largest frequency cannot be right in the middle of the distribution, but it is as close to the middle as it can be, under the circumstances.

Real data is not always as clear as contrived textbook examples for determining if data is approximately normal. There are rigorous tests that can be performed regarding the normality of data. At this point, just have this general concept of how approximately normal data looks.

By the way, what was the total number of students in our assessment data collection? What was the total number of bolts we measured? The answer to each question is simply the total of the frequencies. Total number of students is 3+12+22+14+1 = 52. Total number of bolts is 22+49+116+78+36+11 = 312. For any frequency distribution, the sum of the frequencies tells you the total number of values in the data set.

Let's return to the exercise time data as we continue our discussion of organizing and displaying data.

Example: Create a relative frequency distribution for the exercise time data.

Answer:

exercise time	relative frequency
41-57	2%
58-74	60%
75-91	24%
92-108	6%
109-125	5%
126-142	0%
143-159	3%

To get the relative frequency for a class, take the frequency of the class divided by the total of all the frequencies. The sum of all the exercise time frequencies is 62. So, to get the relative frequency of the first class, which has a frequency of 1, we evaluate $\frac{1}{62}$ = .01613. Relative frequencies can be in decimal form or in percentage form. To motivate a review of how to convert from decimal form to percentage form, which we will need to do a lot in this course, I decided to use whole number percentages. Remember that we just move the decimal point two places to the right (the shortcut for multiplying by 100) to do the conversion. Since I am using whole number percentages, after getting .01613x100% = 1.613% I round to 2%. For the second class, which has a frequency of 37, the relative frequency is $\frac{37}{62}$ = .5968x100% = 59.68%, rounded to 60%. The other relative frequencies were found using this same technique.

The sum of all the relative frequencies, in percentage form, should be 100%. The sum of all the relative frequencies, in decimal form, should be 1. Occasionally, due to some rounding oddities, you may get 99% or 101% in percentage form, or .99 or 1.01 in decimal form. The sum of exact (not rounded) relative frequencies is always 100% or 1.

These are the relative frequency distributions for the scores and bolts data:

scores	relative frequency
0-19	5.8%
20-39	23.1%
40-59	42.3%
60-79	26.9%
80-99	1.9%

tensile strength	relative frequency
21300-21309.9	.0705
21310-21319.9	.1571
21320-21329.9	.3718
21330-21339.9	.2500
21340-21349.9	.1154
21350-21359.9	.0353

Purely for variety's sake, non-whole percentages were used for one and decimals for the other.

Example: Create a cumulative frequency distribution for the exercise time data.

Answer:

exercise time	cumulative frequency
less than 58	1
less than 75	38
less than 92	53
less than 109	57
less than 126	60
less than 143	60
less than 160	62

The class designations are now "less than" statements, with the first number being the second lower class limit from the frequency distribution, and then successive lower class limits are used, with one being added at the end in order to capture all the data. From the frequency distribution, all of the exercise times in the first class (quantity, 1) were less than 58. The exercise times in the first two classes (quantity, 1+37=38) were less than 75. The exercise times in the first three classes (quantity, 1+37+15=53) were less than 92, and so on. We are accumulating the frequencies as we progress through the distribution, which is why this frequency distribution is called cumulative.

We don't care about the sum of the cumulative frequencies because that does not tell us anything useful. However, the last cumulative frequency listed must be the same as the sum of all the frequencies because, by the time we get to the last line of the distribution, we have accumulated all the frequencies.

These are the cumulative frequency distributions for the scores and bolts data:

scores	cumulative frequency
less than 20	3
less than 40	15
less than 60	37
less than 80	51
less than 100	52

tensile strength	cumulative frequency
less than 21310	22
less than 21320	71
less than 21330	187
less than 21340	265
less than 21350	301
less than 21360	312

Example: Create a histogram of the exercise time data.

Answer:

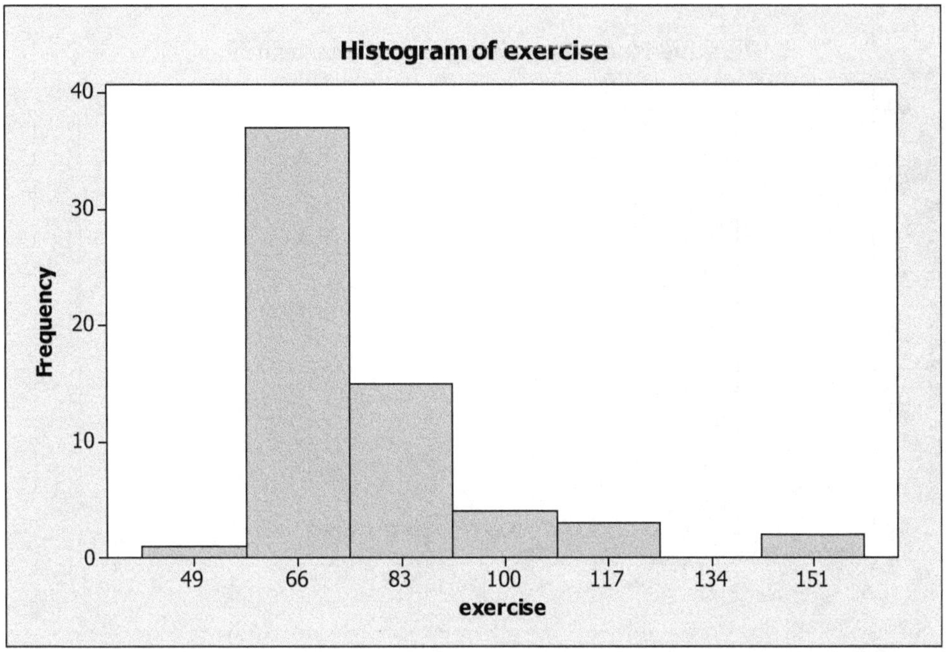

A histogram is a display of classed (or grouped) quantitative data that represents each class with a bar, the height of which corresponds to the frequency of the class. The first class had a frequency of one, so the height of the bar is one. The second class had a frequency of 37 so its bar's height is 37, and so on. The frequency is labeled on the vertical axis. For labeling the horizontal axis, one of three strategies can be used: lower class limits, class boundaries, or class midpoints. Since the directions did not indicate which of those to use, the arbitrary decision was made to use midpoints. You'll recognize those as the midpoints we calculated previously. When using midpoints to label the horizontal axis, each midpoint sits in the middle of its bar.

Notice that the bars are touching in this display. That is always true for histograms. An analogous display for categorical data is called a bar graph (or bar chart). With that display, the categories of the categorical data are shown on the horizontal axis, the frequency for each category is represented by the height of its respective bar, but the bars do not touch.

Example: Create a relative frequency histogram of the exercise time data.

Answer:

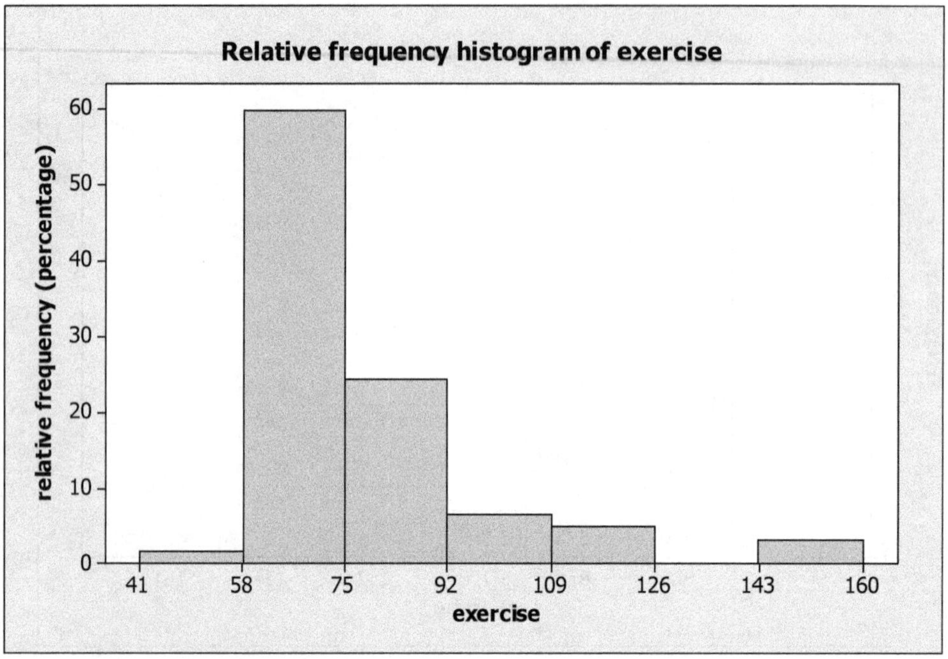

As the name suggests, the relative frequency histogram has bar heights determined by relative frequencies. The relative frequencies are plotted on the vertical axis. For the horizontal axis, there are the same options for labeling: lower class limits, class boundaries, or class midpoints. Since the directions did not indicate which of those to use, the arbitrary decision was made to use lower class limits. You'll recognize those as the lower class limits from the frequency distribution. We needed to add another lower class limit to the diagram, keeping the same class width, in order to have a place to close the last bar. When using lower class limits to label the horizontal axis, the lower class limits sit on the edges of the bars. If we labeled the horizontal axis using class boundaries, they would also sit on the edges of the bars (that's why we need a class boundary before the first class and after the last class; so we will be able to have complete bars in the histogram).

Example: Create a frequency polygon of the exercise time data.

Answer:

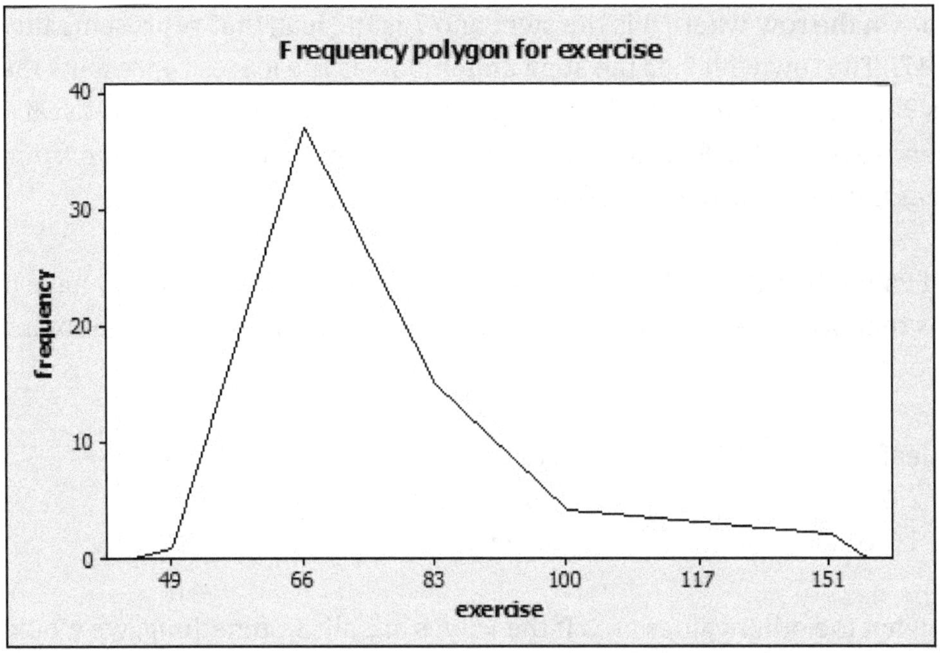

The frequency polygon has class midpoints on the horizontal axis. Over each class midpoint we place a dot, the height of which corresponds to the frequency for the respective class. All of the dots are connected to create the figure shown. [In some books and software, the figure is not closed at the ends; it just hovers over the horizontal axis.] The frequencies are indicated by the vertical axis.

Example: Create a stemplot for the exercise time data.

Answer:

stem	leaf
4	7
5	
6	00000011111122233333345568
7	000112222334667
8	011457788
9	00058
10	05
11	0
12	00
13	
14	4
15	0

When using a stemplot (also called a stem-and-leaf plot) we first physically split each data value into two pieces: a stem and a leaf. How this is done depends on what seems to make sense for the data. For the exercise times data, the last digit is the leaf and the first digit(s) make the stem. On the row where 4 is the stem and 7 is the leaf, that represents the exercise time 47. The row with 9 as the stem and 00058 as the leaves represents the exercise times 90, 90, 90, 95, and 98. The row with 14 as the stem and 4 as the leaf represents exercise time 144. Each leaf represents a different individual value from the data. The leaves must be in order from 0 to 9.

If our data had values like .302, .312, .304, .316, .311, etc. we would obviously have to use different stems than what we used for the exercise time data. We would have something like:

stem	leaf		stem	leaf
.30	2 4	OR	.3	02 04 11 12 16
.31	1 2 6		etc.	
etc.				

depending on what the other values are. If the values are all .3-something, we would probably go with the display on the left. If there are values that are .4-something, .5-something, and so on, we would probably use the setup on the right. Do what makes sense for the data (after following any stated directions).

Example: Randomly selected Statsville College students were asked how many units they have completed so far. Their responses were:

32 38 44 51 59 38 4 32 47 38 61 54

Create a dotplot for this data.

Answer:

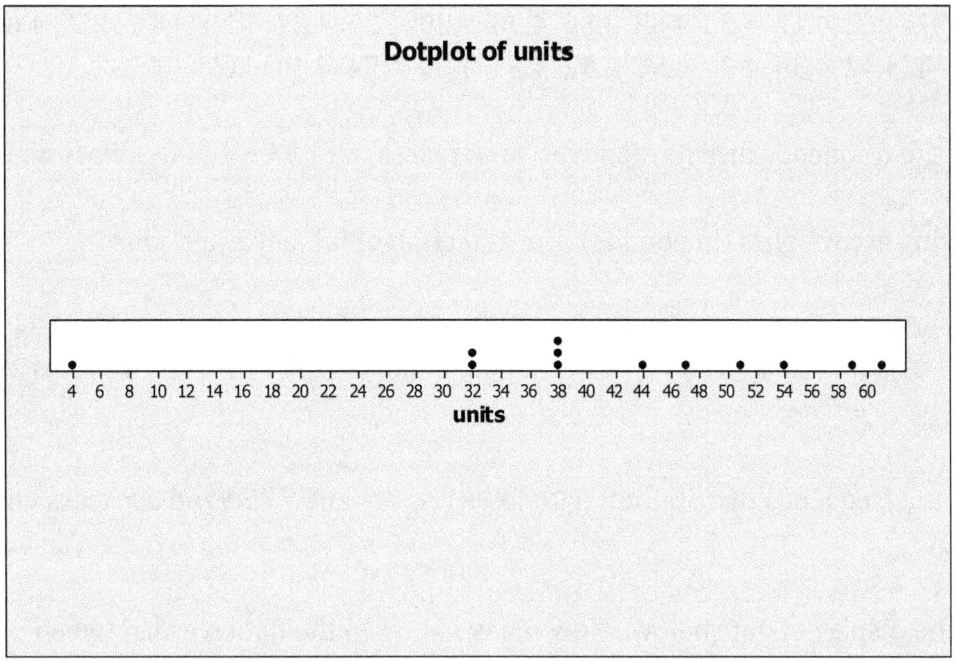

Like a stem-and-leaf plot, a dotplot is a display of data for the individual data values. The data values are plotted on the number line. For each occurrence of a data value, a dot is placed over that value. By looking at the dotplot, we can see there is one 4, two 32's, three 38's and so on in the data set.

As mentioned, there are many other displays that can be made for a set of data. Those discussed here are some common displays. You can easily understand other displays based on this basic background.

Exercises

1) Look ahead at the wind chill data in Exercise 1) of Section 3.1. Construct a dotplot for that data.

2) Look ahead at the elevation data in Exercise 2) of Section 3.1. Construct a dotplot for that data.

For each set of data given in 3) and 4), a) create a frequency distribution, b) create a relative frequency distribution, c) create a cumulative frequency distribution, d) create a histogram, e) create a relative frequency histogram, f) create a frequency polygon, g) create a stemplot, h) use the frequency distribution and histogram to answer the question, "Is the data approximately normal?"

3) Randomly selected pH readings of rooftop rain are shown below:

4.38 4.42 4.27 4.62 4.17 4.20 4.53 4.36 4.30 4.29 4.26 4.31 4.20 4.46 4.50 4.60 4.49 4.35 4.28 4.44 4.42 4.39 4.34 4.51 4.52 4.55 4.60 4.54 4.40 4.07 4.45

(start the frequency distribution with lower class limit 4.06 and use class width .14)

4) The following are weights (in pounds) of a collection of full beverage cans:

.7762 .7855 .7960 .7887 .7938 .7962 .7827 .7797 .8077 .7914 .7841 .7941 .7773 .7927 .8170 .7804 .7803 .7817 .7950 .7784 .8102 .7814 .7823 .8123 .7819

(start the frequency distribution with lower class limit .7760 and use class width .0072)

5) Consider the display of data below. How many values in the data were between 130 and 139?

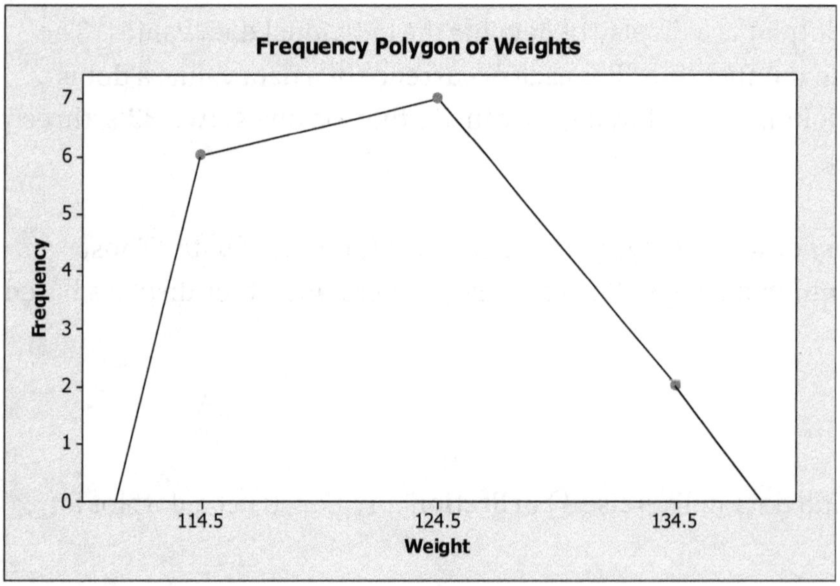

3 Summarizing Data

3.1 Measures of Center

Now that we have considered the organizing and displaying data portion of descriptive statistics, let's turn our attention to summarizing data.

Our summary typically begins with measuring the center (or central tendency) of the data. There are different ways to conceptualize the "center" of a set of data. **Measures of center** quantify these different conceptualizations.

One way to think about the center of data is in terms of the center of mass. If you were to place all the data values in their positions on a balance beam that was labeled with a number line, what number would be the balancing point of the data? The **mean**. When all the data values are plotted on a number line, there is just as much space between the mean and the numbers to its left, as there is between the mean and the numbers on its right. The mean is what you think of when you hear the word "average". To find the mean, simply add all the numbers in the data and divide by how many numbers there are. [In the world of mathematics there are several kinds of means. The one we use in Statistics is the *arithmetic mean*. Since that is the only mean used in Statistics, we just refer to it as the mean.]

Example: Find the mean for this data: 4 7 16

Answer: $\frac{4+7+16}{3} = 9$

Piece of cake. Notice that, if you placed 9 on a number line, 4 and 7 would be to its left and 16 would be to its right. The space between 9 and 4 is five units, and the space between 9 and 7 is two units, for a total of seven units of space to the left of 9. The space between 9 and 16 is seven units, for a total of seven units of space to the right of 9. There is just as much space between 9 and the numbers to its left, as there is between 9 and the numbers on its right. So 9 is the balance point, the center of mass, or the "center" of the data in terms of space.

Another way to think of center is in terms of the middle of an ordered list. When you list a set of data, which number is in the middle? The **median**. To find the median, just arrange the numbers in order (ascending or descending will work, but ascending is better for procedures that will be performed later), locate the number in the middle of the list, and that is the median. If there is an even number of values in the data, there will be two numbers in the middle of the list. The mean of those two numbers constitutes the median.

Example: Find the median of each set of data:
1) 4 7 16
2) 4 7 16 19
3) 7 16 4

Answers: 1) 7 2) 11.5 3) 7

Set 1) is arranged in order, there is an odd quantity of values, the median is the one in the middle.

Set 2) is arranged in order, there is an even quantity of values, the median is the mean of the two numbers in the middle, so we have the median is $\frac{7+16}{2} = 11.5$.

Set 3) does not have ordered data. The "middle" of the data has no significance unless the values are in order. Upon ordering the data, we realize it is actually just Set 1) again: 4 7 16. So, again, the median is 7. Be sure to order the data first whenever you are finding the median!

Another way to think of the center is in terms of the center of attention. Which value in the data is most common? The **mode**. To find the mode just look for the value in the data that is most frequently occurring. If there is a tie between two equally frequent values, both numbers are the mode and the data is said to be *bimodal.* If there are more than two values that are tied for most frequently occurring, all of those numbers are the mode and the data is said to be *multimodal.* If all data values appear once, there is no mode.

Example: Find the mode of each set of data:
1) 4 7 7 16
2) 4 7 7 16 16
3) 4 7 7 16 16 23 23
4) 4 7 7 16 16 23 23 23
5) 4 7 16

Answers: 1) 7; 2) 7, 16 (bimodal); 3) 7, 16, 23 (multimodal); 4) 23 5) no mode

Remember, we are looking for the most frequently occurring. In Set 4) only 23 is the mode. We are not looking for what appears more than once, but what appears the most. [By the way, if you had a data set like: {4 4 7 7 16 16} most sources consider that multimodal.]

The units on measures of center will be the same as the original data. For example, if the data was measured in feet, the measures of center would have units of feet as well.

Example: The monthly electricity consumption, measured in Kilowatt Hours (KWH), of randomly selected households are shown below. Calculate the measures of center for this data.

565 892 736 370 517 831 736

Answer: mean = 663.8571 KWH, median = 736 KWH, mode = 736 KWH

This is pretty straightforward. If you are wondering why the median is not 370 KWH, remember that the data must be ordered when finding the median. The ordered version of the data is {370 517 565 736 736 831 892} and, clearly, 736 is in the middle.

This is a good place to introduce some notation (symbols) that will be used throughout the course. It is imperative that you learn, know, and memorize this notation. Whenever you are introduced to notation in the course, begin committing it to memory immediately. Notation gives us a way to communicate lots of information in a small amount of space, which is what people who work with mathematics and Statistics prefer to do.

notation	general description
Σ	the sum of a set of values
x	individual data values
n	sample size (quantity of values in a sample)
N	population size (quantity of values in a population)
$\bar{x}$	sample mean (mean of a set of sample values)
μ	population mean (mean of the set of population values)

The descriptions provided, as stated, are the general descriptions. In any research context, which we will demonstrate through the use of application problems throughout the text, these symbols take on a specific meaning for that situation. It is very important to memorize the general meaning and then tailor that to the specific meaning for each problem. This will be necessary in order to correctly use formulas and state conclusions. In different books and statistical software, there will be variations in the notation but, once you are told what notation is being used, you simply adjust.

MEMORIZE THE NOTATION!!!!!

Σ is the Greek capital letter "sigma" that you have no doubt seen before on a shirt, sticker, or building. In a mathematical context, it signals us to add values. You should have seen that notation in your algebra class.

μ is the Greek lower case letter "mu".

$\bar{x}$ is referred to as "x bar".

x is the variable commonly used to represent individual data values. Suppose we have values 4, 7, and 16 in a set of data. We can list them as:

x
4
7
16

As they are listed here, $x_1 = 4$, $x_2 = 7$, and $x_3 = 16$. To formally use sigma notation to communicate adding these values, that would look like:

$$\sum_{i=1}^{3} x_i$$

As it turns out, in Statistics, the starting value (below the sigma) is always 1 and the ending value (atop the sigma) is always the quantity of values we have. So, in many introductory Statistics books, the notation is simplified to this:

$$\Sigma x$$

We will follow this convention in this text. So when you see Σ just think, "add up all the". For example, Σx is telling me to "add up all the x (values)".

As we know, to get the mean we add all the data values and then divide by how many we have. If the data is from a sample, we can write the mean and how it is calculated as:

$$\bar{x} = \frac{\Sigma x}{n}$$

If the data is from the whole population, we have: $\mu = \frac{\Sigma x}{N}$

The process for finding the mean is the same whether dealing with a sample or a population, but the notation is different. We were told that the electricity consumption data was for randomly selected households. So, when reporting the mean, we could write $\bar{x} = 663.8571$ KWH. If the whole population consisted of just those seven households, then we would have μ = 663.8571 KWH. As the problem is stated, the electricity consumption example has n = 7. If the data represented the whole population, we would have N = 7.

One more thing... **MEMORIZE THE NOTATION!!!!!**

If we have a set of data, finding the mean is easy—simply add all the data values and divide by how many you have. Suppose we are given a frequency distribution and we need to find the mean? Consider the frequency distribution of the student assessment scores from before:

scores	number of students
0-19	3
20-39	12
40-59	22
60-79	14
80-99	1

Suppose we want to find the mean assessment score for these students. Well, we can't add all of the assessment scores and divide by how many there are because we don't know the individual scores. We have to take a different approach. It can be shown, and is also quite intuitive, that the best way to estimate the individual values that are represented using a frequency distribution is to use the class midpoint for each individual value. For example, there are three assessment scores in the 0-19 class. Our best estimate for each score is $\frac{0+19}{2} = 9.5$. There are 12 assessment scores in the 20-39 class. Our best estimate for each of those scores is $\frac{20+39}{2} = 29.5$, and so on. Since we cannot add all the individual assessment scores and divide by how many we have, we will use the respective class midpoint as a substitute for each individual assessment score, add all of those, and divide by how many we have. If we were going to add three 9.5's, that would be the same as multiplying 3x9.5. If we were going to add twelve 29.5's, that would be the same as multiplying 12x29.5. When we add all of the class midpoints that are substituting for each individual value, and we do so in terms of multiplication as just described, and we divide by how many values there are, we can communicate these steps with the following formula:

$\bar{x} = \frac{\Sigma(fx)}{\Sigma f}$ where x stands for the class midpoint and f is the frequency

$\Sigma(fx)$ is telling us to "add up all the f times x". The first (f times x) is 3x9.5. The next (f times x) is 12x29.5, and so on. Σf is telling us to add all the frequencies. Remember that the sum of the frequencies is the same as how many values we have. So that formula, in a very concise way, describes the procedure explained just before the formula was introduced. This is the formula we will use when we need to find the mean of a set of data, but all we have is the frequency distribution for that data. Let's try it.

Example: Assessment scores for a sample of students are represented in the frequency distribution below. Find the mean of the assessment scores.

scores	number of students
0-19	3
20-39	12
40-59	22
60-79	14
80-99	1

Answer: To get the needed information for the formula, we will add to the table we were given:

scores	number of students (f)	class midpoint (x)	fx
0-19	3	$\frac{0+19}{2} = 9.5$	3(9.5) = 28.5
20-39	12	$\frac{20+39}{2} = 29.5$	12(29.5) = 354
40-59	22	$\frac{40+59}{2} = 49.5$	22(49.5) = 1089
60-79	14	$\frac{60+79}{2} = 69.5$	14(69.5) = 973
80-99	1	$\frac{80+99}{2} = 89.5$	1(89.5) = 89.5
total	52		2534

$$\bar{x} = \frac{\Sigma(fx)}{\Sigma f} = \frac{2534}{52} = 48.7308$$

The mean assessment score for students in this sample is 48.7308

Technically, this is the best estimate of the mean. It's probably not exactly equal to the mean for the raw data but, without having the raw data, this is the best we can do. So we just consider this to be the mean of the data.

In the expression $\Sigma(fx)$, we don't need the parentheses because, by the order of operations, we would multiply before we add. Including the parentheses, though, is not a problem.

Let's take a moment to consider whether or not this answer makes sense. It is reasonable to conclude that the mean of the scores is 48-something because many of the values are in the 40-59 class, with a considerable amount of values on either side (greater than or less than) of scores in that class. What if you got an answer of 192.8 for the mean? Well, while we don't know any of the individual scores, the largest possible score is 99, based on the upper class limit for the last class. How could the mean—the average—be larger than the biggest possible value? What about a final answer of 2534? That also makes no sense for the same reason. An answer of 2534 would suggest that the person forgot to divide by Σf. An answer of 192.8 just suggests some other error. You should always take a moment, after you get an answer, and see if the answer makes sense. For some types of problems you won't really have a feel for what answers would or would not make sense. For other types of problems, like the one we just did, we can certainly rule out some answers as wrong. Just because an answer makes sense does not guarantee that it is correct (suppose we got 49.1238, for example—makes sense, but it's wrong), but if an answer makes no sense or is totally impossible, it is likely wrong. If you discover that, you can go back and find your mistake.

Exercises

1) Consider the following wind chill readings (^{o}F) for randomly selected cities on a winter day:

0 -16 -3 -3 3 28 0 -7 26 5 -34 3 2

Find the measures of center for this data.

What would be the measures of center if this were a population instead of a sample?

2) The elevations (above sea level, measured in meters) for a sample of cities are shown below:

98 76 34 60 63 3 105 40 -2 37

Find the measures of center for this data.

3) Return to the elevations data from problem 2). Multiply each value by 2. Find the new measures of center for this data. Compare and contrast the new answers with your answers for problem 2).

4) Return to the elevations data from problem 2). Add 2 to each value. Find the new measures of center for this data. Compare and contrast the new answers with your answers for problem 2).

5) The display of data below gives the tensile strength (in foot pounds) of a selection of bolts. Find the mean tensile strength for these bolts.

tensile strength	number of bolts
21300-21309.9	22
21310-21319.9	49
21320-21329.9	116
21330-21339.9	78
21340-21349.9	36
21350-21359.9	11

3.2 Measures of Variation

Example: Find the measures of center for each sample:

1) 20 21 22 23 23 24 25 26

2) 47 −1 23 68 −10 56 23 −22

Answer: Sample 1) $\bar{x} = 23$, median = 23, mode = 23
Sample 2) $\bar{x} = 23$, median = 23, mode = 23

Hmmm...wait, what?! The two samples have the same measures of center. By inspection, though, I can see that the samples are quite different. The first sample has only values that are 20-something. The second sample has values that are far more dispersed. Those values are spread from −22 all the way up to 68. The numbers in the second sample vary much more than those in the first sample. For that reason, when we summarize data, we are not only interested in measuring the center, because that does not tell us the whole story. We also want to capture how varied, or dispersed, or spread the values are.

Measures of variation (or **measures of dispersion** or **measures of spread**) are used to quantify how different the values in a set of data are.

The simplest of these measures is the **range**. In everyday use, the word "range" is often used to indicate an interval ("the incomes range from $25,000 to $64,000"). However, in Statistics, the range is referring to a single number calculated by taking the largest value in a data set, minus the smallest number.

Example: Find the range of Sample 1) and Sample 2) provided earlier.

Answer: Sample 1) range = 6 Sample 2) range = 90

For Sample 1), range = 26 – 20 = 6

For Sample 2), range = 68 – (−22) = 90

Another measure of variation is the **variance**. When working with sample data the variance, s^2, of a set of data is found by:

$$s^2 = \frac{\sum(x-\bar{x})^2}{n-1} \qquad \text{OR} \qquad s^2 = \frac{n\sum x^2-(\sum x)^2}{n(n-1)}$$

These are algebraically equivalent formulas. The first formula is a more conceptual formula because it shows explicitly that we are measuring variation in terms of how far individual data values are from the sample mean. The second formula is a better computational formula because it is often easier to use when performing the calculation.

Example: The exit velocities (in MPH) for randomly selected batted balls of major league baseball players were as shown:

96.0 61.7 108.5 107.2 120.1 118.3 106.4 108.9

Find the variance for this data. Use both variance formulas.

Answer:

Using first formula:

x	$x-\bar{x}$	$(x-\bar{x})^2$
96.0	96.0−103.3875 = −7.3875	$(-7.3875)^2 = 54.57515625$
61.7	61.7−103.3875 = −41.6875	$(-41.6875)^2 = 1737.847656$
108.5	108.5−103.3875 = 5.1125	$(5.1125)^2 = 26.13765625$
107.2	107.2−103.3875 = 3.8125	$(3.8125)^2 = 14.53515625$
120.1	120.1−103.3875 = 16.7125	$(16.7125)^2 = 279.3076563$
118.3	118.3−103.3875 = 14.9125	$(14.9125)^2 = 222.3826563$
106.4	106.4−103.3875 = 3.0125	$(3.0125)^2 = 9.07515625$
108.9	108.9−103.3875 = 5.5125	$(5.5125)^2 = 30.38765625$
	total	2374.24875

$$s^2 = \frac{\sum(x-\bar{x})^2}{n-1} = \frac{2374.24875}{8-1} = 339.1784\ (\text{MPH})^2$$

Using second formula:

	x	x^2
	96.0	9216
	61.7	3806.89
	108.5	11772.25
	107.2	11491.84
	120.1	14424.01
	118.3	13994.89
	106.4	11320.96
	108.9	11859.21
total	827.1	87886.05

$$s^2 = \frac{n\sum x^2-(\sum x)^2}{n(n-1)} = \frac{8(87886.05)-827.1^2}{8(8-1)} = 339.1784\ (\text{MPH})^2$$

We get the same answer using either formula because, again, the formulas are algebraically equivalent. The algebra needed to show that the formulas are equivalent is relatively sophisticated, especially for students with only an intermediate algebra background.

Notice that there is a huge difference between Σx^2 and $(\Sigma x)^2$. Σx^2 tells us to "add up all the x^2". Σx tells us to "add up all the x" and, since that is in parentheses with exponent 2 outside, we square that sum. The order of operations will always apply. As it relates to these terms, we take care of what is inside parentheses first, then exponents, then adding. In Σx^2 we have to perform the squaring before the adding. In $(\Sigma x)^2$ we have to evaluate the expression inside the parentheses first (the adding) before performing the squaring.

In the table needed to use the first formula, each $(x-\bar{x})$ is called a *deviation*. If a data value is less than the mean (which we needed to calculate for this data, and found it to be 103.3875), its deviation will be negative. If a data value is greater than the mean, its deviation will be positive. A data value that is equivalent to the mean has a deviation of zero. The sum of the deviations is always zero. [This is a good check to perform to make sure you calculated the deviations correctly.] That makes sense because the mean is the center of mass, as we discussed in Section 3.1.

For this and all calculations for the rest of the course, you should retain all decimal places during intermediate steps of a calculation and not round until you have the final answer. This will save you tons of headache from having answers that are slightly off but not finding any errors in your steps. If you are able to enter huge chunks of calculations into your calculator without stopping, that is great, but if you have to stop and write numbers during a calculation, keep all decimal places shown on the calculator display and do not round until the very end.

The second formula is nice because we do not have to calculate the mean. If we calculate the mean incorrectly, then all of the deviations would be wrong, as would the entire rest of the calculation. Also, there are generally less hassles with decimal places when the second formula is used.

When using the first formula, each deviation has units MPH. When each deviation is squared, the resulting value has units $(MPH)^2$. After adding all those squared deviations, the sum has units $(MPH)^2$. When we divide by the unit-less quantity $(n-1)$, the final answer has units $(MPH)^2$.

Using the second formula, when we square each data value, each result has units $(MPH)^2$. The sum of those $(MPH)^2$ values also has units $(MPH)^2$. The term $[n\Sigma x^2]$ has units $(MPH)^2$, and the term $[(\Sigma x)^2]$ has units $(MPH)^2$. The difference (subtraction) of those terms would retain the units $(MPH)^2$. Finally, dividing by the unit-less quantity $[n(n-1)]$ would give a final answer in $(MPH)^2$.

The variance always has units that are the square of the units for the original data.

If the data is measured in feet, the variance has units square feet (or ft^2). OK, that's fine. If the data is measured in days, the variance has units days-squared. Huh?! What is a squared day? If the data is measured in dollars the units for the variance are dollars-squared, if the data is measured in people the variance has units people-squared, if the data is measured in doughnuts the units for the variance are doughnuts-squared (which actually sounds delicious), etc. Very strange. Even $(\text{MPH})^2$ squared—what's that?!

A measure of dispersion that is often more useful for reporting is the **standard deviation**. The standard deviation is simply the square root of the variance. When working with sample data, the notation of the standard deviation is s.

Example: Find the standard deviation of the exit velocity data.

Answer: $s = \sqrt{s^2} = \sqrt{339.1784(\text{MPH})^2} = 18.4168 \text{ MPH}$

It's that easy. To get the standard deviation for a set of data, if you already have the variance, just take the square root of the variance. [Conversely, if you have the standard deviation for a set of data, and you want the variance, just square the standard deviation.]

Observe that the standard deviation has units MPH. If the data is measured in feet, the standard deviation has units feet. If the data is measured in days, the standard deviation has units days. If the data is measured in dollars the units for the standard deviation are dollars, if the data is measured in people the standard deviation has units people, if the data is measured in doughnuts the units for the standard deviation are doughnuts (hmmm...this is somehow disappointing...), etc.

The standard deviation has the same units as the original data.

This quality makes the standard deviation a preferred measure for reporting purposes. We can always wrap our heads around the units. The variance, though, has useful mathematical properties that the standard deviation does not have. The variance's properties help us understand distributions of data, which allows us to perform the statistical procedures that we use later in the text. For these reasons, both the variance and standard deviation are important.

We can get formulas for the standard deviation of a sample by taking the square root of the formulas used for the variance:

$$s = \sqrt{\frac{\sum(x-\bar{x})^2}{n-1}} \qquad \text{OR} \qquad s = \sqrt{\frac{n\sum x^2 - (\sum x)^2}{n(n-1)}}$$

Often, you will be asked to calculate a sample standard deviation without first having the variance.

Example: Using both formulas, find the standard deviation for this sample: 1 5 8 2

Answer:

x	x−x̄	(x−x̄)²
1	1−4 = −3	(−3)² = 9
5	5−4 = 1	(1)² = 1
8	8−4 = 4	(4)² = 16
2	2−4 = −2	(−2)² = 4
	total	30

	x	x²
	1	1
	5	25
	8	64
	2	4
total	16	94

$$s = \sqrt{\frac{\Sigma(x-\bar{x})^2}{n-1}} = \sqrt{\frac{30}{4-1}} = 3.1623 \qquad s = \sqrt{\frac{n\sum x^2-(\sum x)^2}{n(n-1)}} = \sqrt{\frac{4(94)-16^2}{4(4-1)}} = 3.1623$$

So far, we have only dealt with measures of dispersion—range, variance, and standard deviation—for sample data. In the rare case that you have data for the entire population, the formulas and notation change. The range is still found in the same manner: largest value in the data minus the smallest value in the data. However, there are changes in the other two measures.

The variance of a population has notation σ^2, (Greek lower case letter "sigma", squared) and is found using:

$$\sigma^2 = \frac{\Sigma(x-\mu)^2}{N} \qquad \text{or} \qquad \sigma^2 = \frac{1}{N^2}\left(N\sum x^2 - \left(\sum x\right)^2\right)$$

The same relationship exists between a population variance and population standard deviation as there is between a sample variance and sample standard deviation. The notation for the standard deviation of a population is σ, with formulas:

$$\sigma = \sqrt{\frac{\Sigma(x-\mu)^2}{N}} \qquad \text{or} \qquad \sigma = \frac{1}{N}\sqrt{N\sum x^2 - \left(\sum x\right)^2}$$

Again, the first version of each formula shows conceptually what is happening, but the second is usually easier to use for calculations. The formulas now involve notation (μ and N) indicative of population values.

Example: Using both formulas, find the standard deviation for this population: 1 5 8 2

Answer:

x	x−μ	(x−μ)²
1	1−4 = −3	(−3)² = 9
5	5−4 = 1	(1)² = 1
8	8−4 = 4	(4)² = 16
2	2−4 = −2	(−2)² = 4
	total	30

	x	x²
	1	1
	5	25
	8	64
	2	4
total	16	94

$$\sigma = \sqrt{\frac{\Sigma(x-\mu)^2}{N}} = \sqrt{\frac{30}{4}} = 2.7386 \qquad \sigma = \frac{1}{N}\sqrt{N\sum x^2 - \left(\sum x\right)^2} = \frac{1}{4}\sqrt{4(94) - 16^2} = 2.7386$$

It is extremely rare for you to have an entire population of data, so it is rare that you would use one of these population formulas. However, you will make great use of the notation throughout the course! When you have a set of data values, if it is not clear to you whether it is population data or sample data, assume it is sample data.

As you review the formulas for standard deviation and variance, it should be clear to you that neither quantity could ever be negative. By virtue of the arithmetic you are performing, you cannot get a negative answer. This is more clearly seen from the conceptual formulas. When you square quantities, add them all, and divide by a positive number, there is no way that can be negative. Since the computational versions of the formulas are algebraically equivalent, they likewise can never give you a negative answer. A common mistake when using the computational formula for a sample is forgetting to multiply n times Σx^2, but only using the Σx^2 part of that term. That mistake often gives you a negative variance. Hopefully, if you make that error, you will say to yourself, "Hey, this can't be negative!" and then find your mistake.

What do you think would happen if all the data was identical? Suppose your data was all 23's. The mean of that data would be 23. All of the deviations would then be zero. So variance and standard deviation, whether we are talking about sample or population, would be zero. So, unless the data is all identical, the variance and standard deviation will both be positive.

If we want to see which of two data sets is more dispersed (or varied, or spread), we can compute and compare their measures of variation.

Example: Which set of data is more dispersed? Justify your answer.

Set 1) 12 45 93
Set 2) 15 39 86

Answer: Set 1) range = 81, s = 40.7308, s^2 = 1659
Set 2) range = 71, s = 36.1156, s^2 = 1304.3333

Set 1) is more dispersed because its measures of dispersion are larger.

We assumed that this was sample data since the problem did not specify.

Notice that we did not just eyeball the data and make some arbitrary conclusion: "Hmmm...well, let me see now. It kinda looks like the first ones may be further apart. But, then, again, I don't know. Uh, yeah, well..." Any question having to do with dispersion, variation, or spread will require us to employ one or more of the measures of dispersion, variation, or spread. In some cases, one measure (like the range) may lead to a different conclusion than another (like the standard deviation) when comparing the variation of two sets of data. If that happens, simply report that fact, along with the conflicting measures.

There is a big difference between the dispersion of data and the center of data. Look at these two sets:

Set 1) 1000 1000 1000
Set 2) 3 91 106

Without doing any calculations, we know Set 1 has range, variance, and standard deviation of zero, while its measures of center are 1000. In the second set, we don't know the measures of dispersion off hand but, whatever they are, they are more than zero. Set 2)'s measures of center, whatever they are, are less than 1000. Set 1) has larger center, but Set 2) has larger variation. The numbers in Set 2) are "more different from each other" than are the numbers in Set 1).

Just a quick note on why the variance and standard deviation formulas for sample and population are different. Let's focus on the conceptual formulas. For a sample the denominator is n-1, but for a population it is N. When we divide by n-1, the sample variance gives us what's called an *unbiased estimator* for the population variance. That means that, on average, the sample variance does not systematically overestimate or underestimate the population variance (which would be the case if we divided by n), but estimates it exactly (again, on average). This topic of biased and unbiased estimators is one of many topics alluded to in this course, but explored in great detail in more advanced Statistics courses.

Know that in other texts or formula sources, you will see formulas that look slightly different from those presented in this section (throughout the text, in fact). All of those are algebraically equivalent alternatives to the formulas you are given. When we say algebraically equivalent, it's like these expressions:

A) $(x-3)(x+4)$
B) $x^2 + x - 12$
C) $x^2 + 37 + x - 49$

At first glance, these all look different. However, if you take a value for x, plug it into each expression, and evaluate each expression, you get the same result.

At this point, students sometimes have the question, "What exactly is the standard deviation?" If you need some sort of easy practical sense of the standard deviation in order to make this material more palatable, you can think of the standard deviation as (virtually) the average distance each value in the data is from the mean. The variance can be thought of as the average squared distance each value in the data is from the mean. [There are less-practical but more complete and technically accurate definitions, which you can ascertain by simply putting the conceptual formulas into words.]

From now until the day you close this book for the final time, you will need to calculate standard deviations and variances for sets of data. Every single time you have to do so, you will make use of these same formulas. So, when those occasions present themselves, start setting up your table and get to it! There is no need to be confused, flummoxed, or befuddled because you have already learned how to perform those calculations in this section. If the directions do not indicate which formula to use (conceptual or computational), it will be your choice. When you complete the exercises for this section, though, you should use both formulas so you are comfortable with them.

Remember, when you are given new notation, immediately work on memorizing it. In this section you got:

s	standard deviation of a sample
s^2	variance of a sample
σ	standard deviation of a population
σ^2	variance of a population

Exercises

1) Consider the following wind chill readings (°F) for randomly selected cities on a winter day:

0 -16 -3 -3 3 28 0 -7 26 5 -34 3 2

Find the measures of dispersion for this data.

What would be the measures of dispersion if this were a population instead of a sample?

2) The elevations (above sea level, measured in meters) for a sample of cities are shown below:

98 76 34 60 63 3 105 40 -2 37

Find the measures of variation for this data.

3) Return to the elevations data from problem 2). Multiply each value by 2. Find the new measures of spread for this data. Compare and contrast the new answers with your answers for problem 2).

4) Return to the elevations data from problem 2). Add 2 to each value. Find the new measures of variation for this data. Compare and contrast the new answers with your answers for problem 2).

5) A sample of dieters who used the Looz Now diet were weighed (in pounds) before beginning the diet program and then again two weeks later. The weight differences (weight before minus weight after) were:

10.23 8.71 -3.18 6.01 4.59

Find the measures of dispersion for this data.

3.3 Introduction to z-scores

A z-score (or standardized score) can be calculated for an individual value, x, in a set of data. The magnitude (absolute value) of the score tells how many standard deviations from the mean the individual data value is. If a value from the data has a z-score of 3.71, that tells us that the individual data value is 3.71 standard deviations from the mean. More specifically, since the z-score is positive, the individual data value is 3.71 standard deviations above the mean. If a value from the data has a z-score of −2.76, that tells us that the individual data value is 2.76 standard deviations from the mean. More specifically, since the z-score is negative, the individual data value is 2.76 standard deviations below the mean. If an individual data value has a z-score of zero, that tells us that the individual data value is equal to the mean.

The information in the previous paragraph is clarified when we consider how z-scores are calculated:

$$z = \frac{x-\bar{x}}{s} \qquad \text{OR} \qquad z = \frac{x-\mu}{\sigma}$$

A z-score is calculated by taking the individual data value, x, and subtracting the mean. That difference is divided by the standard deviation. If we are dealing with a value from a sample, the formula on the left is applicable. If we have the entire population available to us, and we take a value from it, the formula on the right is applicable. I am sure you recognized the notation in the left formula as referring to a sample, while the notation in the right formula refers to a population. In most cases, though, you don't have to worry about the look of the formula because the process of the calculation is the same either way.

There are several uses of z-scores in Statistics. A z-score can be used to determine if the individual data value used to calculate it is "unusual" or an *outlier*. By convention, an individual data value is considered unusual, or an outlier, if its z-score is less than −2 or greater than 2. Otherwise, the data value is not considered unusual.

Example: Scores on the College Preparedness Exam (CPE) have a mean of 84.2 and standard deviation of 11.6. Would it be unusual for a test-taker to earn a score of 51.3? Explain.

Answer: $z = \frac{51.3-84.2}{11.6} = -2.8362$. Yes, that would be an unusual (specifically, unusually low) CPE score because its z-score is less than −2.

On the other hand, a CPE score of, say, 101.1 would not be unusual because its z-score, 1.4569, is between −2 and 2, inclusive.

How many standard deviations from the mean is the CPE score of 51.3? 2.8362. We get that from the magnitude of its z-score.

Another use of z-scores is for making comparisons of values that come from different scales.

Example: Scores on the College Preparedness Exam (CPE) have a mean of 84.2 and standard deviation of 11.6. Two years ago, the format and scoring on the CPE were changed to its current version. Before the change, CPE scores had a mean of 243.2 and standard deviation 39.1. An admissions and scholarships officer for Statsville College is trying to determine which is a better: an applicant's score of 305.7 on the old CPE or an applicant's score of 108.6 on the current CPE. What decision do you think should be made? Explain.

Answer: old CPE: $z = \frac{305.7-243.2}{39.1} = 1.5985$

new CPE: $z = \frac{108.6-84.2}{11.6} = 2.1034$

The score of 108.6 on the new CPE is better because its z-score is bigger.

It would not be appropriate for us to compare the raw scores, 305.7 and 108.6, directly because they mean different things. They are from different scales. To make a fair comparison, we put them on the same scale, a standard scale, the "how many standard deviations from the mean" scale, by converting to z-scores. As the saying goes, we can then compare "apples to apples" and make a decision.

This sort of process is used all the time when different tests are supposed to measure the same thing (like college readiness, or potential for college success). Or, as in the example, it could be the same test, but the scoring of it has changed. Or, from one period to the next (like fall of one year versus spring of the next) the test versions administered had different levels of difficulty so a lower score on the harder test would be equivalent to a higher score on the easier test version. z-scores can be used for contexts not involving exams. If a situation requires comparing values from different sets, standardization (i.e., converting to z-scores) can facilitate a fair comparison.

Example: 1000 employed Wudmoor residents were surveyed and asked how long (in miles) they commute to work (one way). After the data was entered into a statistical software program, a list of z-scores was generated based on each respondent's answer. Here is a partial list:

respondent	z-score
Claire	.65
Cliff	.22
Denise	−2.31
Rudy	3.16
Theodore	−.49
Vanessa	−1.43

a) Who has a below average commute?

b) Who has an unusual commute?

c) How many standard deviations from the mean is Theodore's commute?

Answers: a) Denise, Theodore, and Vanessa, because their z-scores are negative.

b) Denise and Rudy because their z-scores are outside of the interval from −2 to 2.

c) .49 standard deviation from the mean because that is the magnitude (absolute value) of his z-score.

Recall that negative z-scores are indicative of data values that are below the mean (below average) and positive z-scores are indicative of data values that are above the mean (above average). We could add that Theodore's commute is specifically .49 standard deviation *below* the mean, since the z-score is negative.

Note that z-scores never have units. If the data was measured in feet, the difference calculated in the numerator of the z-score would be in feet, the denominator (the standard deviation) would be in feet so, when dividing, those feet units would cancel. This always happens with z-scores which is why there are no units (like feet, hours, cm^3, bananas, etc.). One could say that z-scores have "standard units" which just means that a z-score is in z-score units, but not any general unit of measurement.

Exercises

1) A certain contractor licensing exam was administered in September and then again in the following March. The licensing commission experimented with a more challenging exam in March in an ongoing effort to improve the quality of contractors. The September exam had a mean score of 101.6 with standard deviation 28.3. The March exam had a mean score of 83.6 with standard deviation 31.7. Which is better: a score of 124.2 on the September exam or a score of 129.4 on the March exam? Which exam score is unusual?

2) Suppose the average golf club distance for mid-hitting men using a 3-iron is 180 yards, with standard deviation 51.2 yards. For mid-hitting women the mean is 125 yards, with standard deviation 62.9 yards. Which is more noteworthy: Martin's (a man) 3-iron distance of 163.0 yards or Gina's (a woman) 3-iron distance of 134.2 yards? (Both of them are mid-hitters.)

3.4 Quartiles and Boxplots

Quartiles are used to divide a set of quantitative data into four sections. There are a couple different ways to find quartiles, but we will use the most common method when finding them by hand. (Another method, using percentiles, is used by some technology. We will consider percentiles later.) Quartiles are labeled Q_1, Q_2, and Q_3.

Q_2: the median of all the data
Q_1: the median of the first half of data
Q_3: the median of the second half of data

Example: The monthly electricity consumption, measured in Kilowatt Hours (KWH), of randomly selected households are shown below. Find the quartiles for this data.

565 892 736 370 517 831 736

Answer: Ordered data: 370 517 565 736 736 831 892

370	517	565	736	736	831	892
	Q_1		Q_2		Q_3	

The quartiles are 517, 736, 831 (KWH)

Since we are working with medians, we know the data must be ordered. We find Q_2 first, in the usual fashion for finding the median. Then, excluding Q_2, we look at the data to the left of Q_2 and find its median. That's Q_1. Then, excluding Q_2, we look at the data to the right of Q_2 and find its median. That's Q_3. What if we had an even quantity of data values? I'm glad you asked.

Example: The exit velocities (in MPH) for randomly selected batted balls of major league baseball players were as shown below. Find the quartiles for this data.

96.0 61.7 108.5 107.2 120.1 118.3 106.4 108.9

Answer: Ordered data: 61.7 96.0 106.4 107.2 108.5 108.9 118.3 120.1

Q_1	Q_2	Q_3
101.2	107.85	113.6

The quartiles are 101.2, 107.85, 113.6 (MPH)

Again, we find Q_2 first from the ordered data. That is the mean of the two numbers in the middle (107.2 and 108.5). Then, excluding Q_2, we look at the data to the left of Q_2 and find its median. That's Q_1. 107.2 was NOT Q_2 so it is part of the first half of data that is strictly to the left of Q_2. For those four numbers to the left of Q_2, their median is the mean of 96.0 and 106.4, which is 101.2. Then, excluding Q_2, we look at the data to the right of Q_2 and find its median. That's Q_3. 108.5 was NOT Q_2 so it is part of the second half of data that is strictly to the right of Q_2. For those four numbers to the right of Q_2, their median is the mean of 108.9 and 118.3, which is 113.6.

As mentioned, quartiles divide data into four regions. Approximately 25% of the data is less than Q_1, approximately 50% of the data is less than Q_2, and approximately 75% of the data is less than Q_3. [The exact percentages will vary depending on the exact size of a set of data.]

The five-number summary for a set of data is: the smallest value, Q_1, Q_2, Q_3, the largest value.

Example: Write the five-number summary for the exit velocity data.

Answer: 61.7, 101.2, 107.85, 113.6, 120.1 (MPH)

The interquartile range (IQR) is $Q_3 - Q_1$. We can identify outliers using the IQR.

A value from the data, x, is considered an extreme outlier if:

$$x < Q_1 - 3\ IQR \qquad \text{or} \qquad x > Q_3 + 3\ IQR$$

A value from the data, x, that is not an extreme outlier is considered a mild outlier if:

$$x < Q_1 - 1.5\ IQR \qquad \text{or} \qquad x > Q_3 + 1.5\ IQR$$

Example: Identify outliers in the electricity consumption data, using the IQR.

Answer: $IQR = Q_3 - Q_1 = 831 - 517 = 314$

$3\ IQR = 3(314) = 942$

An extreme outlier is a value in the data that is

less than $Q_1 - 3\ IQR = 517 - 942 = -425$ or

greater than $Q_3 + 3\ IQR = 831 + 942 = 1773$

No values in the data are less than −425 or greater than 1773 so there are no extreme outliers.

$1.5\ IQR = 1.5(314) = 471$

A mild outlier is a value in the data that is

less than $Q_1 - 1.5\ IQR = 517 - 471 = 46$ or

greater than $Q_3 + 1.5\ IQR = 831 + 471 = 1302$

No values in the data are less than 46 or greater than 1302 so there are no mild outliers.

There are no outliers in the electricity consumption data.

We learned previously how to find outliers based on z-scores, but the directions specifically asked us to use the IQR. [The two approaches to finding outliers do not always give the same outliers. In practice, you can decide beforehand which method seems most sensible for your situation or, if desired, you can use both methods and compare results.]

We simply applied the definitions of extreme and mild outliers. Based on the definitions we found the cutoffs (or fences) for each type of outlier and then looked to see if any of our data values fell outside of the cutoffs. Notice that our calculations did not give us the outliers; they gave us the cutoffs for defining outliers.

Example: Identify outliers in the exit velocity data, using the IQR.

Answer: IQR = $Q_3 - Q_1$ = 113.6 – 101.2 = 12.4

3 IQR = 3(12.4) = 37.2

An extreme outlier is a value in the data that is

less than Q_1 – 3 IQR = 101.2 – 37.2 = 64 or

greater than Q_3 + 3 IQR = 113.6 + 37.2 = 150.8

61.7 is less than 64 so it is an extreme outlier.

1.5 IQR = 1.5(12.4) = 18.6

A mild outlier is a value in the data (besides the extreme outlier) that is

less than Q_1 – 1.5 IQR = 101.2 – 18.6 = 82.6 or

greater than Q_3 + 1.5 IQR = 113.6 + 18.6 = 132.2

Besides the extreme outlier, no values in the data are less than 82.6 or greater than 132.2 so there are no mild outliers.

There is one outlier in the data, 61.7 MPH, and it is extreme.

Example: Construct a boxplot for the electricity consumption data.

Answer:

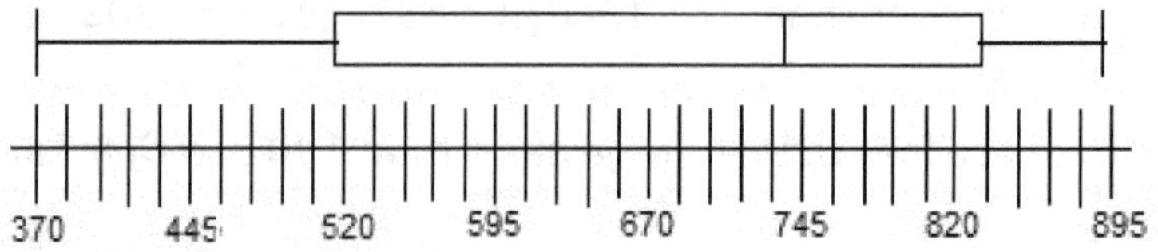

A boxplot, also called a box-and-whisker plot, is another display of data. The "box" extends from Q_1 (517) to Q_3 (831). Inside the box we draw a segment over Q_2 (736). The horizontal segments extending from the box in either direction are called *whiskers*. The whiskers end at the smallest and largest non-outlier values in the data. Since this data has no outliers, the smallest non-outlier value is just the smallest value (370) and the largest non-outlier value is just the largest value (892). The diagram hovers over a number line to give it a scale. [In some software, the boxplots are drawn to perfect scale so there is no need for a number line but, when drawing these by hand, a number line is need to keep the diagram to scale.]

Example: Construct a boxplot for the exit velocity data.

Answer:

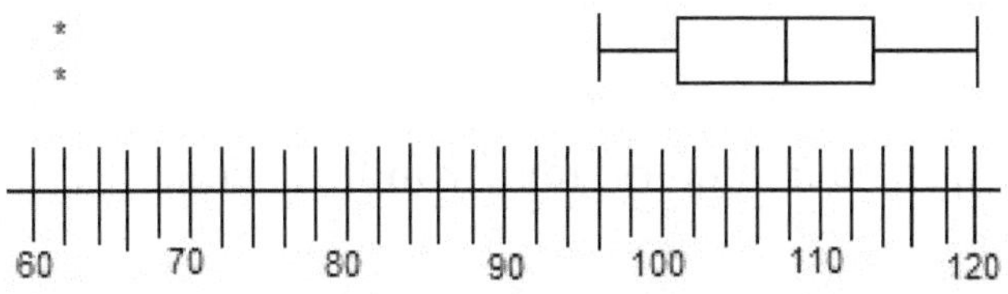

Again, the box extends from Q_1 (101.2) to Q_3 (113.6), with a segment inside at Q_2 (107.85). We already identified 61.7 as an outlier so the smallest non-outlier value in the data is 96.0, and that is where the left whisker ends. There are no outliers on the right side of the data so the largest non-outlier value in the data is just the largest value (120.1). Asterisks are often used to indicate outliers. One asterisk indicates a mild outlier and two asterisks indicate an extreme outlier. So, over our extreme outlier 61.7, you see two asterisks. [Various software may have different ways of indicating outliers.]

Why do we care about outliers? They could be mistakes in measuring or recording. Suppose your data consists of recorded heights, in inches, for randomly selected people. You later look at your data and notice one entry is 647 inches. There's no way that is correct. If you still have access to the person represented by that number, perhaps you can measure that person again. If you recorded the data yourself, and you are reasonably sure as to the reason for the error (say, that the value should actually be 64.7) you can correct it before doing the sorts of tests and analysis we will perform later in the book. Sometimes outliers are not mistakes at all. It is not uncommon for a population to have outliers. So, a sample you collect could also have outliers. The point is that we identify and investigate them before moving on. Depending on the situation, you may remove the value(s) completely (if known to be irremediable errors), or perform statistical tests and procedures with the outlier(s) and without, and then compare results. This is one example of how Statistics is part science but also part art.

Exercises

1) Consider the following wind chill readings (°F) for randomly selected cities on a winter day:

0 -16 -3 -3 3 28 0 -7 26 5 -34 3 2

a) Find the 5-number summary for this data.

b) Find the outliers for this data, using the IQR.

c) Create a boxplot for this data.

2) The elevations (above sea level, measured in meters) for a sample of cities are shown below:

98 76 34 60 63 3 105 40 -2 37

a) Find the 5-number summary for this data.

b) Find the outliers for this data, using the IQR.

c) Create a boxplot for this data.

3.5 Descriptive Statistics: A Mixture

We will begin this section by looking at Chebyshev's (the spelling of this name varies) theorem. It is not of much value to us in this course, but it is usually mentioned in an introductory Statistics text. According to the theorem, which is proven in higher level courses and texts, the proportion of data lying within k standard deviations of the mean (regardless of the shape of the distribution) is at least $1 - 1/k^2$, where k is bigger than one. So, for the CPE scores that have a mean of 84.2 and standard deviation of 11.6,

for k = 2, at least $1 - 1/2^2$ (or 75%) of the scores are within two standard deviations of the mean (so that's between 61 and 107.4)

for k = 3, at least $1 - 1/3^2$ (or 89%) of the scores are within three standard deviations of the mean (so that's between 49.4 and 119)

for k = 4, at least $1 - 1/4^2$ (or 94%) of the scores are within four standard deviations of the mean (so that's between 37.8 and 130.6)

The rest of the section is devoted to putting together some of the material considered in previous sections. The exercises contain various elements of the material that has been considered previously, plus Minitab output relating to that material.

Consider the following output from Minitab:

Descriptive Statistics: rain pH

Variable	N	N*	Mean	SE Mean	StDev	Minimum	Q1	Median	Q3
rain pH	31	0	4.3942	0.0248	0.1379	4.0700	4.2900	4.4000	4.5100

Variable	Maximum
rain pH	4.6200

This output from the statistical software Minitab shows descriptive statistics from randomly selected pH readings of rooftop rain. Minitab shows us that the information refers to the "**rain pH**" data. Minitab uses N to indicate the number of values in the set of data. We see have 31 pH readings in our sample. N* shows how many missing pieces of data we have. In this course that will always be zero, but need not be in practical situations. The mean of the sample of pH readings is 4.3942. SE Mean is the Standard Error of the Mean which we will use later, but will never need to read it from output (in this text). StDev is showing the standard deviation of the sample. The five-number summary for the data is 4.0700, 4.2900, 4.4000, 4.5100, and 4.6200 which is easily ascertained from the labels.

We will practice a lot with Minitab output in this course. Minitab is one of many statistical software packages. No matter which package you wind up using later, the output is always labeled. Once you are skilled at reading statistical output, you will be able to transfer that skill to any statistical package you encounter in the future.

Exercises

1) There are four furniture stores in a certain city. One is in the north, one is in the south, one is in the east, and one is in the west. We collected random samples of prices (in dollars) from each store and used the data for the Minitab output below.

Descriptive Statistics: North

Variable	N	N*	Mean	SE Mean	StDev	Minimum	Q1	Median	Q3
North	91	0	1508.9	4.84	46.2	1416.5	1470.8	1512.4	1545.9

Variable	Maximum
North	1597.8

Descriptive Statistics: South

Variable	N	N*	Mean	SE Mean	StDev	Minimum	Q1	Median	Q3
South	82	0	1514.6	5.35	48.4	1397.5	1484.4	1517.1	1547.0

Variable	Maximum

```
South       1607.6
```

Descriptive Statistics: East

```
Variable   N  N*    Mean  SE Mean  StDev  Minimum      Q1  Median      Q3
East      91   0  2005.9     3.90   37.2   1903.6  1977.2  2011.4  2033.8

Variable  Maximum
East       2095.1
```

Descriptive Statistics: West

```
Variable   N  N*    Mean  SE Mean  StDev  Minimum      Q1  Median      Q3
West      73   0  1601.8     9.26   79.1   1510.1  1538.9  1586.5  1661.0

Variable  Maximum
West       2002.3
```

a) State all known measures of center for the sample from the North store.

b) State all known measures of spread for the sample from the South store.

c) See if the biggest value in the East store sample is an outlier.

d) How many prices are in the West store's sample?

2) The stem-and-leaf plot below shows run times (to the nearest whole minute) for randomly selected movies. What is the mode of this data?

stem	leaf
8	7
9	3 3 5 6 8 8 9
10	0 0 1 1 1 2 2 5 7 8
11	0 1 2 2 3 3 3 3 4 4 4 8 9 9 9
12	1 1 2 6

3) Suppose people keep their cars an average of 11.4 years with variance 24.2064 years^2. The amount of time Dexter kept his car before trading it in for something new was 1.3 standard deviations below the mean. How long did Dexter keep his car?

4) We chose 28 residents from Quahog and asked each person how long he/she slept the night before. (They knew they would be involved in this research so they used a device to accurately record the time.) The total amount of hours of sleep for the sample was 180.6.

The greatest amount of time provided by one respondent was 9.4 hours. The standard deviation of the sample was 1.58 hours.

In this research we also sampled residents from Springfield and asked them how long they slept the night before. These were the responses (in hours):

5.6	8.1	4.8	7.3	6.5	4.8	6.2	5.3	4.9	7.5	6.6	5.0
3.5	6.4	5.2	8.3	5.6	8.2	4.7	4.4	8.2	4.5	5.5	

a) Provide a measure of center for the Quahog sample.

b) Create a frequency distribution for the Springfield data starting with lower class limit 3.4, using class width .9.

c) What is the variance for the Quahog sample?

5) Weights of elephants (in tons) are shown below:

3.5	4.3	5.2	4.6	5.1	5.2	3.8	4.7	5.0	4.6

a) How many standard deviations from the mean is the weight 5.1 tons?

b) Give the five-number summary for this data.

4 Probability Basics

Probability is a world unto itself. There are entire books and graduate courses on that topic. The foundations of pure probability are rather abstract, not easily relatable to real-world situations, and require math at a level much higher than what is required for a course using this book. However, pure probability has very specific and exact properties that the real world does not. In Statistics we try to best approximate real-world situations, with all their unknowns and inexactness, using the well-known and mathematically sound structure in probability theory. Probability provides the framework and the tools that we use in Statistics to best understand and frame our real world. To use Statistics, however, it is not necessary to be an expert in probability, just like a person who uses a car need not be an expert on engines. Many introductory Statistics books and instructors spend too much time on probability (in this author's opinion) which confuses and sidetracks students because the thinking used in probability is different from the thinking employed in Statistics. With that said, some exposure to concepts of probability is necessary for students because we will be referring to probability in the rest of the book.

To fully understand probability, one must be familiar with set theory and measure theory. Since students using this book are not expected to have that experience, the following will provide some basic background and simplified explanations.

A sample space (often indicated with the letter S or Ω) contains all possible outcomes for a procedure. Each individual outcome is indicated with s or ω. Individual outcomes and groups of outcomes that are of interest are called events. Those events are organized in a mathematical structure called a sigma algebra. A probability is a value between 0 and 1 (inclusive) assigned to each event. For example, I could randomly select a number between 200 and 300 and consider questions like, What is the probability of selecting a number between 203 and 229? What is the probability of selecting 284? As another example I could consider tossing a two-sided coin infinitely many times and assigning probabilities to events like, "exactly 37 heads are face up" or "37 tosses until the first head is face up". Probabilities could be assigned in different ways to events but, as a practical matter, it is helpful to think of a probability as a measure of how likely an event is, with higher numbers representing more likely events. (As a side note, an event with probability zero, depending on the situation, may or may not mean that the event is impossible).

Probabilities are assigned to these events based on what appears to be true about how frequently the events can occur or how frequently the events actually occurred empirically. Many widely-accepted formulas and rules are used to assign probabilities to events in a wide variety of situations that seem appropriately described (or reasonably approximated) by those formulas and strategies. A researcher who is using Statistics often begins by finding the probability assignment framework that is generally used for his type of research situation based on the assumptions and conditions associated with that framework. This book focuses on the practical application of basic probability frameworks. You can certainly take more advanced courses to study probability frameworks (distributions) in detail in more advanced calculus-based classes.

Here are a couple axioms about probabilities:

a) A probability is always between 0 and 1 (inclusive).

b) The sum of the probabilities for all separate events of a particular procedure is 1.

If we want to get the probability of one event OR another event, we add the probabilities of the two events (being sure to consider only once any outcomes common to both events).

As these features of general probability become relevant in later sections, we will highlight them.

<u>Discrete Probability Distribution</u>

A discrete probability distribution is a framework that assigns a probability to each outcome of a sample space. Suppose we have 1000 people in the quaint town of Statsville. 512 of them prefer dogs, 304 of them prefer cats, 126 of them prefer birds, and 58 of them prefer snakes. We are going to randomly select and interview a resident and assume that all residents would be equally likely for selection (which is a huge assumption since some residents might be harder to reach than others due to work schedules or general unwillingness to be selected). We could setup a discrete probability distribution as follows:

Animal preference	Probability
dog	512/1000 = .512
cat	304/1000 = .304
bird	126/1000 = .126
snake	58/1000 = .058

The sample space, Ω, would contain 1000 outcomes. Each outcome, ω, would be an animal preference of a resident (ω_1 is the preference for the first resident, ω_2 is the preference for the second resident, and so on). Again assuming equally likely outcomes, the probability for each ω is 1/1000. We have four events in this problem, one for each animal preference. The group (subset) of 512 ω's in Ω that are "dog" constitute one event, the group of 304 ω's in Ω that are "cat" constitute another event, and so on. With this, and discrete probability distributions in general, we can calculate the probability of an event by simply adding the probabilities of the outcomes that comprise that event. This is how we got the probabilities in the table.

What is the probability of selecting a resident who prefers cats? P(cat) = .304

What is the probability of selecting a resident who has a preference that ends with a vowel? P(ends with a vowel) = .058 (snake ends with a vowel)

What is the probability of selecting a resident whose preference is bird or dog? P(bird or dog) = .126 + .512 = .638

What proportion of the residents prefers cats? .304

What percentage of the residents prefers cats? 30.4%

(Notice the strong relationship among probability, proportion, and percentage. If you know one of those you can very quickly determine the other two!)

Now suppose that, instead of considering the type of animal each Statsville resident prefers, we consider the number of stocks each resident owns. Suppose we get the following results:

Number of stocks owned	Probability
0	512/1000 = .512
1	304/1000 = .304
2	126/1000 = .126
3	58/1000 = .058

Based on this, what is the average (mean) number of stocks owned by Statsville residents? Well, we know that 512 residents own no stocks, 304 own one stock, 126 own two stocks, and 58 own three stocks. As we learned in Section 3.1, we get the mean by adding all the numbers and dividing by how many we have. So, we could get the average number of stocks owned by adding 512 zeros plus 304 ones plus 126 twos plus 58 threes and dividing by 1000. Using multiplication for those repeated sums (for example, adding 58 threes is the same as multiplying 58 times three) we have:

$$\frac{512(0)+304(1)+126(2)+58(3)}{1000}$$

Using arithmetic we have: $\frac{512(0)}{1000}+\frac{304(1)}{1000}+\frac{126(2)}{1000}+\frac{58(3)}{1000}$

$$= 0\left(\frac{512}{1000}\right)+1\left(\frac{304}{1000}\right)+2\left(\frac{126}{1000}\right)+3\left(\frac{58}{1000}\right)$$

$$= 0\ P(0) + 1\ P(1) + 2\ P(2) + 3\ P(3)$$

$$= 0(.512) + 1(.304) + 2(.126) + 3(.058)$$

$$= .73$$

So Statsville residents own an average of .73 stock.

Notice that, to get the mean, we simply took the product of each number times the probability of that number and then found the sum of those products. The mean is also called the expected value, E. So, we have:

E = expected value = mean = $\sum x\text{P}(x)$
where x is a number and P(x) is the probability of that number

In Section 3.2 you learned how to calculate the variance of a population using:

$$\sigma^2 = \frac{\Sigma(x-\mu)^2}{N}$$

In similar fashion we could get the variance of this discrete probability distribution using multiplication in place of repeated addition (like, adding 512 of the quantity $(0-.73)^2$ is the same as 512 times the quantity $(0-.73)^2$) as we did to find the mean:

$$\sigma^2 = \frac{\Sigma(x-\mu)^2}{N}$$

$$= \frac{512(0-.73)^2+304(1-.73)^2+126(2-.73)^2+58(3-.73)^2}{1000}$$

$$= \frac{512(0-.73)^2}{1000} + \frac{304(1-.73)^2}{1000} + \frac{126(2-.73)^2}{1000} + \frac{58(3-.73)^2}{1000}$$

$$= (0-.73)^2\frac{512}{1000} + (1-.73)^2\frac{304}{1000} + (2-.73)^2\frac{126}{1000} + (3-.73)^2\frac{58}{1000}$$

$$= (0-.73)^2P(0) + (1-.73)^2P(1) + (2-.73)^2P(2) + (3-.73)^2P(3)$$

$$= \sum[(x-\mu)^2P(x)]$$

$$= .7971 \text{ stock}^2$$

From this same discrete probability distribution we could consider questions like:

What is the probability that a randomly selected resident owns no stocks? P(0) = .512

What is the probability that a randomly selected resident owns at least 2 stocks? P(2 or 3) = P(2) + P(3) = .184

Consider another example. In a bag we have three blue marbles, four green marbles, and two yellow marbles, all of which have the same texture, size, and weight so we assume that each would have an equal probability of being selected. Our sample space looks like:

b1 b2 b3 g1 g2 g3 g4 y1 y2

where b1 is one blue marble, b2 is another blue marble, and so on. We want to consider the probability of the event: "selecting a blue marble on the first trial and then a green marble on the second trial." There are a couple ways we could approach this. We could create a new sample space based on the two trial procedure:

b1b1 b1b2 b1b3 b1g1 b1g2 b1g3 b1g4 b1y1 b1y2
b2b1 b2b2 b2b3 b2g1 b2g2 b2g3 b2g4 b2y1 b2y2
b3b1 b3b2 b3b3 b3g1 b3g2 b3g3 b3g4 b3y1 b3y2
g1b1 g1b2 g1b3 g1g1 g1g2 g1g3 g1g4 g1y1 g1y2
g2b1 g2b2 g2b3 g2g1 g2g2 g2g3 g2g4 g2y1 g2y2
g3b1 g3b2 g3b3 g3g1 g3g2 g3g3 g3g4 g3y1 g3y2
g4b1 g4b2 g4b3 g4g1 g4g2 g4g3 g4g4 g4y1 g4y2
y1b1 y1b2 y1b3 y1g1 y1g2 y1g3 y1g4 y1y1 y1y2
y2b1 y2b2 y2b3 y2g1 y2g2 y2g3 y2g4 y2y1 y2y2

where b1b1 means selecting marble b1 on the first trial and then selecting marble b1 on the second trial, y2g4 means selecting marble y2 on the first trial and marble g4 on the second trial, and so on. Notice that this setup allows me to select the same marble for both trials (like b1b1) so I must replace the first marble before making my second selection (called a "with replacement" procedure). There are 81 possible outcomes of which, 12 (b1g1, b1g2, b1g3, b1g4, b2g1, b2g2, b2g3, b2g4, b3g1, b3g2, b3g3, b3g4) comprise our event in which a blue marble is selected first, then a green marble second. So, again making the assumption of equal probability outcomes, the probability of a blue marble on the first trial and a green marble on the second trial is 12/81 = .1481. Notice that we would get the same result by using the original sample space and taking the probability of selecting a blue marble on the first trial (3/9) multiplied by the probability of selecting a green marble on the second trial, given that a blue marble was selected on the first trial (4/9), which is also .1481.

How would the situation have changed if we did not replace the first marble prior to making the second selection (called a "without replacement" procedure)? This time the new sample space based on the two-trial procedure would look like:

	b1b2	b1b3	b1g1	b1g2	b1g3	b1g4	b1y1	b1y2
b2b1		b2b3	b2g1	b2g2	b2g3	b2g4	b2y1	b2y2
b3b1	b3b2		b3g1	b3g2	b3g3	b3g4	b3y1	b3y2
g1b1	g1b2	g1b3		g1g2	g1g3	g1g4	g1y1	g1y2
g2b1	g2b2	g2b3	g2g1		g2g3	g2g4	g2y1	g2y2
g3b1	g3b2	g3b3	g3g1	g3g2		g3g4	g3y1	g3y2
g4b1	g4b2	g4b3	g4g1	g4g2	g4g3		g4y1	g4y2
y1b1	y1b2	y1b3	y1g1	y1g2	y1g3	y1g4		y1y2
y2b1	y2b2	y2b3	y2g1	y2g2	y2g3	y2g4	y2y1	

Now there are only 72 possible outcomes, still 12 (b1g1, b1g2, b1g3, b1g4, b2g1, b2g2, b2g3, b2g4, b3g1, b3g2, b3g3, b3g4) of which comprise our event in which a blue marble is selected first, then a green marble second. With our assumption of equal probability outcomes, the probability of our event is now 12/72 = .1667. Again, we could use the original sample space and take the probability of selecting a blue marble on the first trial (3/9) multiplied by the probability of selecting a green marble on the second trial, given that a blue marble was selected on the first trial (4/8) which is also .1667. As you can imagine, it is generally preferred to use the original sample space and this *multiplication rule*, but either method works.

Binomial Distribution

There are many discrete probability distributions that are described by formulas (called probability distribution functions) instead of listing each outcome or event with its probability. Formulas are nice because they allow for easier application of those distributions to many different real world situations. One such distribution is the Binomial Distribution.

A Binomial distribution is applicable for a procedure when that procedure:
1) has a fixed number of trials
2) has only two possible outcomes on each trial
3) is such that the probability of each outcome remains the same on each trial
4) has independent trials

(In probability the term "independent" has a very specific mathematical definition, but you can think of independent trials as trials that have no effect on each other.)

Suppose we have a procedure as just described and we want the probability of exactly r of a certain outcome (called a "success") when performing n trials. Using the strategy for how to count the possible number of ways to take n things r at a time (when order does not matter), along with strategies similar to that described previously we have:

$P(r) = \frac{n!}{r!(n-r)!}p^r(1-p)^{n-r}$ where p is the probability of a success on each trial

$\mu = np \quad \sigma^2 = np(1-p)$

(If this were a probability text, we would detail how these formulas are obtained.)

For example, we know that 19% of people in Statsville have back problems. We are going to randomly select six Statsville residents. What is the probability that exactly two of them have back problems?

$P(2) = \frac{6!}{2!(6-2)!}.19^2(1-.19)^{6-2} = .2331$

There are many other discrete probability distributions that have formulas, each with its own set of conditions about the experimental procedure.

Continuous Probability Distribution

A continuous probability distribution is a framework that does not assign a probability to each outcome in the sample space but, rather, to intervals of outcomes. In these cases, the events are intervals. For example, suppose our sample space is all real numbers between 200 and 300 and we are going to randomly select a number. There are uncountably infinitely many values in the sample space (we are not limited to just integers but all real numbers) so we are not able to make a list of each outcome and assign it a probability. Instead we would assign probabilities to intervals like the interval from 202.3 to 209.83, from 237.06 to 256.1, etc., based on the length and location of the interval. There are many ways this could be done, based on what is known or assumed about the procedure. (However this is done, the probability for any single value is zero because the length of an interval from a value to that same value [like, from 206.1 to 206.1] is zero.) A probability density function (pdf) gives the probability density (amount [mass] of probability per unit) for a given outcome in the sample space.

Like functions you learn in algebra, each pdf can be graphed. A practical way to visualize a pdf graph (or curve) is to start by thinking about a dotplot with billions of dots. For example, imagine a dotplot representing the weights of every human on earth, in pounds, to ten decimal places. Now imagine drawing a smooth curve over the top of the dots. Based on the function producing that curve, and calculus methods, we could figure out (or at least get a really good estimate) what proportion of dots lie in any intervals we select (from 202.3 pounds to 209.83 pounds, from 237.06 to 256.1 pounds, etc.). This would be much easier than trying to count how many dots are in the intervals of interest! [A pdf and its curve get a lot more technical than discussed here, but this is a good way to initially wrap your mind around how we will use probability density functions.] Each pdf is constructed in such a way that the area under a pdf curve on a certain interval corresponds to the probability of selecting a value in that interval (area = probability). The total area under a pdf curve is 1.

In practical (as opposed to theoretical) problems, we try to find known probability density functions that best approximate our situations (data) because that makes it much easier to find probabilities and make research conclusions. The most commonly used pdf is the normal (or Gaussian) pdf.

Normal Distribution

Major beliefs about a situation that would justify using a normal distribution is that most of the values in the sample space are near the mean, the mean median and mode are all equal, and values are less likely to appear the further away they are from the mean.

$$f(x) = \frac{1}{\sigma\sqrt{2\pi}} e^{-\frac{1}{2}\left(\frac{x-\mu}{\sigma}\right)^2}$$

That is the normal pdf. The graph of this function is a bell curve. The *standard* normal pdf has $\mu = 0$ and $\sigma = 1$. When using the standard normal pdf the input is a z-score. While it is possible to find the area under a curve using calculus, we have a table of areas (or probabilities) that we can use to get these values since calculus is not a prerequisite for this course. There are many real-world situations in which the data is approximately normally distributed so the normal distribution is used to obtain probabilities of selecting values in intervals.

Again, there is much more to know about probability! This text is a Statistics text, not a probability text (there are many of those available). This is sufficient general background for you to be able to grasp the statistical procedures and methods that will follow. Here are the most important takeaways from this material that will be needed as we study statistical procedures:

1) Any probability is between zero and one, inclusive.
2) The sum of the probabilities for all disjoint (no shared outcomes) events is one.
3) The smaller the probability of an event, the less likely the event is (an event having probability less than .05 is generally considered an unusual event), and the larger the probability of an event, the more likely the event is considered to be.

We will now practice finding probabilities in the normal distribution because this will be necessary for our statistical work.

5 Normal Probability Distribution

5.1 Standard Normal Probability Distribution

As mentioned in Chapter 4, the normal probability distribution is described by the function

$$f(x) = \frac{1}{\sigma\sqrt{2\pi}} e^{-\frac{1}{2}\left(\frac{x-\mu}{\sigma}\right)^2}$$

This actually represents a family of infinitely many different distributions, each specified by its mean (μ) and standard deviation (σ). The area under the curve for an interval of x values provides the probability of selecting a value in that interval (or, the proportion of the values that lie in that interval). If you had calculus, you know that finding the area under a curve on an interval requires integration. As it turns out, the normal pdf does not have an antiderivative that can be used with the Fundamental Theorem of Calculus. Numerical integration must be used.

It is not expected that you learned calculus before entering your current Statistics course, so the last few sentences likely went right over your head. Good news: we won't need to perform the calculus steps because someone else did that for us and preserved many useful results in tables. We can use these tables to get needed probabilities without calculus.

We will begin by considering a special normal distribution: the standard normal distribution. Normal distributions have a symmetric bell-shaped curve, and the standard normal distribution in particular has a mean of zero and standard deviation of one. The independent variable is z-scores, and those are plotted on the horizontal axis. Zero, the mean, is under the peak of the curve and the numbers to the left and right of zero are negative and positive, respectively, like you find on a typical number line.

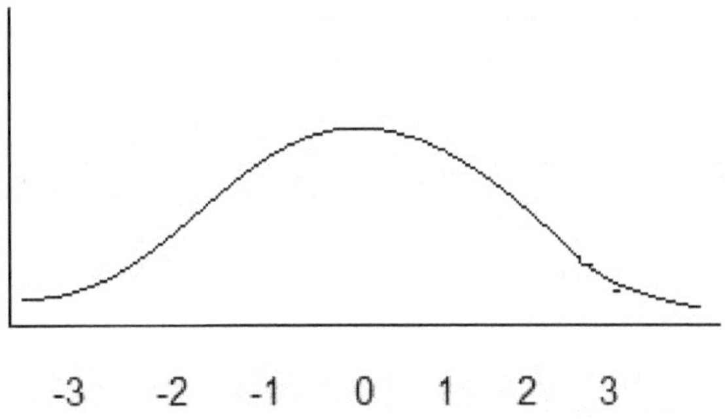

Sketches in this section, and those that you draw by hand, will not necessarily be to scale, but your negative z-scores should always be on the left and the positive z-scores to the right.

Table A (pages 331 and 332) provides probabilities to the LEFT (that's important to remember) of z-scores. We will practice using that table.

Example: In the standard normal distribution, find...

a) the area left of z = −1.28

Answer: Whenever we complete a problem requiring a probability (or area) or a z-score, we will draw a diagram to help us visualize the answer. Probabilities are labeled by the curve, and z-scores are along the horizontal axis. We start by labeling what we know (what we are given in the problem):

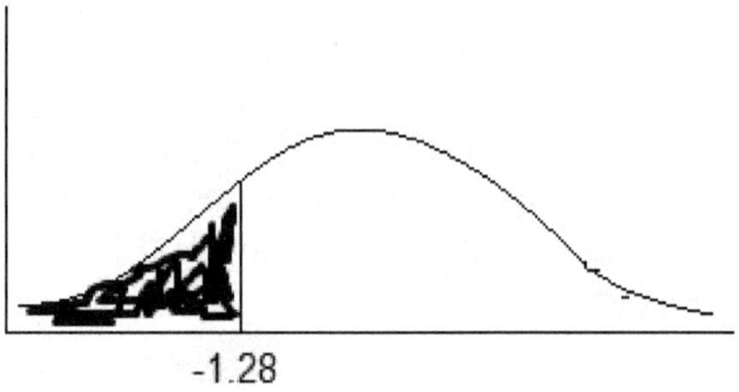

So, we labeled the given z-score. It's negative so it's on the left of the diagram. Since we want the area to the left of that z-score, we shade to its left. Now we look in Table A for the probability. By the way, remember that the total area under the curve is one, and that the curve is symmetric. Just by looking at the diagram we can see that the area must be less than .5, since less than half of the diagram is shaded. When we go to Table A, we find the −1.2 row of the z-scores, and look for its intersection with column 8 to get the second decimal place for our z-score:

second decimal place of z-score

z	0	1	2	3	4	5	6	7	8	9
-3.5 and below	.0001									
-3.4	.0003	.0003	.0003	.0003	.0003	.0003	.0003	.0003	.0003	.0002
⋮										
-1.3	.0968	.0951	.0934	.0918	.0901	.0885	.0869	.0853	.0838	.0823
-1.2	.1151	.1131	.1112	.1093	.1075	.1056	.1038	.1020	.1003	.0985
-1.1	.1357	.1335	.1314	.1292	.1271	.1251	.1230	.1210	.1190	.1170

<u>The answer is .1003</u>

Our completed diagram would look like:

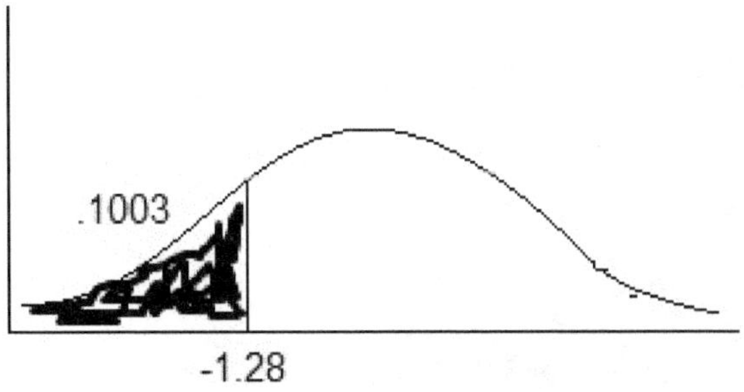

So, the area to the left of z = −1.28 is .1003. Said another way, the probability of randomly selecting a z-score less than −1.28 is .1003. Said another way, the proportion of z-scores that are less than −1.28 is .1003. Said another way, 10.03% of z-scores are less than −1.28. Remember from Chapter 4 that probability, proportion, and percentage are all closely related (if you know one, you know the other two) and probability is the same as area when working with continuous probability distributions.

b) the probability left of z-score 2.07

Answer:

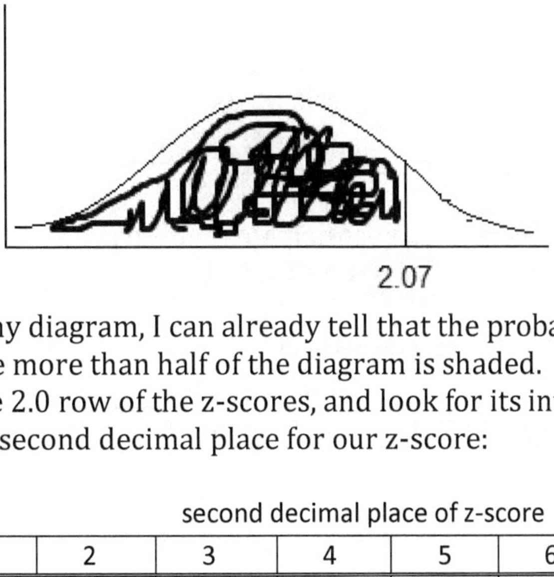

From my diagram, I can already tell that the probability will be greater than .5, since more than half of the diagram is shaded. When we go to Table A, we find the 2.0 row of the z-scores, and look for its intersection with column 7 to get the second decimal place for our z-score:

second decimal place of z-score

z	0	1	2	3	4	5	6	7	8	9
0.0	.5000	.5040	.5080	.5120	.5160	.5199	.5239	.5279	.5319	.5359

:

:

1.9	.9713	.9719	.9726	.9732	.9738	.9744	.9750	.9756	.9761	.9767
2.0	.9772	.9778	.9783	.9788	.9793	.9798	.9803	.9808	.9812	.9817
2.1	.9821	.9826	.9830	.9834	.9838	.9842	.9846	.9850	.9854	.9857

<u>The answer is .9808</u>

Our completed diagram looks like:

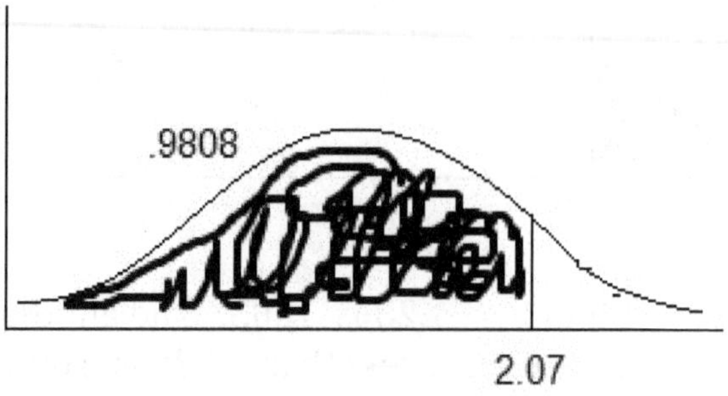

c) the proportion of z-scores above z = .938785

Answer: "Above" is referring to the right side of the z-score. We will need to round the z-score to two decimal places because that is as far as the table represents z-scores. We will use z = .94.

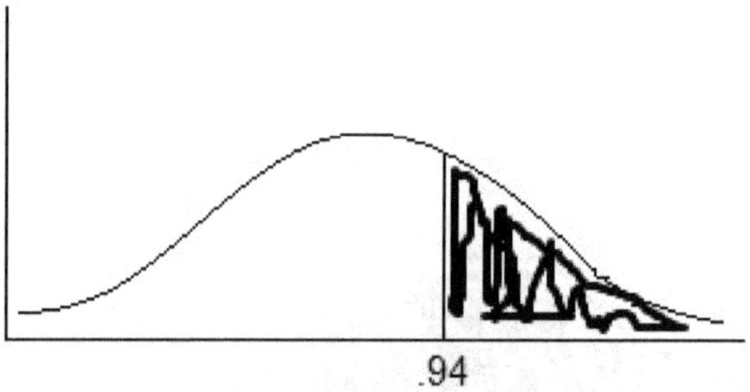

Based on my diagram, I can anticipate that the proportion of my shaded region will be less than .5. Using Table A:

second decimal place of z-score

z	0	1	2	3	4	5	6	7	8	9
0.0	.5000	.5040	.5080	.5120	.5160	.5199	.5239	.5279	.5319	.5359

:
:

0.8	.7881	.7910	.7939	.7967	.7995	.8023	.8051	.8078	.8106	.8133
0.9	.8159	.8186	.8212	.8238	.8264	.8289	.8315	.8340	.8365	.8389
1.0	.8413	.8438	.8461	.8485	.8508	.8531	.8554	.8577	.8599	.8621

As Table A faithfully does, it gave me the area to the LEFT of my z-score. Since I need the area to the right, and since the total area is one, I can get my area by taking 1 – .8264 = .1736

<u>The answer is .1736</u>

Our completed diagram is:

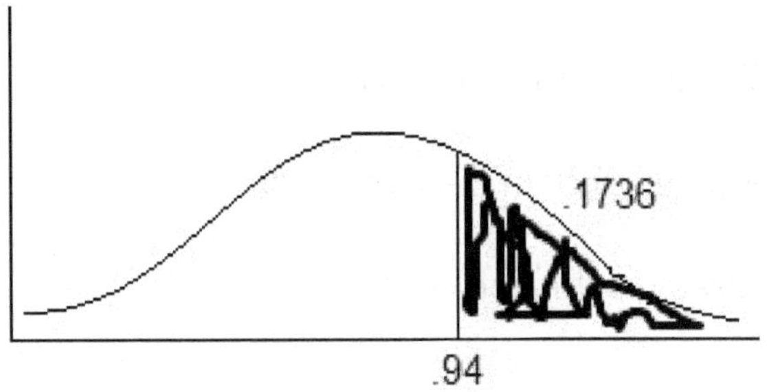

d) the area below z = –.94

Answer: As "above" indicated the right of the z-score, "below" indicates the left. When you draw your diagram, you realize that the final answer has to be less than .5 as you will shade less than half of the diagram. Then you use Table A, look on the –.9 row and see that it intersects column 4 at .1736. That is our answer.

If you are thinking, "Wow, that answer sounds really familiar." Yes!! That is the answer we got for part c). Remember that the distribution is symmetric. That means that the area to the left of a negative z-score is equal to the area to the right of its opposite. The area to the left of z = –1.34 is equal to the area to the right of z = 1.34. The area to the left of z-score –3.21 is equal to the area to the right of z-score 3.21, and so on.

So, what you may do when you need the area to the right of a z-score is to simply look in the table for the opposite z-score. Whatever you find there will be the answer, and you don't have to mess with subtracting from one.

e) the percentage of z-scores to the right of –3

Answer:

When we draw the diagram, we immediately realize that our answer must be more than half (or 50%) because we shade more than half of it.

<u>First method</u>: Look for the –3.0 row of the table and see where it intersects with column 0 (because –3 is the same as –3.00). There you find .0013. That, of course, is the area to the LEFT of the z-score. The area to the right is 1 – .0013 = .9987, giving us a final answer of 99.87%

<u>Second method</u>: Since we are looking for an area to the right, we can

use the symmetry of the distribution to find the area to the left of the opposite z-score, 3.00. If we look for the 3.0 row of the table, we find its intersection with the 0 column at .9987. Our answer is 99.87%.

The answer is 99.87%.

f) the area between z-score −.52 and z-score 1.89

Answer: A diagram of this looks like:

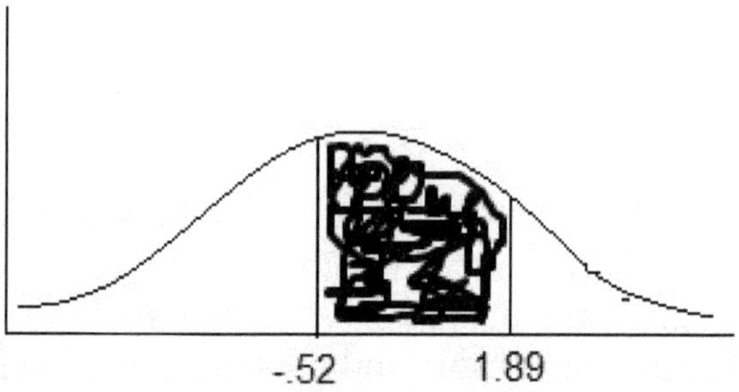

Because the z-scores straddle zero, we don't know for sure if the area will be more than half or less than half just by looking at our sketch. There are a couple ways we could approach this solution, but we will do the following: find the area to the left of the rightmost z-score and subtract the area to the left of the leftmost z-score. From the table, the area to the left of z-score 1.89 is .9706. The area to the left of z-score −.52 is .3015. The area between those z-scores is .9706 − .3015 = .6691 (that's the answer). You don't have to remember "leftmost" and "rightmost" but remember that the area has to be a positive number. So just take the "big area" minus the "little area".

g) the area left of z = −4.56

Answer: When you make your sketch you immediately conclude that the final answer must be less than half. When we look for the −4.5 row of the z-score table, we don't see it. Instead, we see the note: "−3.5 and below" and next to it ".0001". When we need the area to the left of a z-score −3.5 or smaller (or to the right of a z-score 3.5 or larger) we use .0001. In reality, the area is not identical to the left (or right, for the positive extreme) of all those z-scores but, once the area gets that small, that's all we need to know for the statistical procedures we'll use later. So, the answer to this problem is .0001.

If we need the area to the left of z-scores like −3.89, −5.24, −6.35, etc. or the area to the right of z-scores like 3.91, 4.78, 8.49, etc., we would simply use .0001.

h) the percentage of z-scores that are less than .57 or greater than 2.9013

Answer: Using 2.90 (again, we have to round z-scores to two places) for the second z-score, we have this diagram:

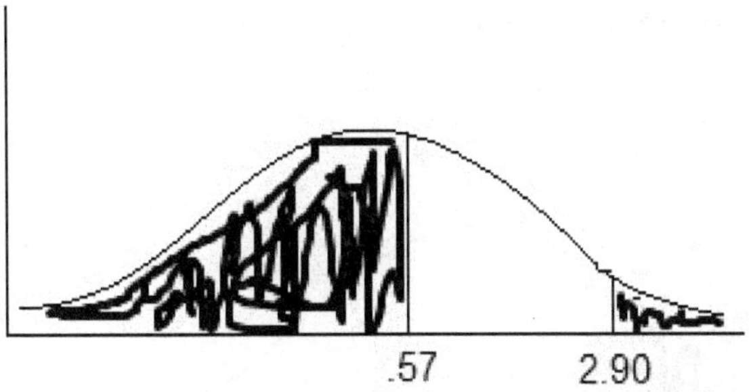

From the diagram we can see that the area to the left of z = .57 is more than half and the area to the right of 2.90 is less than half. Recall from Chapter 4 that, if we want to get the probability of one event OR another event, we add the probabilities of the two events (there are no z-scores that are less than .57 and greater than 2.90 at the same time, so we can just add the individual probabilities with no further considerations). From the diagram, we already know that the final answer will be more than half (50%) because the probability to the left of z =.57 by itself is more than .5.

When we go down to row .5 of the z-score table and over to column 7 we see .7157. So, the area to the left of z-score .57, by itself, is .7157. Suppose we use symmetry to find the area to the right of z = 2.90. We do that by looking for z-score −2.90 and we see that the area is .0019 (the area to the left of z-score −2.90 is .0019, so the area to the right of z-score 2.90 is also .0019).

Our completed diagram looks like this:

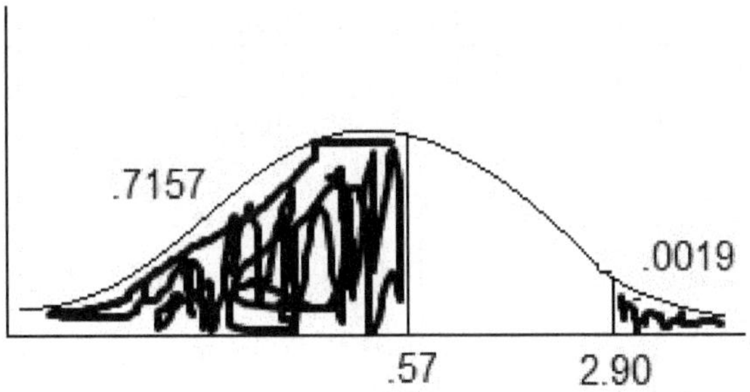

So, the proportion of z-scores that are less than .57or greater than 2.90 is .7157 + .0019 = .7176. Since the question asked for a percentage, we report 71.76% as the answer.

Example: In the standard normal distribution, find the z-score that...

a) has area .1251 to its left

Answer: We are still using Table A, but a little differently. Instead of looking for area based on a z-score, we are looking for a z-score based on area. No problem. As always, we start by diagramming what we are given:

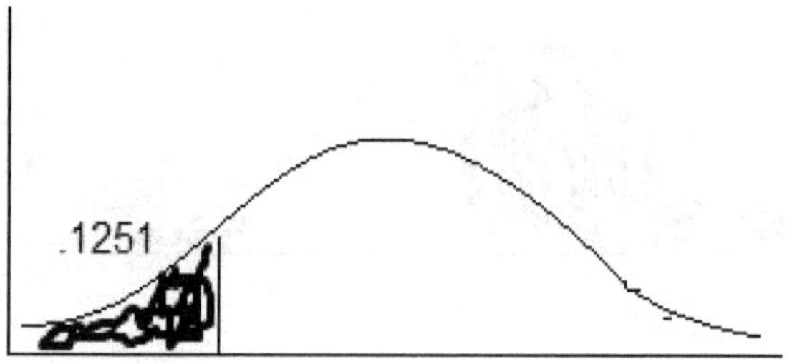

From our diagram we can instantly see that the z-score must be negative because we are on the left of the diagram. We knew we were on the left of the diagram because the area to the left of the yet to be determined z-score is less than .5. This time we will look in Table A for what we have, the area, where the areas are located in the table. Once we find it, we will match it to the appropriate z-score. After a little searching, we find area .1251:

second decimal place of z-score

z	0	1	2	3	4	5	6	7	8	9
-3.5 and below	.0001									
:										
-1.2	.1151	.1131	.1112	.1093	.1075	.1056	.1038	.1020	.1003	.0985
-1.1	.1357	.1335	.1314	.1292	.1271	.1251	.1230	.1210	.1190	.1170
-1.0	.1587	.1562	.1539	.1515	.1492	.1469	.1446	.1423	.1401	.1379

From our area, we trace left and then up to match it to the z-score: −1.15. That is the answer.

The final diagram looks like:

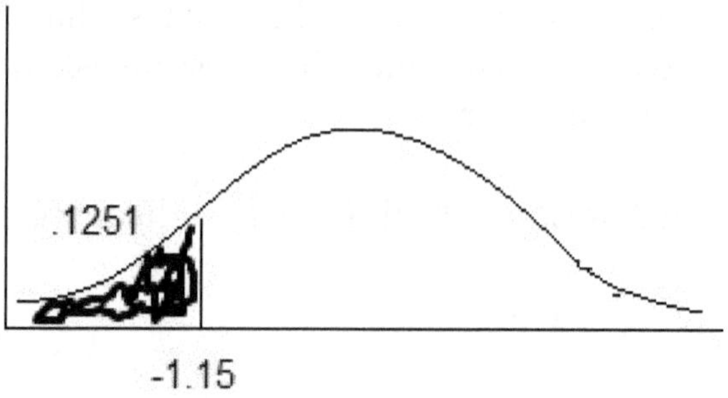

As usual, we label areas at the top and z-scores along the horizontal axis.

b) has probability .2676 to its right

Answer: We are given the area to the right. This is our diagram:

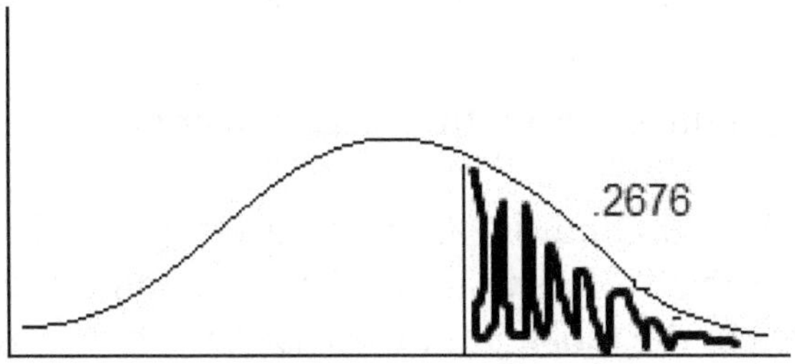

We know we are on the right of the diagram because the area to the right of the yet to be determined z-score is less than .5. So, the z-score is positive.

First method: We can find the area on the left (1 – .2676 = .7324) and look for that in the body of Table A. Once found, we match it to the z-score, .62.

Second method: We can find the given area, .2676, in the body of Table A. Once found, we match it with z-score –.62. That is telling us that z-score –.62 has area .2676 to its left. Using the symmetry of the distribution, we know the z-score we want is .62.

The answer is .62.

Completed diagram looks like:

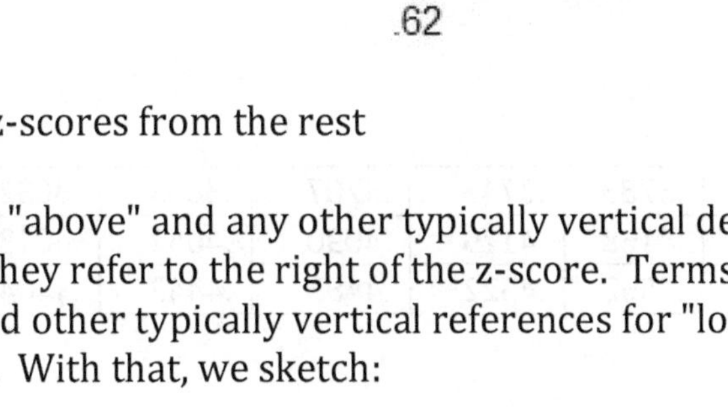

c) separates the top 40% of z-scores from the rest

Answer: "Top" is like "above" and any other typically vertical description referring to "high"—they refer to the right of the z-score. Terms like "bottom", "below", and other typically vertical references for "low" refer to the left of the z-score. With that, we sketch:

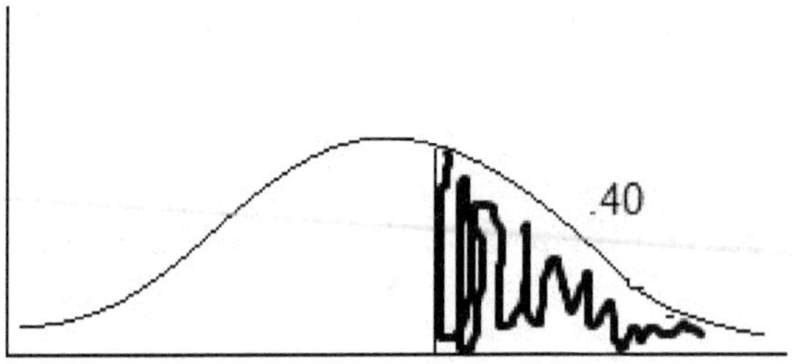

We can see from the diagram that our answer will be a positive z-score.

First method: We can find the area on the left (1 – .40 = .60) and look for that in the body of Table A. It isn't there. This is commonly the case when looking for a given area in the table. In these cases, we have to find the closest area which, for this problem, is .5987.

second decimal place of z-score

z	0	1	2	3	4	5	6	7	8	9
0.0	.5000	.5040	.5080	.5120	.5160	.5199	.5239	.5279	.5319	.5359
0.1	.5398	.5438	.5478	.5517	.5557	.5596	.5636	.5675	.5714	.5753
0.2	.5793	.5832	.5871	.5910	.5948	.5987	.6026	.6064	.6103	.6141
0.3	.6179	.6217	.6255	.6293	.6331	.6368	.6406	.6443	.6480	.6517

.5987 and .6026 are both close, but the closest is .5987. So, we match that to the z-score and report the answer as .25.

Second method: We can find the given area, .40, in the body of Table A. It's not there, but the closest is .4013.

second decimal place of z-score

z	0	1	2	3	4	5	6	7	8	9
-3.5 and below	.0001									
⋮										
-0.3	.3821	.3783	.3745	.3707	.3669	.3632	.3594	.3557	.3520	.3483
-0.2	.4207	.4168	.4129	.4090	.4052	.4013	.3974	.3936	.3897	.3859
-0.1	.4602	.4562	.4522	.4483	.4443	.4404	.4364	.4325	.4286	.4247

.4013 and .3974 are both close, but the closest is .4013. From matching the z-score we find that z-score −.25 has area .4013 to its left so, using the symmetry of the distribution, z-score .25 has area .4013 to its right. We report the answer as .25.

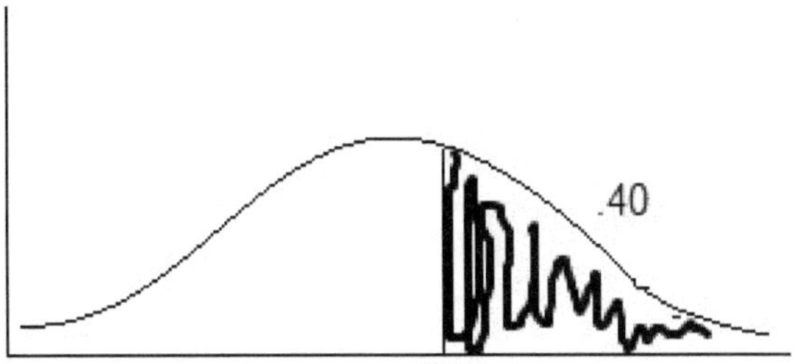

d) separates the bottom 20% of z-scores from the rest (called the 20th percentile)

Answer: [A brief note about percentiles: In a set of values, the kth percentile has k% of the values below it. The 10th percentile has 10% of the values below it. The 82nd percentile has 82% of the values below it, and so on.]

We sketch a bell curve with the dividing line on the left of the diagram, and area .20 shaded to its left. [We already know from the diagram that the z-score must be negative.] Looking at the areas of Table A, the closest is .2005. That corresponds to z-score −.84, our final answer.

The notation, $z_{\alpha/2}$, stands for "the z-score that has area $\frac{\alpha}{2}$ to its right."

(Hey, that was more notation. **MEMORIZE**!)

Example: Find $z_{\alpha/2}$ when α = .05

Answer: We start by calculating the subscript: $\frac{\alpha}{2} = \frac{.05}{2} = .025$. So, we have $z_{.025}$ which is "the z-score that has area .025 to its right." At this point, we employ the same skills used in the recent examples. The area on the left of the z-score is 1 − .025 = .975. Looking for that area and matching to a z-score gives 1.96. Area .025 matches with z-score −1.96 so, using symmetry, the z-score we need is 1.96. Either way, $z_{\alpha/2} = 1.96$.

Example: Find $z_{\alpha/2}$ when α = .10

Answer: We start by calculating the subscript: $\frac{\alpha}{2} = \frac{.10}{2} = .05$. So, we have $z_{.05}$ which is "the z-score that has area .05 to its right." At this point, we employ the same skills used in the recent examples. The area on the left of the z-score is 1 −

.05 = .95. When we look for that area in the table we find a tie between .9495 and .9505. They are equally close. So, because there is no single area that is the closest, we do the following: use the mean of the z-score that matches with area .9495 and the z-score that matches with .9505. The z-score that matches with area .9495 is 1.64, and the z-score that matches with area .9505 is 1.65, so our z-score is $\frac{1.64+1.65}{2} = 1.645$.

We only use this mean of z-scores approach in the event of a tie between two equally close areas.

If we use symmetry, we start by looking for area .05 in the table. We find a tie between area .0495 and .0505. The mean of the z-scores corresponding to those areas is $\frac{-1.65+(-1.64)}{2} = -1.645$. Symmetry tells us to change the sign to 1.645.

Either way, $z_{\alpha/2} = 1.645$

When working with areas to the right of z-scores, you can use either approach (calculating and using the area on the left or using symmetry). However, when we get to other distributions, you will have to use symmetry because of how their tables are organized.

When working with areas to the left of z-scores, no other steps need to be taken. The table is arranged according to areas on the left.

Quick note regarding terminology. In a continuous probability distribution, like the standard normal distribution, "more than" is the same as "at least", and "less than" is the same as "at most". Using mathematical notation:

$z > 2$ is the same as $z \geq 2$

$z < 2$ is the same as $z \leq 2$

The reason is because, as mentioned in Chapter 4, a continuous distribution assigns probabilities to intervals based on their length and location. The probability of a single value is 0. So the probability of the event $z \geq 2$ is equal to $P(z > 2) + P(z = 2)$ [We add because of the "or" in the ≥ symbol].

$$P(z \geq 2) = P(z > 2) + P(z = 2) = P(z > 2) + 0 = P(z > 2).$$

So these problems would all have the same answer:

Find the probability that a randomly selected z-score is greater than 2.
Find the probability that a randomly selected z-score is at least 2.
Find the probability of selecting a z-score that is more than 2.

Find the probability of selecting a z-score that is no less than 2.

Remember that areas (or probabilities or proportions) are always always always between 0 and 1, inclusive. So, something like –.2587 cannot possibly be an area. Neither can something like 1.96.

When working with problems like those in this section, it is paramount that you keep straight what is a z-score and what is an area. As you saw, the areas are in a certain place (the body) of Table A while the z-scores have their place (the left column and the top). Similarly, on the sketches, the areas and the z-scores have their respective places. Most problems dealing with the standard normal distribution are one of two types: 1) you are given z-score(s) and you need to find area, or 2) you are given area and you need to find z-score(s). Start by labeling on your sketch what you are given and then find the requested information.

Exercises

For all exercises, we are in the standard normal distribution.

1) Find the proportion of z-scores to the left of 2.3416.

2) What percentage of z-scores are greater than –.78?

3) If we randomly selected a z-score, what is the probability that our selection would be between –.786 and –.103?

4) What is the probability that a randomly selected z-score would be less than –3.42 or greater than 1.61?

5) What z-score separates the bottom 75% of z-scores from the rest?

6) State the z-score that separates the top 89% of z-scores from the rest.

7) Find $z_{\alpha/2}$, where $\alpha = .12$.

8) For $\alpha = .01$, find $z_{\alpha/2}$.

9) What is the area to the right of z = 7.28?

10) Find the z-score that has probability .0132 above it.

11) What is the probability of randomly selecting a z-score of at least –2.73?

12) What proportion of z-scores are between .2877 and 1.36?

13) Find the percentage of z-scores that are more than 2.6 standard deviations from the mean.

14) How much area in the standard normal distribution is comprised of z-scores that are no more than 1.1098?

15) α = .02. Find $z_{\alpha/2}$.

16) The proportion of z-scores below a certain z-score is .2090. What is the z-score?

17) Find the z-score that is greater than 2% of all z-scores.

18) If a z-score is randomly selected, what is the probability that the selected score will be smaller than −2.65 or at least −1.0713?

19) What z-scores separate the middle 99.4% of z-scores from the rest?

20) Find the z-score that is less than 23% of the other z-scores.

21) What is the probability that a randomly selected z-score would be greater than −3.42 or less than 1.61? [Make your diagram carefully. Compare and contrast this with Exercise 4).]

22) **The Empirical Rule**: This is a rule of thumb in Statistics that estimates the percentage of values within one, two, and three standard deviations of the mean when working with bell-shaped data as 68%, 95%, and 99.7%, respectively. We'll find the motivation behind that rule with these exercises.

a) Find the percentage of z-scores within one standard deviation of the mean (i.e. z-score between −1 and 1).

b) Find the percentage of z-scores within two standard deviations of the mean (i.e. z-score between −2 and 2).

c) Find the percentage of z-scores within three standard deviations of the mean (i.e. z-score between −3 and 3).

23) **For those who have had calculus**: In nearly every introductory Statistics class, there are one or two students who have taken calculus. If you are one, this exercise will give you an opportunity to use skills you learned in that class and see how they relate to what you are now learning.

Recall that the pdf for the normal distribution is $f(x) = \frac{1}{\sigma\sqrt{2\pi}} e^{-\frac{1}{2}\left(\frac{x-\mu}{\sigma}\right)^2}$

For the standard normal distribution in particular, with its mean of zero and standard deviation of one, we have: $f(x) = \frac{1}{\sqrt{2\pi}} e^{-\frac{1}{2}x^2}$, where the "x" values are z-scores. Using this function for the standard normal distribution:

a) Use a Riemann sum with eight subintervals and, using left endpoints, estimate the area under the standard normal curve from z = 1 to z = 2.

b) Using the Trapezoidal Rule with eight subintervals, estimate the area under the standard normal curve from z = 1 to z = 2.

c) Using Simpson's Rule with eight subintervals, estimate the area under the standard normal curve from z = 1 to z = 2.

d) Using Table A, find the area under the standard normal curve from z = 1 to z = 2.

5.2 Non-Standard Normal Probability Distributions

The following display shows graphs of many different normal distributions. As mentioned previously, each normal distribution is distinguished by the value of its two parameters—mean and standard deviation. The value of the mean dictates the location of the curve on the graph, and the standard deviation determines the spread of the curve. For each of the different normal distributions shown, the mean and standard deviation are displayed as well. Observe differences in location and spread based on the values of the parameters.

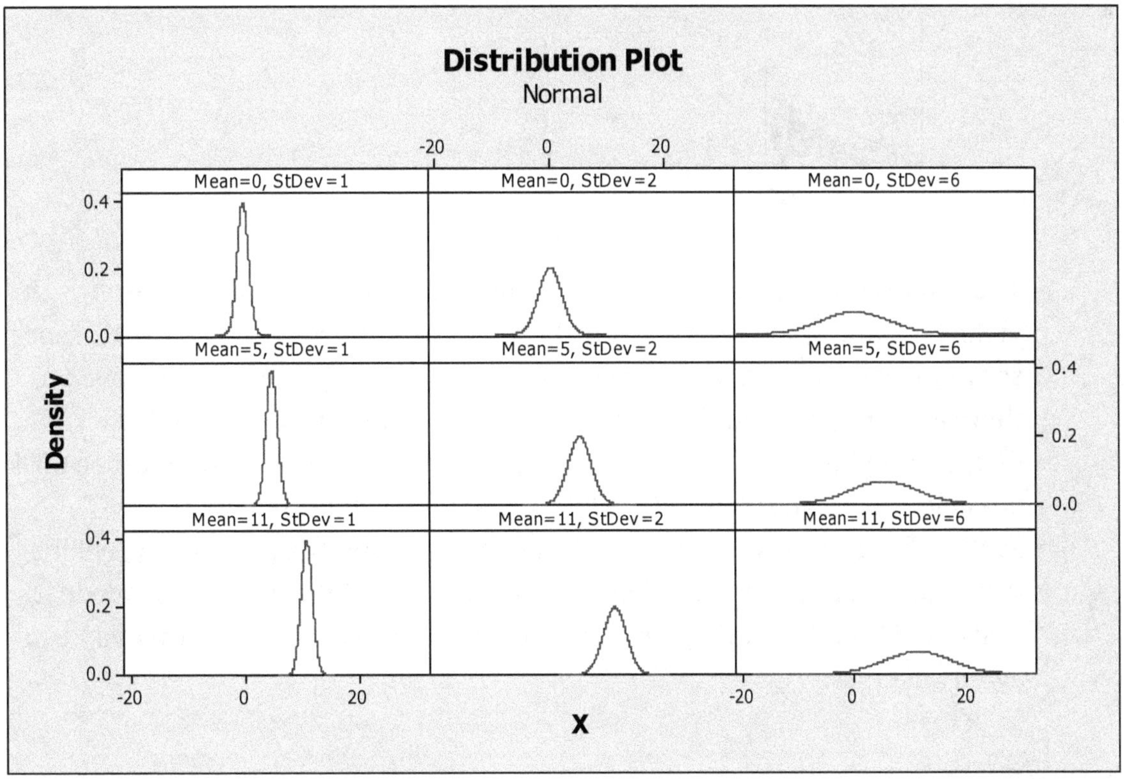

Unfortunately, Statistics books cannot contain tables of probabilities for every different normal distribution (there are infinitely many). As it turns out, any non-standard normal distribution can be transformed to standard normal by subtracting the mean and dividing by the standard deviation. In practical terms, this means we can convert values from any normal distribution into z-scores and then use the z-scores to find needed probabilities.

The basic difference between problems in this section and those in the last is that we will need to convert values to z-scores. Once we do that, the problems become very much like those in the last section. It is assumed that you have mastered Section 5.1 at this point. Solutions will be provided based on that assumption.

Example: Scores on the College Preparedness Exam (CPE) are normally distributed with mean 84.2 and standard deviation 11.6. Find the probability that a randomly selected exam taker has a CPE score...

a) less than 68.4

Answer: If we start with a sketch of the distribution of CPE scores, it looks like this:

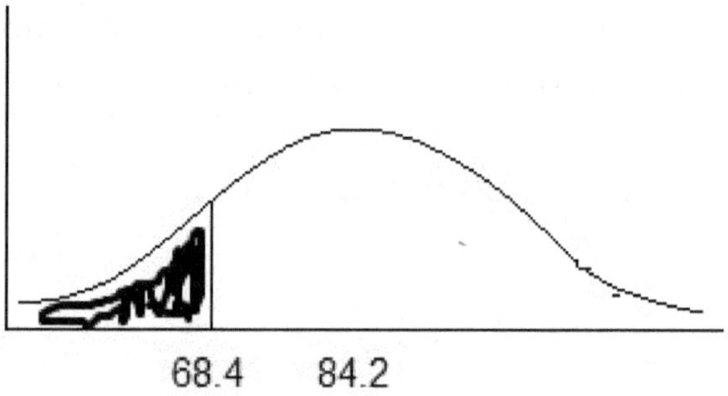

The text does not have a table of probabilities for a normal distribution with mean 84.2 and standard deviation 11.6. We need to convert 68.4 to a z-score so we can use Table A to get the requested probability. Before we move on, there are already a couple of very useful facts we know just from this diagram: 1) The z-score we calculate has to be negative. Why? Because 68.4 is a below average score and we know that data values that are below average have negative z-scores. 2) The final answer, the probability, will be less than .5 because we clearly have less than half of the diagram shaded. It is always helpful to anticipate something about the next step so that you can avoid mistakes.

$$z = \frac{68.4-84.2}{11.6} = -1.3621$$

Once we have the z-score, we proceed just like we did in Section 5.1. The completed diagram after the transformation to the standard normal distribution is:

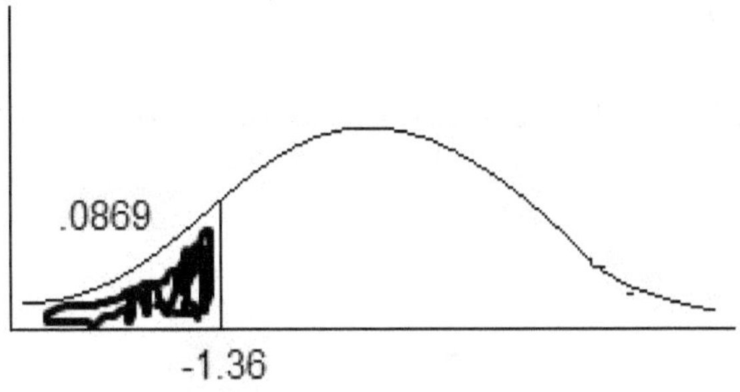

The answer is .0869

What just happened? As we know from the perfect Minitab-generated curves for various normal distributions at the beginning of this section, the CPE distribution and the standard normal distribution have different bell curves. However, the area under the CPE curve to the left of CPE score 68.4 is equal to the area under the standard normal curve to the left of z-score −1.36. You have no doubt observed two different glasses, maybe one is tall and thin and the other is short and wide, but they hold the same amount of a liquid. (That involves volume and we are dealing with area, but it's the same principle.) Similarly, all we are interested in is how much (two-dimensional) space there is—area—at different locations in distributions. We can always get the sought after areas in a non-standard normal distribution by doing this conversion to standard normal.

b) greater than 108.3

Answer: The sketch of the CPE distribution would look like:

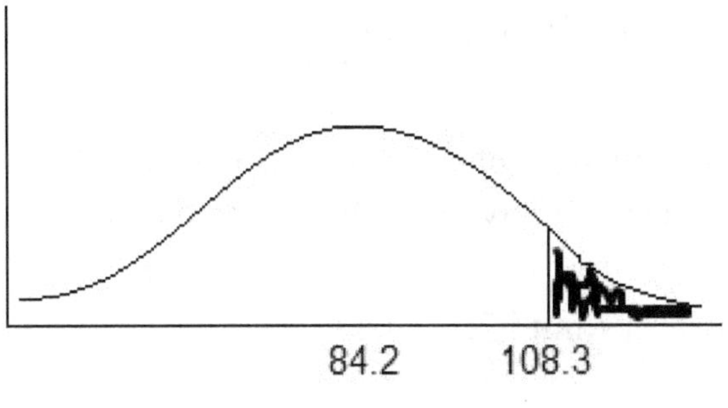

From this we know that the z-score we will momentarily calculate has to be positive. 108.3 is bigger than the mean, and we can see that we are on the right side of the distribution. On the right side of the distribution the z-scores are always positive. Furthermore, we can see that the final answer, the area of the shaded region, is less than .5

$$z = \frac{108.3-84.2}{11.6} = 2.07759$$

Our completed sketch of the standard normal distribution is:

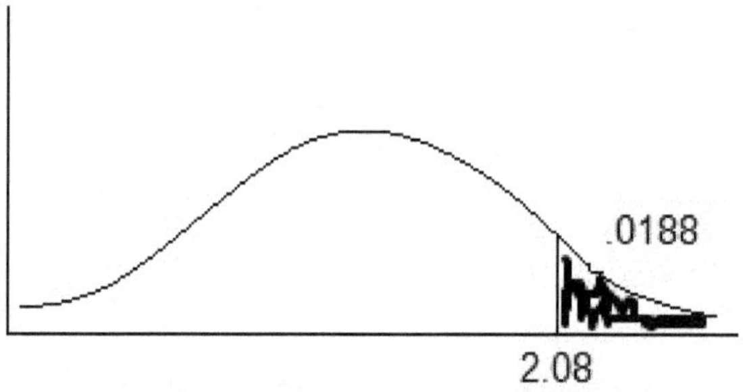

The answer is .0188

The area (remember, area is probability) to the right of CPE score 108.3 in the CPE distribution is the same as the area to the right of z-score 2.08 in the standard normal distribution.

c) between 89.3 and 110.6

Answer: After making your sketch of the CPE distribution you can see that both z-scores we calculate will be positive and the final answer will be less than .5.

for CPE score 89.3: $z = \frac{89.3-84.2}{11.6} = .4397$

for CPE score 110.6: $z = \frac{110.6-84.2}{11.6} = 2.2759$

Using methods from Section 5.1, the area between those z-scores (rounding to .44 and 2.28, respectively), which is our final answer, is: .3187.

d) less than 72.3 or greater than 97.1

Answer: After making your sketch of the CPE distribution you can see that one z-score will be negative and one z-score will be positive. The two areas that we find will individually be less than .5, but the sketch doesn't definitively tell you if the final answer will be more than .5 or less than .5.

$$\text{for CPE score 72.3: } z = \frac{72.3-84.2}{11.6} = -1.02586$$

$$\text{for CPE score 97.1: } z = \frac{97.1-84.2}{11.6} = 1.1121$$

Using methods from Section 5.1, the area to the left of z-score −1.02586 (rounded to −1.03) plus (add because of "or" and the regions have no values in common) the area to the right of z-score 1.1121 (rounded to 1.11) is our final answer: .2850.

e) more than 132.9

Answer: After making your sketch of the CPE distribution you can see that the z-score will be positive and the area of the shaded region is less than .5.

$$z = \frac{132.9-84.2}{11.6} = 4.1983$$

Using methods from Section 5.1, the area to the right of z-score 4.1983 (rounded to 4.20) is our final answer: .0001.

Example: Refer to the information about CPE scores. Would it be unusual for a randomly selected exam taker to earn a score less than 63.1? Explain.

Answer: The first thing to notice is that we are not being asked to assess whether or not one individual CPE score is unusual. If we were, we could simply calculate the z-score for that one individual value and use the [−2, 2] rule. Here, though, we are being asked about all CPE scores less than 63.1. For this, we will need to use the probability definition of "unusual". In Chapter 4, it was mentioned that an event having probability less than .05 is generally considered an unusual event. We will find the probability of the event, "CPE score less than 63.1" and use that rule.

$$z = \frac{63.1-84.2}{11.6} = -1.8190$$

Using methods from Section 5.1, the area to the left of z-score −1.8190 (rounded to −1.82) is .0344. So, yes, it would be unusual for a randomly selected exam taker to earn a score less than 63.1 because the probability of that event, .0344, is less than .05.

Example: Refer to the information about CPE scores.

a) What CPE scores separates the top 10% of CPE scores from the rest?

Answer: Here is our initial diagram of the CPE distribution:

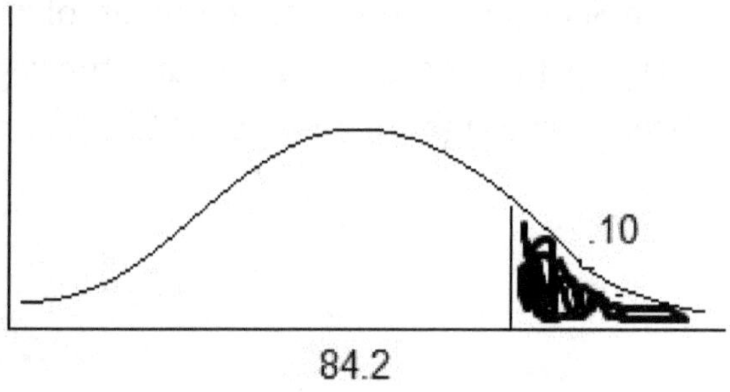

From this we know that the final answer is more than 84.2 because we are on the right of the diagram (since the area in the "top" was less than .5)

Again, we will need to rely on a transformation to standard normal to get the answer. From Table A we will find the z-score that has area .10 to its right which, using methods from Section 5.1, is 1.28. (From the diagram, we knew the z-score would be positive because we are on the right side of the distribution.) What we need to know is the CPE score that corresponds to a z-score of 1.28. This is a job for algebra!

Using the z-score formula with the information we know:

$$1.28 = \frac{x - 84.2}{11.6}$$

From that, 1.28 (11.6) + 84.2 = x = 99.048. That is the answer.

b) What CPE score is the 14th percentile?

Answer: Recall that the 14th percentile is the score that separates the bottom 14% from the rest. With that, our CPE sketch looks like:

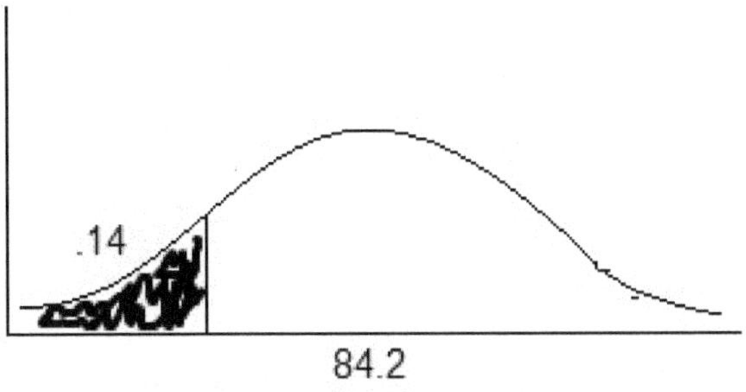

From this we know that the final answer will be less than 84.2 since we are on the left side of the diagram.

From Table A we will find the z-score that has area .14 to its left which, using methods from Section 5.1, is −1.08. (From the diagram, we knew the z-score would be negative because we are on the left side of the distribution.) What we need to know is the CPE score that corresponds to a z-score of −1.08. Using the same approach as in the previous example,

$$-1.08 = \frac{x-84.2}{11.6}$$

$-1.08\ (11.6) + 84.2 = x = 71.672$, the answer.

Example: Suppose the average walking stride length for men is .78m with standard deviation .087m, and that for women the mean is .7m with standard deviation .081m, and that both populations are normally distributed.

a) Suppose a man that has a walking stride length that is greater than at least 95% of men is considered a "long-strider". What is the minimum walking stride length that a man must have to be considered a long-strider?

Answer: The sketch:

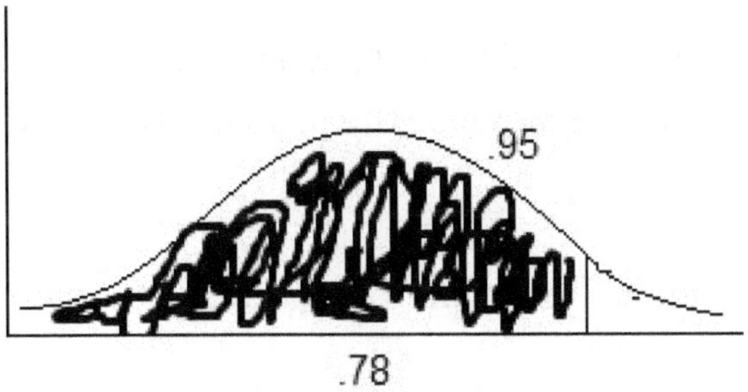

We know the final answer is greater than .78 by looking at the graph (and by thinking about the substance of the problem). We will find the z-score that has area .95 to its left and then use that in the z-score formula to solve for x, the specific male stride length. Altogether, that is:

$$1.645 = \frac{x - .78}{.087}$$

1.645 (.087) + .78 = x = .9231m (the answer)

b) Would it be unusual to find a woman with a walking stride length between .62m and .74m? Explain.

Answer: Our initial sketch:

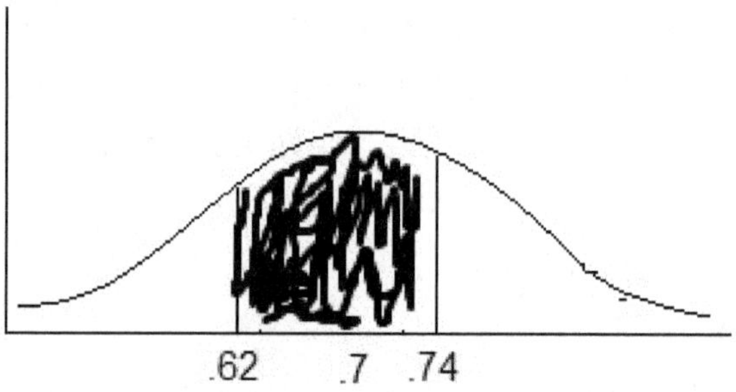

We get the z-score for female stride lengths .62m and .74m, then find the area between them using the methods of Section 5.1. (We anticipate that one z-score will be negative and the other positive.) The respective rounded z-scores are −.99 and .49, the area between them is .6879 − .1611 = .5268. So, this is telling us that 52.68% of women have walking stride lengths between .62m and .74m. Said another way, the proportion of women with walking stride lengths between .62m and .74m is .5268. Said another way, the probability of finding a woman with a walking stride length between .62m and .74m is .5268. To answer the question:

No, it would not be unusual to find a woman with a walking stride length between .62m and .74m because the probability of that event, .5268, is not less than .05.

c) What walking stride length for women separates the top 66% from the rest?

Answer: Sketch:

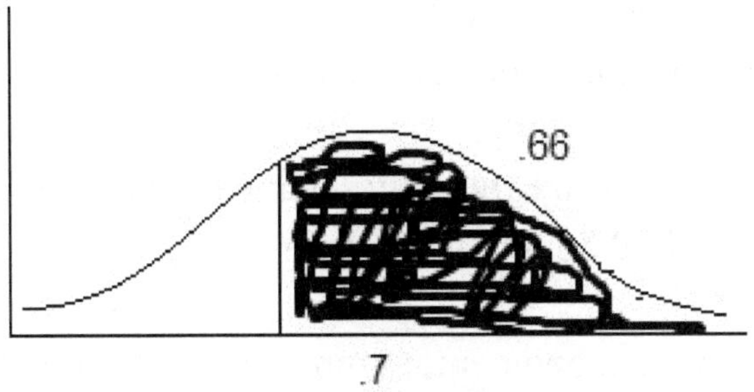

So we know our answer will be less than .7. Notice that even though we were given a percentage for the "top" we are on the left side of the diagram because that percentage is more than 50% (area > .5).

We will get the z-score that has area .66 to its right (or, .34 to its left), which we can see will be negative since we are to the left of the mean, and use that in the z-score formula to solve for the requested walking stride. That looks like:

$$-.41 = \frac{x - .7}{.081}$$

$-.41(.081) + .7 = x = .6668$m (the answer)

d) What proportion of men have walking stride lengths greater than .95m?

Answer: When you make the sketch you immediately realize that the answer will be less than .5 because you shade less than half of the diagram.

We get the rounded z-score, 1.95, and find the area to its right to arrive at the answer: .0256.

As was the case in the previous section, you should sketch a diagram for each problem and start by labeling what you know. Problems like those in this section generally are one of two types: 1) you are given a value from the data and you need to find area, or 2) you are given area and you need to find a value from the data.

Exercises

1) Suppose 8.4 oz. cans of a Kshzow! energy drinks have caffeine contents that are normally distributed with a mean of 80 mg and standard deviation .3 mg.

a) What is the probability that a randomly selected 8.4 oz. can of Kshzow! has between 79.62 mg and 80.26 mg of caffeine?

b) 58% of 8.4 oz. cans of Kshzow! contain more than how much caffeine?

c) Find the proportion of 8.4 oz. cans of Kshzow! that have caffeine contents below 79.22 mg.

d) Among 8.4 oz. cans of Kshzow!, what caffeine content separates the bottom 13% from the rest?

2) Suppose the lengths of softball games are normally distributed with mean 99.4 minutes and standard deviation 24.3 minutes.

a) .9909 is the proportion of softball games that last less than how long?

b) 7% of softball games last no longer than how many minutes?

c) Your friend has asked you to attend his softball game (ugghh...) but you have plans for later that evening. What is the probability that the game will go on for longer than two hours?

d) What percentage of softball games are completed within 90 minutes?

3) **For those who have had calculus**: Recall that the pdf for the normal distribution is

$$f(x) = \frac{1}{\sigma\sqrt{2\pi}} e^{-\frac{1}{2}\left(\frac{x-\mu}{\sigma}\right)^2}$$

Suppose the time it takes to perform the Star Spangled Banner before professional sporting events is normally distributed with mean 116 seconds and standard deviation 8 seconds. The pdf for this specific normal distribution looks like:

$$f(x) = \frac{1}{8\sqrt{2\pi}} e^{-\frac{1}{2}\left(\frac{x-116}{8}\right)^2}$$

We want to get the probability that a Star Spangled Banner performance before a professional sporting event lasts between 96 seconds and 104 seconds.

a) Use a Riemann sum with eight subintervals and, using left endpoints, estimate the desired probability.

b) Using the Trapezoidal Rule with eight subintervals, estimate the desired probability.

c) Using Simpson's Rule with eight subintervals, estimate the desired probability.

d) Using Table A and the methods of this section, find the desired probability.

5.3 Central Limit Theorem

We are now ready to consider a theorem that motivates much of the inferential statistics that we will soon learn—The Central Limit Theorem. A thorough discussion of this theorem requires an advanced mathematics and probability background. Like all theorems, it has a proof (and, as with many theorems, this proof has different versions). At this point, we want a general understanding of the implications of the theorem.

The theorem has different forms based on what conditions are imposed on the data, and all those forms can get quite technical in their presentation. Briefly and generally we can state the theorem in this way:

> Regardless of the distribution of the population, the sampling distribution of $\bar{x}$ (the distribution of all the means of simple random samples of size n) approaches a normal distribution with mean μ and standard deviation $\frac{\sigma}{\sqrt{n}}$ as the sample size, *n*, increases. (Technical side note: The theorem in its general form applies to the cumulative distribution function of the normal distribution but, after certain restrictions are imposed, it also applies to the density function [bell curve]. To make this section simpler, we will assume the density function application and sketch bell curves as we complete examples.) [Simple random samples are defined in Chapter 6.]

Here are guidelines resulting from the Central Limit Theorem:

1) If n > 30 then the simple random samples of size n have a distribution that can be approximated by a normal distribution with mean = μ and standard deviation = $\frac{\sigma}{\sqrt{n}}$.

2) If n ≤ 30 and the population is normally distributed, the simple random samples of size n have a normal distribution with mean = μ and standard deviation = $\frac{\sigma}{\sqrt{n}}$.

[The Central Limit Theorem itself does not specifically state a sample size of 30, but that is a rule of thumb that is widely used. The theorem just says that as the sample size gets bigger the distribution gets closer to normal. 30 has been shown to be enough "gets bigger" to consistently get trustworthy results from the normal distribution. Means from samples having sizes less than 30, sometimes even as little as five, might be well approximated by the normal distribution as long as no outliers are present.]

In Statistics, whenever we say, "approximated by" that means we will be using that distribution to get probabilities. For practical purposes, you could ignore "approximated by" because we will behave as if that was the exact distribution for the data.

If we are dealing with sample sizes less than 30, and we don't have reason to believe the population is approximately normal, then we are not sure that the normal distribution, with the stated values for the mean and standard deviation, would be an appropriate approximation for that distribution of sample means. Other methods would need to be used (non-parametric methods, bootstrapping, etc.) and they are covered in higher level texts and courses.

More notation:

$\mu_{\bar{x}}$ is the mean of the distribution of sample means.
[Under the conditions stated by the Central Limit Theorem, $\mu_{\bar{x}} = \mu$]

$\sigma_{\bar{x}}$ is the standard deviation of the distribution of sample means
[Under the conditions stated by the Central Limit Theorem, $\sigma_{\bar{x}} = \frac{\sigma}{\sqrt{n}}$]
$\sigma_{\bar{x}}$ is called the standard error of the mean

It's probably easiest to see the effect of the Central Limit Theorem, as we have discussed it thus far, with an example. We will revisit the CPE example from the previous section.

Example: Scores on the College Preparedness Exam (CPE) are normally distributed with mean 84.2 and standard deviation 11.6.

a) Find the probability that a randomly selected exam taker has a CPE score less than 85.6.

Answer: This is very much like Section 5.2 (by "very much like" we mean "exactly the same as"). We make our sketch and immediately see that the answer will be more than .5 because more than half of the diagram is shaded. We get the z-score corresponding to CPE score 85.6 (which, rounded to two places, is .12) and find the area to its left, giving a final answer of .5478.

The completed sketch looks like:

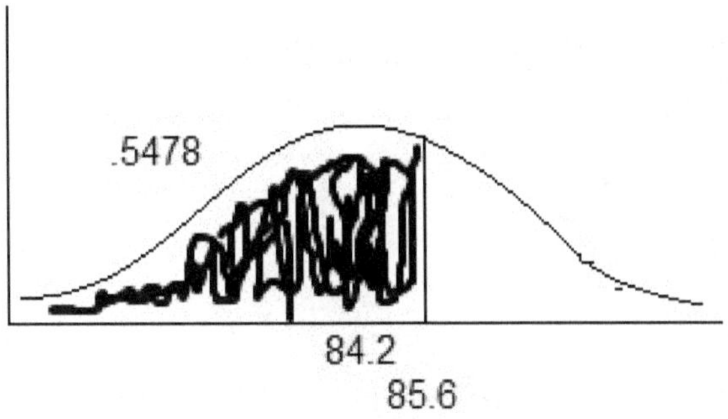

b) Find the probability that a sample of 47 exam takers have a mean CPE score less than 85.6.

Answer: OK, this is different from what we've been doing to this point. In a) of this example, and in all of Section 5.2, we were working in normally distributed populations. We were considering individual values from, or proportions and percentages from, the distribution of the whole population of data. Now we are in a different distribution: the distribution of sample means. Every time we take a different simple random sample of 47 CPE exam takers, its mean CPE score will be different. The distribution of those means, according to the Central Limit Theorem is normal with mean equal to the population mean and standard deviation equal to the population standard deviation divided by the square root of the sample size. We have a big sample size (more than 30) and the underlying population is normal, so the Central Limit Theorem definitely applies.

Please make sure you take a moment to acknowledge how a) and b) are totally different.

We would still begin with a sketch and it would look exactly the same as the first sketch you drew for a). So, again, we anticipate a final answer larger than .5. The big difference comes in how we compute the z-score. The initial formula looks like this:

$$z = \frac{\bar{x} - \mu_{\bar{x}}}{\sigma_{\bar{x}}}$$

We are not in the distribution of x (individual data values) but we are in the distribution of $\bar{x}$. The formula is still "a value from the distribution minus the mean of the distribution all divided by the standard deviation of the distribution", but the symbols are relevant to the $\bar{x}$ distribution.

If you write the formula more explicitly in terms of the numbers we are going to plug into it, you could write:

$$z = \frac{\bar{x} - \mu}{\frac{\sigma}{\sqrt{n}}}$$

since $\mu_{\bar{x}}$ has the same value as the original population mean and $\sigma_{\bar{x}}$ is equal to $\frac{\sigma}{\sqrt{n}}$.

Our z-score for this problem is: $z = \frac{85.6 - 84.2}{\frac{11.6}{\sqrt{47}}} = .8274$

When we use the rounded z-score .83 to get our answer, we find it's: .7967

<u>The answer is .7967</u>

The completed sketch, of our $\bar{x}$ distribution, is:

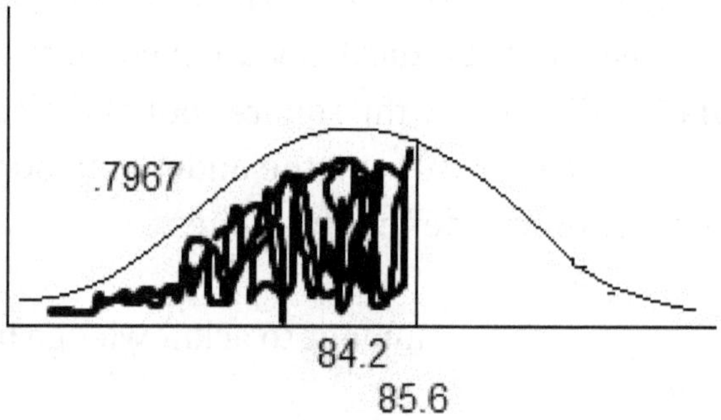

c) Find the probability that a randomly selected exam taker has a CPE score greater than 78.6.

Answer: Back to the Section 5.2 stuff. Your diagram shows you that our final answer will be more than .5.

The z-score is $z = \frac{78.6-84.2}{11.6} = -.4828$ (you knew that was going to be negative...right?!)

Area to the right of z-score −.48 is the final answer: .6844.

d) Find the probability that a sample of 16 exam takers have a mean CPE score greater than 78.6.

Answer: We immediately notice that this question is not about the population of CPE scores, but is about the population of sample mean CPE scores taken from (simple random) samples of size 16 exam takers. Even though the sample size is "small" (using our rule of thumb of 30), we were told that the original population of CPE scores is normal, so the Central Limit Theorem assures us that the population of sample means is also normal for samples of this size. Furthermore, the distribution of sample means has a mean equal to μ and a standard deviation equal to $\frac{\sigma}{\sqrt{n}}$. Armed with that knowledge, we can compute our z-score and go from there.

$$z = \frac{\bar{x}-\mu}{\frac{\sigma}{\sqrt{n}}} = \frac{78.6-84.2}{\frac{11.6}{\sqrt{16}}} = -1.9310$$

The area to the right of that z-score is .9732, our final answer.

Of course, you knew from your diagram that the final answer had to be more than .5.

The final sketch of the $\bar{x}$ distribution would look like:

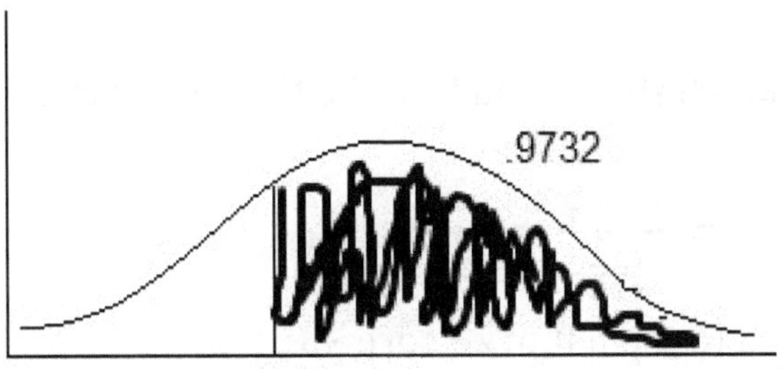

[compare this to the diagram you drew for c)]

Let's revisit the walking strides example.

Example: Suppose the average walking stride length for men is .78m with standard deviation .087m, and that for women the mean is .7m with standard deviation .081m, and that both populations are normally distributed.

a) If we got a sample of 57 men, what is the probability that the mean walking stride for those men does not exceed .748m?

Answer: [Whenever we refer to getting a sample in this book, we assume it's simple random.]

Alright, we see we are in the distribution of sample means. Our sample size is bigger than 30, plus the original population of men's walking strides is normally distributed, so we will answer the question based on the Central Limit Theorem. "Does not exceed" tells me this is a "less than" problem. Upon making my sketch, I see the answer must be less than .5. It's z-score time.

$$z = \frac{.748 - .78}{\frac{.087}{\sqrt{57}}} = -2.7770$$

The area to the left of z-score −2.78 is .0027, our final answer.

b) What percentage of samples of 21 women have a mean walking stride under .64m?

Answer: We are considering the distribution of means of walking strides for samples of 21 women. Our rule of thumb renders our sample size "small" but, not to worry, the population of all women's walking strides is normally distributed, so we can proceed on the basis of the Central Limit Theorem.

Your sketch shows you that the final answer will be less than 50%

$$z = \frac{.64 - .7}{\frac{.081}{\sqrt{21}}} = -3.3945$$

The area to the left of z-score −3.39 gives us the final answer of .03%.

Example: Customers who shop at Inimitable Ingrid's House of Style spend an average of $237.69 per visit with standard deviation $99.75. What is the probability that, in a sample of 136 visits to Inimitable Ingrid's, shoppers would have spent an average of more than \$256.76?

Answer: Notice that the problem does not tell us that the amounts spent at Inimitable Ingrid's are normally distributed. If we were asked a question regarding the probability of an amount spent during a single visit being in a certain interval, or if we were asked about the proportion or percentage for particular intervals of amounts spent at Inimitable Ingrid's on individual visits, we'd be out of luck. For the question we were asked, we do not know, or care, about the shape of the distribution of amounts spent on individual visits. We only care about the distribution of the mean amounts spent for samples of 136 visits to Inimitable Ingrid's. Do we know anything about that? Well, since the sample size is "big" (larger than 30) we know from the Central Limit Theorem that we can use a normal distribution with mean 237.69 and standard deviation $\frac{99.75}{\sqrt{136}}$ to obtain the requested probability.

$$z = \frac{256.76 - 237.69}{\frac{99.75}{\sqrt{136}}} = 2.2295$$

Area to the right of z-score 2.23 is our final answer, .0129.

(Based on your sketch, you are not surprised to get an answer less than .5.)

Thinking Beyond the Formula

As we learned, the standard deviation for the distribution of sample means is $\frac{\sigma}{\sqrt{n}}$, which is smaller than the standard deviation for the distribution of the whole population which is just σ. In other words, there is less variation among the sample means than there is among the individual members of the population. That makes sense. Think about heights. If you randomly go from one person to the next, you can encounter very tall people and very short people. Say you are walking through a shopping mall or a sports arena. You would walk by people of differing heights—there would be a lot of variation. However, if you take a sample of 200 people and calculate their mean height, and then another sample of 200 people and calculate their mean height, and so on, there won't be as much difference among the mean heights as there was among the individual heights. Most of your samples would probably have people near the average height in them. The couple of extremely tall or extremely short people that may appear in a sample would have a minimal effect on the average height for the sample.

Notice also that bigger samples have less variation among their means than smaller samples. (As n gets bigger, the denominator of $\frac{\sigma}{\sqrt{n}}$ gets bigger, making the whole quantity smaller.) That also makes sense. If you have an extremely tall or short person in a sample of size two, that would more greatly affect the sample mean than if that person was in a sample of size 200. So the larger samples will not vary as much in their means because extreme values generally have less of an effect on the mean in larger samples than in smaller samples.

Note that the Central Limit Theorem is predicated on the simple random samples being taken with replacement and/or from an infinite population. If the sample size is less than 5% of the population size, use of the theorem provides good results. When these conditions do not hold, an adjustment can be made to $\sigma_{\bar{x}}$, which we won't detail here, but just be aware that it exists.

Exercises

1) Suppose 8.4 oz. cans of a Kshzow! energy drinks have caffeine contents that are normally distributed with a mean of 80 mg and standard deviation .3 mg.

a) What is the probability that a randomly selected 8.4 oz. can of Kshzow! has more than 80.0597 mg of caffeine?

b) What is the probability that 118 randomly selected 8.4 oz. cans of Kshzow! have an average of more than 80.0597 mg of caffeine?

c) What percentage of samples of 118 8.4 oz. cans of Kshzow! have an average of more than 80.0597 mg of caffeine?

d) Among samples of 22 8.4 oz. cans of Kshzow!, what mean caffeine content separates the bottom 13% of sample means from the rest?

2) Suppose the lengths of softball games are normally distributed with mean 99.4 minutes and standard deviation 24.3 minutes.

a) If you attend 10 softball games at random, what is the probability that they have a mean length less than 94.95 minutes or more than 102.97 minutes?

b) If you attend 102 softball games at random, what is the probability that they have a mean length less than 94.95 minutes or more than 102.97 minutes?

3) Suppose commuters' one-way distances to work have a mean of 15.41 miles with standard deviation 17.3 miles.

a) If 235 commuters are randomly selected, what is the probability that their mean commute distance is between 15.67 miles and 18.84 miles?

b) If three commuters are randomly selected, what is the probability that their mean commute distance is less than 36.78 miles?

6 Inferential Statistics Ideas

Suppose you wanted to know the mean age of all people who own a car. To get that mean, you could visit all people on earth. For those who own a car, record their age. Add all those ages, divide by how many people on earth own a car, and that would be the mean you wanted!

Suppose you wanted to know the standard deviation of the weights of all water bottles. You could find every water bottle, weigh it, use one of the standard deviation formulas (for a population) that you learned previously, and that would be it!

Suppose you want to know if there is a different rate of depression among Americans compared to Canadians? Well, simply visit all Americans and all Canadians, test each person for depression, and compare the results.

Are certain jobs associated with greater anxiety? We could easily answer that question by interviewing everyone who has a job. In the interview we will determine what kind of job each person has and we will also administer an anxiety assessment. When we are done, we can compare all of our numbers and perhaps make some histograms and other displays of data.

Does one drug generally do a better job of treating headaches than another drug? To address this question, every time any person anywhere gets a headache, you can give the person one drug. The next time each person gets a headache you can give him or her the other drug. You can keep track of all people and when you gave them which drug, how much you gave, how long it took for headaches to become 75% as bad, 50% as bad, 25% as bad, and then completely gone. After you get all those numbers you can make comparisons.

Wait a minute.

While it is easy to describe hypothetically how we would address these and infinitely many other research questions, in practice, though, it would be quite difficult, if not impossible, to actually perform the work described. With most research questions of any significance and real interest, the populations being studied are too big to get the information you want from each individual member. In other classes you will learn about case studies in which one member of the population, hopefully a stellar example of a typical member of the population, is studied and observed in great detail. While this approach has its merits, there are significant limitations. Another approach is to use the inferential statistics methods that comprise the rest of this text.

The basic idea behind inferential statistics is this: We take a sample from a population, get the information we need for our research question from the members of the sample, and then use that information to make an inference about our research question for the whole population.

We can't visit everyone on earth who owns a car so we get a sample of people who own a car, calculate the mean age of those people, and then use that information to make an inference about the mean age of all people who own a car.

We can't weigh all water bottles so we get a sample of water bottles, calculate the standard deviation of the weights of those bottles, and then use that information to make an inference about the standard deviation of the weights of all water bottles.

We can't test all Americans and all Canadians for depression so we get a sample of Americans and a sample of Canadians, get the rate of depression for each sample, and then use that information to make an inference about a comparison of the rate of depression for all Americans and Canadians.

We can't interview everyone who has a job so we get a sample of people who have a job, observe the type of job for each of those people and measure the anxiety of each of those people, and then use that information to make an inference about the relationship between job type and anxiety for all employed people.

We can't administer drugs to all people each time they get a headache so we get a sample of people, administer the test drugs to them when they experience headaches and measure the reduction of their headache, and then use that information to make an inference about the reduction of all people's headaches who use the drugs.

As you can imagine, we can't take the results from the sample and conclude that the whole population is EXACTLY like the sample we got. If the mean age of our sample of people who own a car is 36.12 years, it's unreasonable to expect that the mean age of all people who own a car is EXACTLY 36.12 years. There is sampling error (the discrepancy between the sample value and the true population value). To get our inference, we will take the 36.12 and do stuff with it (hmmm...that doesn't sound very official—let's make that, "perform some mathematically sound procedures on it"—yeah, much better) in order to get a well-supported conjecture about the entire population from which the sample was taken. The rest of this text and the Statistics course you are taking are all about the stuff (mathematically sound procedures) we perform to get those inferences about populations.

Inferences come in two forms: hypothesis tests and confidence intervals. Strictly speaking, confidence intervals are for the purpose of making estimates of parameters while hypothesis tests are for the purpose of testing ideas about parameters. Recall from an earlier section that parameters are numbers that describe whole populations.

Note that, if you have the complete information from the entire population, you don't need the statistical methods you are about to learn. As mentioned though, it is extremely rare to get the information from the whole population so these methods that you will learn are used daily by researchers all over the world. You may become one of those researchers yourself one day.

Research questions from business, economics, social science, psychology, political science, administration of justice, life science, physical science, health science, information technology, education, and many other areas are addressed using inferential statistics. That is why students are required to take the Statistics class in which you are currently enrolled. Try to master the material in this course because, to the extent you do, you will find later classes so much easier and actually more enjoyable. I personally observed many students (including fellow classmates of mine when I was a student) struggle in upper division and graduate level classes because they did not get the solid foundation available to you in your current Statistics course.

In order to get useful inferences, we need to start with useful sample data. If we start with a bad sample, no amount of statistical procedures will fix that problem. There is a whole separate field with textbooks and bodies of published research that deals with sampling techniques as well as research design, some of which you will likely learn in later classes. Also, there are conventions and preferences in different fields. For now, though, we will briefly consider basic methods of acquiring data.

Obtaining Data

Your research is usually either an observational study or an experiment. In an observational study, you simply examine the members of your sample and record what is relevant to your research. With an experiment, you are attempting to change something about the elements of your sample. If we survey a bunch of people and ask how long they have been at their current job, how they enjoy their job, etc. that is an observational study. If we are testing a program that we hope will improve the reading ability of first graders and we test their initial ability, put some through the program while giving some regular instruction for a month, and then test their ability again after the program, then compare the reading ability for the groups, that is an experiment. With an experiment we manipulate the conditions but, with an observational study, we merely record the conditions.

Studies can be cross-sectional, retrospective, or longitudinal. In a cross-sectional study, data is collected at one point in time. For example, you might find a bunch of elementary school children and measure their current academic achievement. In a retrospective study, data is collected from the past. For example, for a group of 30-year-olds you may go back to the elementary school they attended and retrieve records regarding their academic achievement. In a longitudinal study, you collect data from subjects over time. For example, you collect academic achievement data from a group of elementary school children and continue to do so each year until they graduate from college or turn 30 (whichever comes first). In this sort of study you often follow multiple groups (say, students from different cities or school districts), called *cohorts*, in order to compare outcomes.

There are different ways to choose which elements from the population will be members of your sample. In other words, there are different types of samples:

probability sample: each member of the population has a known probability of being selected

random sample: each member of the population has an equal chance of being selected

simple random sample: every sample of a given size, n, has an equal chance of being selected

Suppose there are five first grade classes at a school, each with the same number of students, and they all take recess at the same time. For your sample, you (an education researcher) will select a sample of 10 of this school's first graders from the playground. That would be a simple random sample, where n is 10. Every sample of size 10 first-graders from the school has the same chance of being selected. Now suppose that the five classes all have recess at different times. You randomly select during which recess time you will get your sample and, at that time, you get a sample of size 10 first-graders. Well, each student still has the same chance of being in the sample, but not all samples are equally likely (there is no chance of getting a sample of students from different classes). This time, suppose that again all the first-graders have recess at the same time, but some students have to use some recess time to catch up on class work they didn't finish before recess. Only when their work is done can they go out to recess. Now we have a probability sample because, depending on when you select your sample, some students have a smaller chance of being selected for the sample. The gold-standard for research is a simple random sample but practical considerations do not always make this possible. The second best is a good random sample. Notice that a random sample is a specific type of probability sample, and a simple random sample is a specific type of random sample. (In some texts, no distinction is made between random samples and simple random samples. Truly random samples are generally simple.)

Within those descriptions of types of samples, we can make further distinctions:

systematic sampling: after selecting a starting point, every k^{th} element is selected for the sample

convenience sampling: select people who are easy to get

stratified sampling: split the population into subgroups (usually of like elements), then randomly select elements from within each subgroup

cluster sampling: split the population into subgroups (usually of like elements), then randomly select which of the entire subgroups will be your sample

Let's now consider getting a sample of first graders from the whole state of Maryland. If we're using a systematic sample we could list all first graders in Maryland and then select, say, every 78th first grader in the list to be in our sample. Suppose you personally know a first grade teacher in Maryland. As a favor, you ask her to let her class be your sample—that's a convenience sample. Generally speaking, a convenience sample is a horrible choice of sampling technique due to its non-randomness, but the limitations of real-life sometimes result in a research structure that is less than ideal. (Another non-random sampling method is snowball sampling. Here, one member of the sample suggests to the researcher, or personally recruits, another member, and so on. When the individuals being studied are rare or hard to locate this is sometimes used.) We could consider all the elementary schools in Maryland and randomly select five first graders from each school. That would be a stratified sample (as you probably know, a "stratum" is a level or layer). Another choice would be to randomly select five of the elementary schools in Maryland and then include every first grader in each of those schools to be in our sample. That would constitute a cluster sample.

There are advantages and disadvantages to each of these standard methods of obtaining data. The specifics vary by situation. Some may not give you samples that are likely to be representative of the population (like, your friend might teach an advanced first grade class that is not representative of the abilities of typical first graders in Maryland). If you use clustering, and your clusters are quite different and you choose a relatively small number of them, your sample may completely miss certain features of your population. For a particular situation, a stratified sample might give you a good sample, but reaching the different strata (like, the different schools) might require a large investment of time and other resources. The practical considerations of the specific research being done and the conventions of the research field will often dictate which sampling method is used and, if you are going into any research field, you will spend lots of time discussing very important issues relating to research methods. At the very least, you want a sample that is representative (includes all the types of characteristics found in your population) and is chosen through some random selection process.

Another important issue in research is how to know the right data to get for a research question. Suppose a sociologist wants to know what makes good families. What kind of data should be acquired to answer that question? Is that even a good question to ask? What is a "good family"? What data would indicate someone has a "good family"? How would we know that the data would show what is responsible for a family being good? These sorts of issues would need to be addressed far before data collection began. Again, upper division classes shed light on the formulation of good research questions and how to collect appropriate data for those questions.

As indicated, this discussion can get far more detailed and sophisticated, but this is just to give you some background into acquiring data for studies. As is the case with our coverage of probability, this is just enough for you to be able to grasp the rest of this course. There are scores of other books that cover sampling techniques and research design. In the applications presented in the book, just assume that the data was correctly acquired for the situations presented.

Before We Go

As we move into inferential statistics, here are a few things to keep in mind. The interpretations and conclusions we get from confidence intervals and hypothesis tests are always about populations, NEVER samples. When we get a sample, we are able to get all the information from it that we want. There is nothing unknown or uncertain about the sample for which we need to make an estimate or inference. What is unknown is what is happening in the entire population. In fact, that was our sole motivation for getting a sample—to use the information we got from it to make an inference about the unknown population information. For that reason, your inferences should never have the word "sample" in them (this is for you as entry-level Statistics students). When beginning Statistics students have the word "sample" in their inferences they are usually indicating that the statistical process they just used was to make an inference about the sample, and that is always completely wrong.

Applications in the text will always be greatly oversimplified so we can concentrate on illustrating and practicing the relevant basic statistical method being discussed.

Our discussion of inferential statistics begins with confidence intervals.

7 Confidence Intervals

7.1 Confidence Interval Foundations

A confidence interval is an estimate. That's it.

Suppose we wanted to know the mean age of all people who own a car. Since we aren't able to get the ages of all people who own a car, we decide to get a sample of people who own a car, calculate the mean age of those people, and then use that information to make an estimate about the mean age of all people who own a car. In the end, we could state our estimate as something like, "The mean age of all people who own a car is between 30.851 and 41.389 years." Assuming that estimate came from using the correct mathematical procedures, that would be a confidence interval. Using notation we learned in algebra, we could write the interval in the form: $30.851 < \mu < 41.389$. We learned already that μ always stands for the mean of a population. For this specific situation, μ stands for "the mean age of all people who own a car."

In the previous example, the mean age for the people in the sample is called a point estimate. A point estimate is a statistic calculated from sample data and it is the best single estimate of the analogous parameter. Due to sampling error, we are not satisfied with this single estimate so we build an interval around it to increase the chance that our estimate contains the true value of the parameter.

It's important to keep in mind that all of the values in a confidence interval are considered equally likely values of the parameter. In our mean age of car owners example, any age strictly between 30.851 and 41.389 is a likely value of the parameter. 30.89, 35.87, 37.21, 41.30 are all considered reasonable estimates for the mean age of all car owners. If we place no limitations on decimal places, there are infinitely many values that are included in our interval. Values like 30.851, 29.74, 21.22, 41.389, 47.65, 55.23 are NOT plausible values of the mean age of all car owners because they are not included in our interval. (Confidence intervals involve strict inequalities; there's no "or equal to" included with the "less than" symbols.)

Each process that is used to create a confidence interval has a certain success rate. The process correctly captures the true value of the parameter a certain percentage of the time. That success rate is called the confidence level. Confidence levels are usually in the 80's or 90's as percentages (like, 89%, 92%, 95%, for example). Alpha, "α", is the complement of the confidence level as a decimal. For example, if the confidence level is 93%, $\alpha = 1 - .93 = .07$.

The process used for making a confidence interval for a parameter starts with the distribution of its point estimate. The distribution of the point estimate is known as long as the data meets certain conditions. [These conditions will be mentioned when each process is discussed, but you can assume that all problems in this book meet the necessary conditions. In practice, the processes we will discuss generally provide useful results as long as all necessary conditions are close to being met, if not perfectly. When doing legitimate research the conditions should be verified and, when data grossly fails to meet necessary conditions, there is a whole body of other methods—bootstrapping and nonparametric statistics—that should be used. There are upper division and graduate level classes and textbooks for these special topics.] Based on the relevant distribution and the confidence level we are using, we can employ algebra to help us create the interval. How this is done for specific types of confidence intervals will be demonstrated in detail for a couple types of intervals.

Sometimes, a confidence interval is just used to make an estimate, and that is all. Other times, a confidence interval can be used to informally address a research question (we will look at formal methods of testing hypotheses about parameters later). For example, suppose someone claimed that the average age of car owners is 45.6 years. Based on our confidence interval, we would conclude that the claim appears to be too high because it is larger than all of the numbers in our interval, $30.851 < \mu < 41.389$. In addition to making confidence intervals, we will also practice informally testing claims using them.

Like most introductory Statistics books, this one focuses only on two-sided confidence intervals, but more advanced books and statistical software packages deal with one-sided confidence intervals as well.

Let's dive in to the first type of confidence interval we will consider: confidence intervals for proportions.

7.2 Confidence Intervals for Proportions

Conditions: The data is from a simple random sample. The requirements for the binomial distribution (given in Chapter 4) are met. There are at least five successes and five failures. (Assume all conditions are met for each example and exercise.)

Use the methods of this section when...

A) creating a confidence interval for a proportion

B) determining the minimum sample size needed for a specified estimate of a proportion

Important formulas: $E = z_{\alpha/2}\sqrt{\frac{\hat{p}\hat{q}}{n}}$ $\qquad \hat{p} - E < p < \hat{p} + E$

$$n = \frac{\left(z_{\alpha/2}\right)^2(.25)}{E^2} \qquad n = \frac{\left(z_{\alpha/2}\right)^2\hat{p}\hat{q}}{E^2}$$

We need to begin by introducing new notation. As usual, when new notation is introduced, start to **memorize** it right away because it will keep reappearing throughout the course!

p proportion of successes in the population

x number of successes in the sample

$\hat{p}$ proportion of successes in the sample (pronounced "p-hat")

$\hat{q}$ proportion of failures in the sample (pronounced "q-hat")

A "success" is whatever is the topic of the research and a "failure" is anything else. As is true with most notation, it has a general meaning but it takes on a specific meaning for each problem. You must be clear what the specific meaning is for the problem in front of you in order to understand what you are doing.

Example: In a sample of 219 Statsville drivers, 181 had a used car as their first car. Identify p, x, $\hat{p}$, and $\hat{q}$ for this problem.

> **Answer**: p is the proportion of all Statsville drivers who had a used car as their first car (value unknown). x is the number of Statsville drivers in the sample who had a used car as their first car, which equals 181. $\hat{p}$ is the proportion of Statsville drivers in the sample who had a used car as their first car, which is $\frac{181}{219} = .8265$. $\hat{q}$ is the proportion of Statsville drivers in the sample who did not have a used car as their first car, which is $\frac{38}{219} = .1735$.

Since this research is about Statsville drivers who had a used car as their first car, and p is the proportion of successes in the population, here p stands for "the proportion of all Statsville drivers who had a used car as their first car". A "success" in this research is having a used car as the first car. It's not that we love used cars, or that anything is wrong with new cars but, since we're studying having used cars as the first car, that constitutes a success. So, we had 181 successes in the sample. To get the proportion of successes, we just take the number of successes divided by the sample size. Since we had 181 successes in the sample, the number of failures is 219−181 = 38. The proportion of failures would be the number of failures divided by the sample size.

Notice that there is a huge difference between the *number* of something and the *proportion* of something.

As is always the case when we are using inferential statistics, we do not know the population value (if we knew that, there would be nothing to infer).

If you were told that, for all Statsville drivers, 181000 out of 219000 of them had a used car as their first car, p would be $\frac{181000}{219000} = .8265$. Obviously, if we knew that, there would be no inferential statistics needed because we would already know the parameter (the number describing the whole population). For that reason, the value of p will not be known in our problems but we will have to describe it in words.

As you learned previously, the sum of all proportions has to be one. So, since we only have two groups, successes and failures, the proportion for one group could be obtained by subtracting the proportion for the other group from one. So,

$$1 - \hat{p} = \hat{q}$$

Example: In a sample of 219 Statsville drivers, 82.65% had a used car as their first car. Identify p, x, p̂, and q̂ for this problem.

> **Answer**: p is the proportion of all Statsville drivers who had a used car as their first car (value unknown). x is the number of Statsville drivers in the sample who had a used car as their first car, which is (.8265)(219) = 181.0035, rounded to 181. p̂ is the proportion of Statsville drivers in the sample who had a used car as their first car, which (converting the given percentage) is .8265. q̂ is the proportion of Statsville drivers in the sample who did not have a used car as their first car, which is 1−.8265 = .1735.

As you no doubt observed, this is the same as the previous example but with the information provided in a different way. This time we were given the percentage of successes in the sample instead of the number of successes. As you learned previously, if we know a percentage we can easily convert that to a proportion by moving the decimal two places to the left (that's a shortcut for dividing by 100). So this time we did not have to calculate p̂ because it was essentially given in the problem. To get q̂ we used the $1 - \hat{p} = \hat{q}$ formula. Since x stands for a number of people, it clearly could not be a decimal. When we multiplied the proportion by the total number of people (as you learned to do in previous math classes) we knew to round to the nearest whole number. The reason we did not get a whole number from the multiplication calculation itself is because the given 82.65% was rounded. In most problems, once we have p̂ we won't care about x.

Rationale for creating a (1−α)100% confidence interval for p

As you can imagine, whenever you take a different sample from a population, and calculate the relevant proportion of successes for that sample, p̂, there is a different p̂ that comes from each sample. When the stated conditions for this section are met, those p-hats are approximately normal with a mean of p and standard deviation of $\sqrt{\frac{pq}{n}}$. (Whenever we say something "approximately" follows a certain distribution, like the normal distribution, it means we are going to use that distribution. For all practical purposes, you can ignore the "approximately" and just use the distribution. This will occur throughout the text.)

To create a (1−α)100% confidence interval for p, we start by visualizing the distribution of p-hats as follows:

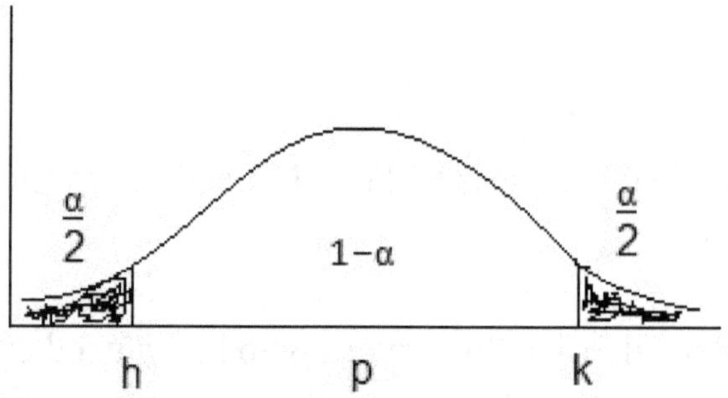

We focus our attention on the middle (1−α) proportion of the p-hat distribution, where α is the small proportion of unusually small and unusually large p-hats. p is the mean of the distribution with h and k being the p-hats separating the middle (1−α) proportion of p-hats from the rest. The z-score that corresponds to h has the symbol $-z_{\alpha/2}$ and the z-score for k would be $z_{\alpha/2}$, based on the z-score notation you learned previously (like, if $z_{\alpha/2}$ is z-score 1.96, then $-z_{\alpha/2}$ would be z-score −1.96). Using the z-score formula we have:

$$-z_{\alpha/2} = \frac{h-p}{\sqrt{\frac{pq}{n}}} \quad \text{and} \quad z_{\alpha/2} = \frac{k-p}{\sqrt{\frac{pq}{n}}}$$

solving for h and k gives:

$$h = p - z_{\alpha/2}\sqrt{\frac{pq}{n}} \quad \text{and} \quad k = p + z_{\alpha/2}\sqrt{\frac{pq}{n}}$$

So, any p-hat in the middle (1−α) proportion of all the p-hats is between h and k which also means for any such p-hat:

$$p - z_{\alpha/2}\sqrt{\frac{pq}{n}} < \hat{p} < p + z_{\alpha/2}\sqrt{\frac{pq}{n}}$$

Use the algebra that you learned in previous classes to get p in the middle of the compound inequality by itself. After those steps, you get:

$$\hat{p} - z_{\alpha/2}\sqrt{\frac{pq}{n}} < p < \hat{p} + z_{\alpha/2}\sqrt{\frac{pq}{n}}$$

So, at this point we know that p, the population proportion of successes, is between the quantity on the left of the compound inequality and the one on the right of the compound inequality. Those are not values we could calculate, though, because the radicals contain p which is the unknown value we are attempting to estimate. In their stead we need to use the best point estimate of p, which is $\hat{p}$, to get:

$$\hat{p} - z_{\alpha/2}\sqrt{\frac{\hat{p}\hat{q}}{n}} < p < \hat{p} + z_{\alpha/2}\sqrt{\frac{\hat{p}\hat{q}}{n}}$$

To make our confidence interval for p, we will get a sample and use the indicated information from it to fill in that formula. That's our confidence interval!

[In this process we are hoping that the sample we select has a p-hat that comes from the middle $(1-\alpha)$ proportion of all the p-hats. That will usually be the case because we use pretty large $(1-\alpha)$ proportions. When we happen to get a sample with a p-hat not in the middle $(1-\alpha)$ proportion of all the p-hats, the confidence interval fails to contain the true value of p. $(1-\alpha)100\%$ confidence intervals contain the true value of p $(1-\alpha)100\%$ of the time. If $(1-\alpha) = .95$ then $(1-\alpha)100\% = 95\%$, so 95% confidence intervals for p contain the true value of p 95% of the time.]

The process outlined above is simplified by using these steps:

<u>Steps for creating a $(1-\alpha)100\%$ confidence interval for p</u>

1) Find the critical value, $z_{\alpha/2}$, that corresponds to the desired confidence level ($z_{\alpha/2}$ is a positive value).

2) Evaluate the margin of error, $E = z_{\alpha/2}\sqrt{\frac{\hat{p}\hat{q}}{n}}$

3) Form the interval $\hat{p} - E < p < \hat{p} + E$

4) Interpret the interval in real world terms.

5) Respond to any given research question.

As used in this text, "research question" does not necessarily indicate an interrogative followed by a question mark. The research question is what we are trying to investigate, test, or determine. It's the research-motivated reason for the statistical procedure we are using.

[There are other more tedious (and generally better performing) methods for making confidence intervals for proportions which are used in some software. Also, there are special procedures when dealing with cases in which the "five successes five failures" condition is not met. When making confidence intervals by hand in this text, the method in this section using the normal distribution will be used as is typical for introductory Statistics classes.]

Example: Randomly selected employed Americans were asked if they are employed by the government or a private employer. 62.2047% of the 127 people surveyed were employed by a private employer. The mean age of the sample was 37.5. 13.4% of respondents reported being "very dissatisfied" with their current employment and 20 respondents reported having searched for a job in the past year. Create a 90% confidence interval that can be used to investigate the claim that the majority of employed Americans are employed by private employers. Respond to the claim.

Answer:

1) First, we need $z_{\alpha/2}$ and we get that as follows:

$$\alpha = 1 - .90 = .10 \qquad \frac{\alpha}{2} = \frac{.10}{2} = .05 \qquad \text{from Table A } \; z_{\alpha/2} = z_{.05} = 1.645$$

2) Next, we get the margin of error:

$$E = z_{\alpha/2}\sqrt{\frac{\hat{p}\hat{q}}{n}} = 1.645\sqrt{\frac{(.622047)(.377953)}{127}} = .07078$$

(recall that $\hat{q} = 1 - \hat{p}$ so here $\hat{q} = 1 - .622047 = .377953$)

3) Now, we form the interval: $\hat{p} - E < p < \hat{p} + E$

$$.622047 - .07078 < p < .622047 + .07078$$

$$.5513 < p < .6928$$

4) Interpretation: With 90% confidence we can say that the proportion of all employed Americans who are employed by private employers is between .5513 and .6928.

5) Respond to research question: Since the entire interval is greater than .50, this supports the claim that the majority of employed Americans are employed by private employers.

Notice that, using the interval notation you learned in your algebra class, this interval could also be expressed as (.5513, .6928). This notation is used in other books and also in some statistical software, like Minitab.

This interval could also be written as .622047 ± .07078.

What exactly is this result communicating? Well, we don't know the exact proportion of all employed Americans who are employed by private employers, but we estimate that proportion to be a number between .5513 and .6928. If we surveyed every employed American, counted the number employed by private employers, divided that number by the total number of employed Americans, the answer to that division problem would be somewhere between .5513 and .6928. Any of the numbers in that interval are equally likely estimates for the true population proportion. Is it possible that the true population proportion is outside of the interval? Yes, that is possible, but we consider that to be unlikely and, for all intents and purposes, we rule out any values not inside the interval. For this problem specifically, since we were investigating a "majority", the whole interval would need to be above .5 for the claim to be supported. If any part of the interval fell at or below .5, the "majority" claim would not be supported.

There was some information provided in the problem that we did not use to get the specific confidence interval that was requested. Ignore it and move on. That other information could certainly be used to perform other statistical analysis, so we would need to know what is needed to answer certain research questions and what is not. There is always "extra information" when you answer a research question using data and statistical methods. Get used to it.

Whenever you use a formula, each symbol means something very specific for that problem and you should take a moment to identify what each symbol means. This makes it easier to know which formula to use, how to use it correctly, and what the results mean. For this problem,

$\hat{p}$ is the proportion of the employed Americans in the sample who are employed by private employers

$\hat{q}$ is the proportion of the employed Americans in the sample who are not employed by private employers

n is the number of employed Americans in our sample

p is the proportion of all employed Americans who are employed by private employers

$\hat{p}$, $\hat{q}$, and n are all known and are used to estimate the unknown parameter p. Why is p unknown? Because we did not go to all employed Americans and ask their employer status (private or government) so the best we could do is get the estimate shown in the confidence interval.

Not unique to this problem, E is the margin of error and α is the probability of the process we used to get this interval resulting in an interval that fails to include the true value of the parameter. α is not a measure of this specific interval but, rather, of the process used to create the interval. We discussed $z_{\alpha/2}$ in detail in Section 5.1. The meaning of the notation discussed in this paragraph is the same for every confidence interval.

Suppose we selected a different sample of 127 employed Americans. Likely we would find a slightly different proportion in that sample who are employed by a private employer. Therefore, we would get a slightly different 90% confidence interval than what our original sample gave us. Suppose we repeated this process of getting a sample and making a 90% confidence interval until we had 100 confidence intervals. How many of those 90% confidence intervals would contain the true proportion of all employed Americans who are employed by private employers? About 90. Suppose we repeated the process until we had 200 confidence intervals. How many of those would have the true population proportion? About 180. In other words, about 90% of those 90% confidence intervals would contain the true population parameter. That's what we mean by the "success rate" of the procedure. [Note: We say "about" because the population undoubtedly does not perfectly meet the conditions for this interval creation process, plus the confidence level describes the success rate for confidence intervals as a whole not just for this one research situation, so we may not get exactly 90 or 180 intervals as stated. Also, note that in a practical research situation we would not get 100 or 200 different samples, but the largest single sample we could get—this paragraph was merely to illustrate what the confidence level communicates.]

Example: Randomly selected current Statsville College students were asked how many units they have completed so far. Their responses were:

32	38	44	51	59	38	54	32	47	38	61	54

Of the students surveyed, 42% intend to major in Statistics and the standard deviation of the GPAs for everyone in the sample is .8017 with mean 2.4583. At Wudmoor College, 65% of the students have completed less than 50 units. 10 years ago, Statsville College students completed an average of 55 units with standard deviation 15.5 units. Create a 91% confidence interval for the percentage of current Statsville College students who have completed less than 50 units. Use your confidence interval to see if there is a difference between the percentage of Statsville College students who have completed less than 50 units and the percentage of Wudmoor College students who have completed less than 50 units.

Answer:

1) First, we need $z_{\alpha/2}$ and we get that as follows:

$\alpha = 1 - .91 = .09$ $\quad \frac{\alpha}{2} = \frac{.09}{2} = .045$ $\quad$ from Table A $z_{\alpha/2} = z_{.045} = 1.7$

2) Next, we get the margin of error:

$$E = z_{\alpha/2}\sqrt{\frac{\hat{p}\hat{q}}{n}} = 1.7\sqrt{\frac{\left(\frac{7}{12}\right)\left(\frac{5}{12}\right)}{12}} = .2419$$

(seven of the data values are less than 50 so $\hat{p} = \frac{7}{12}$)

3) Now, we form the interval:

$$\hat{p} - E < p < \hat{p} + E$$
$$\frac{7}{12} - .2419 < p < \frac{7}{12} + .2419$$
$$.3414 < p < .8252$$

4) Interpretation: With 91% confidence we can say that the percentage of all Statsville College students who have completed less than 50 units is between 34.14% and 82.52%.

5) Respond to research question: Since 65% (the percentage of students at Wudmoor College who have completed less than 50 units) is in our interval, there does not appear to be a difference between the percentage of Statsville College students who have completed less than 50 units and the percentage of Wudmoor College students who have completed less than 50 units.

$\hat{p}$ is the proportion of current Statsville College students in the sample who completed less than 50 units, x is the number of current Statsville College students in the sample who completed less than 50 units, n is the total number of current Statsville College students in the sample, $\hat{q}$ is the proportion of current Statsville College students in the sample who completed 50 or more units, and p is the proportion of all current Statsville College students who completed less than 50 units. Notice that the information provided for Wudmoor is for the whole population—if that information was pertinent to only a sample there would be some indication of that. When reading problems and, more importantly, when assessing real world situations, it will be really important to know if the information you have is for a population or a sample.

Suppose the confidence interval was $.1324 < p < .2281$. Since the entire interval contains only values that are less than .65, we would be confident that the proportion (and, therefore, the percentage) of current Statsville College students who completed less than 50 units is less than the proportion for Wudmoor. Suppose the confidence interval was $.7033 < p < .8843$. Since the entire interval contains only values that are greater than .65, we would be confident that the proportion (and, therefore, the percentage) of current Statsville College students who completed less than 50 units is greater than the proportion for Wudmoor. As long as .65 is somewhere (anywhere) within the interval, there is no evidence that there is any difference in the proportions for the two schools.

Notice that there are different ways that $\hat{p}$ can be obtained. If you are given the percentage of successes in the sample, convert that to a proportion (move the decimal two places to the left). If you are given the raw data, you count the number of successes then divide that number by the sample size. A problem may give you the number of successes in the problem description as well as the sample size in which case you divide the number of successes by the sample size to get $\hat{p}$. No matter what, like any proportion in Statistics, $\hat{p}$ is always between 0 and 1 inclusive. A common mistake is to use the number of successes in formulas instead of the proportion of successes. p and $\hat{p}$ always represent <u>proportions</u>—not percentages, and not counts ("how many").

What would be wrong with this interpretation? "With 91% confidence we can say that the percentage of Statsville College students in the sample who have completed less than 50 units is between 34.14% and 82.52%." The confidence interval is estimating the percentage for the entire population, not the sample. We know the exact percentage for the sample—it's 58% ($\frac{7}{12} \times 100$). What we are estimating is something we don't know—the percentage for the whole population. The sample percentage is one of the numbers in the interval for the population percentage because of the procedure used to create the interval, but the interval itself is making an estimate for the whole population.

Be sure to avoid a few common mistakes when making an inference for this type of problem. Notice that we did not compare $\frac{7}{12}$ to the interval for p to get our response to the research question. The sample proportion of success is always in the interval for the population proportion of successes (because of how the interval is made) so that observation has no usefulness when trying to make an inference. Also, we did not compare $\frac{7}{12}$ to .65. $\frac{7}{12}$ is a sample number but .65 is a population number so they cannot be compared in order to make an inference.

Example: A dietician interviewed 192 people and asked if they find the texture of cottage cheese to be gross. 115 answered "yes".

a) State the meaning of p for this problem.

b) State the meaning and value of x, $\hat{p}$, and $\hat{q}$.

c) Make a 96% confidence interval using this sample data.

Answer:

a) p stands for the proportion of all people who find the texture of cottage cheese to be gross.

b) x is the number of people in the sample who find the texture of cottage cheese to be gross and that value is 115. $\hat{p}$ is the proportion of people in the sample who find the texture of cottage cheese to be gross and that value is $\frac{115}{192} = .5990$. $\hat{q}$ is the proportion of people in the sample who do not find the texture of cottage cheese to be gross and that value is $1 - \frac{115}{192} = \frac{192}{192} - \frac{115}{192} = \frac{77}{192} = .4010$.

c) 1. First we need $z_{\alpha/2}$ and we get that as follows:

$\alpha = 1 - .96 = .04$ $\frac{\alpha}{2} = \frac{.04}{2} = .02$ from Table A $z_{\alpha/2} = z_{.02} = 2.05$

2. Next, we get the margin of error:

$$E = z_{\alpha/2}\sqrt{\frac{\hat{p}\hat{q}}{n}} = 2.05\sqrt{\frac{\left(\frac{115}{192}\right)\left(\frac{77}{192}\right)}{192}} = .07251$$

3. Now, we form the interval:

$$\hat{p} - E < p < \hat{p} + E$$
$$\frac{115}{192} - .07251 < p < \frac{115}{192} + .07251$$
$$.5264 < p < .6715$$

4. Interpretation: With 96% confidence we can say that the proportion of all people who find the texture of cottage cheese to be gross is between .5264 and .6715.

5. Respond to research question: N/A

(We were not asked a research question.)

These answers are all based on the limited information given in the problem. Suppose we were also told that 35 of the people in the sample exercise daily and 89 of them have no siblings. Well, we really wouldn't be able to identify p, x, $\hat{p}$, and $\hat{q}$ without knowing the research question. The meaning and value of the symbols, as well as the interpretation of the resulting confidence interval, would all depend on the research question. Based on the limited information given in this problem, the only research question that could be answered is something about the proportion (or percentage) of all people who find the texture of cottage cheese to be gross. Of course, we don't know the value of p (the proportion of all people who find the texture of cottage cheese to be gross) because, if we did, we would not be estimating it with a confidence interval. Confidence intervals are used to estimate something we do not already know.

Remember that a confidence interval is always estimating something we don't know about the population. The sample tells you what the population is: if we interview some people, the population is all people; if we observe some registered voters in Statsville, the population is all registered voters in Statsville; if our sample is some third graders in Miami, the population is all third graders in Miami; etc.

Suppose we wanted an interval that is not so wide. For example, suppose you made a confidence interval for the proportion of all residents in your city who wish they could travel to another planet, and your interval was: $.0001 < p < .9987$. That interval does not provide helpful information because it is so wide (it's essentially 0 to 1, which you know is true of any proportion!). Suppose that the nature of our research is such that our interval, $.5264 < p < .6715$ is too wide to provide useful information. Mathematically, there are two ways to resolve this problem. One solution is to get a larger sample. Consider the radical in the margin of error. If, instead of 192, we had 19200 the margin of error would become much smaller (assuming the number of successes also changed so that $\hat{p}$ and $\hat{q}$ retain approximately their original respective values) which would make the new endpoints of the confidence interval closer to each other (i.e., the interval is not so wide). Another solution is to decrease the confidence level. Suppose we used 90% confidence instead of 96%. That would make the critical value ($z_{\alpha/2}$) 1.645. If that number was used in place of the 2.05 previously used, again the margin of error would decrease (without touching the sample size) resulting in an interval that is not so wide. Though either option works, getting a bigger sample is the preferred solution (assuming the situation allows that) because this will give us more actual information about our population as opposed to playing with the confidence level which is just a superficial change (and may make your results less convincing to consumers of your research).

So far we have seen confidence levels like 90%, 91%, and 96%. Why don't we use 100% confidence so we are sure to get the correct estimate every time? First, since we are estimating a value for a population based on a sample we could never do so correctly 100% of the time (unless, of course, we made up a silly interval like $-5000 < p < 256$). Second, if you tried to make a 100% confidence interval, α and $\frac{\alpha}{2}$ would both equal zero. Because of the asymptotic nature of the standard normal distribution, there is no critical value $z_{\alpha/2}$ that has zero area to its right.

When we make a confidence interval, the margin of error is a function of the sample size. However, what if we wanted to predetermine our margin of error before ever getting a sample and making a confidence interval? Well, we would have to get a minimum sample size to make sure that our estimate remained within the desired margin of error. But what minimum sample size would we need?

We call on our good friend algebra again. Start with the margin of error formula and perform the necessary algebra to solve for n. That gives:

$$n = \frac{\left(z_{\alpha/2}\right)^2 \hat{p}\hat{q}}{E^2}$$

Now, n is a function of E. Using that formula gives the minimum sample size needed to make an estimate of p within predetermined margin of error, E . This formula requires that you have some outside source (pilot study, similar research from a trusted source, etc.) to use for p-hat since you obviously don't have your own p-hat (you don't have a sample yet).

In cases where no p-hat is available from some outside source, we replace $\hat{p}\hat{q}$ with the biggest number that product could ever equal, .25. (Go ahead, try to find values of $\hat{p}$ and $\hat{q}$, each of which is between zero and one, where $1 - \hat{p} = \hat{q}$, and the product $\hat{p}\hat{q}$ is bigger than .25. I'll wait... You can actually use algebra and a quadratic function to prove .25 is the biggest possible product). The formula becomes:

$$n = \frac{\left(z_{\alpha/2}\right)^2 (.25)}{E^2}$$

Using the biggest possible value for $\hat{p}\hat{q}$ gives us the safest (i.e., largest) minimum sample size to make sure our estimate stays within the margin of error we are willing to tolerate.

When we determine a minimum sample size, we have to get a whole number. This usually requires rounding the answer. Since we are calculating a minimum we specifically need to round UP whenever the calculation results in a non-whole number.

Example: The dietician mentioned earlier feels that the margin of error, .07251, used for her estimate is too large. Actually, she wants to make sure that her estimate of the proportion of all people who find the texture of cottage cheese to be gross is within 1.5 percentage points of the true proportion, while keeping a 96% confidence level.

a) How many people must be surveyed so that she accomplishes her goal?

b) Suppose she relies on a pilot study in which 60% of respondents found the texture of cottage cheese to be gross. Revise the answer you found in a).

Answer:

a) $\alpha = 1 - .96 = .04$ $\quad \frac{\alpha}{2} = \frac{.04}{2} = .02$ $\quad$ from Table A $z_{\alpha/2} = z_{.02} = 2.05$

$$n = \frac{\left(z_{\alpha/2}\right)^2(.25)}{E^2} = \frac{2.05^2(.25)}{.015^2} = 4669.4444$$

answer: 4670 people must be surveyed

b) $\alpha = 1 - .96 = .04$ $\quad \frac{\alpha}{2} = \frac{.04}{2} = .02$ $\quad$ from Table A $z_{\alpha/2} = z_{.02} = 2.05$

$$n = \frac{\left(z_{\alpha/2}\right)^2 \hat{p}\hat{q}}{E^2} = \frac{2.05^2(.60)(.40)}{.015^2} = 4482.6667$$

answer: 4483 people must be surveyed

Remember that we must round UP because the calculation is giving us a minimum sample size. 4669 is smaller than the minimum 4669.4444 so the dietician needs to survey (at least) 4670 people in part a). We rounded UP in part b) for the same reason.

We knew this was a sample size question because of, "how many people must be surveyed..." Every problem doesn't have those exact words, of course, but the situation is that we are trying to collect a sample with the correct minimum size so we stay within the predetermined margin of error. We were NOT asked to make the confidence interval, but merely to determine the size of the sample she will need so that, when her estimate (the confidence interval) is later made, it will stay within the predetermined margin of error.

In part a) we were not given any information to use for $\hat{p}$, so we used the formula that has .25 in it. For b) we were given the 60% figure so we used .60 for $\hat{p}$ and $1 - \hat{p} = 1 - .60 = .40$ for $\hat{q}$.

The "within 1.5 percentage points" information provided the predetermined margin of error, .015 (in proportion form—E is always expressed as a proportion in the formulas).

It should come as no surprise that our answer for a) was larger than the answer for b). The difference in the formulas used is that the .25 in the numerator for a) is replaced by the product (.60)(.40) in b). We established earlier that, even without performing the calculations, we know .25 is larger than (.60) multiplied by (.40) so the result for a) has to be bigger than the result for b). In practical terms this highlights the fact that getting a trusted value for $\hat{p}$ to use in the sample size calculation is ideal, if at all possible, because usually a smaller sample means less resources (time, money, researchers, etc.) will be needed to accomplish the research.

Example: Forty Statsville residents were interviewed and asked whether or not they are employed. Using the Minitab output below, respond to the claim that over 60% of Statsville residents are employed.

Test and CI for One Proportion: Employed?

```
Event = y


Variable    X   N  Sample p          95% CI
Employed? 25   40  0.625000  (0.458015, 0.772737)
```

Answer:
With 95% confidence we can say that the proportion of all Statsville residents who are employed is between .458015 and .772737. Since the whole interval is not above .60, these results do not support the claim that over 60% of Statsville residents are employed.

The column where the employment status responses are entered in Minitab is labeled "**Employed?**" Each respondent's entry for this question was "n" (no) or "y" (yes). Minitab considers "y" to be a success as indicated by the "`Event = y`" portion of the output. By default, whichever occurs later alphabetically (or numerically, if numbers are used) is considered a success. "y" occurs later than "n" alphabetically, so "y" is considered a success. In the sample, 25 people answered "y". The sample size was 40. The proportion of successes in the sample was .625. The 95% confidence interval is shown using interval notation under the "95% CI" label. Take a moment to identify all these parts of the output. We will look at lots of Minitab output for inferential statistics. Everything is labeled so just read the labels and you can easily understand the output!

General word of caution: Often the results that are shown in Minitab (or any other statistical software) output are different from what you get if you do the problem by hand. There are various reasons for this relating to rounding and estimating. The software result is always the most accurate. If you are particularly motivated and you complete an exercise by hand and then decide to try it in Minitab and you notice a difference do not be alarmed. In practice, though, when you have access to Minitab output you just use it. Notice that I did not perform any calculations when I reported my answer. Why would I?????! I have Minitab output! The whole purpose for using statistical software is so that we do NOT have to do calculations by hand. We do need to be able to interpret the output, though, because the software is not able to make inferences for us about our research questions.

What's wrong with this conclusion?: "The proportion of Statsville residents who are employed is .625 which is the same as 62.5%. That verifies the claim that over 60% of Statsville residents are employed." Ouch, that hurts. .625 is just the proportion for THE SAMPLE not the entire population. We can never take a single sample statistic and use it to make an inference about an entire population (if we could do that, there would be no Statistics classes!). Just citing the sample statistic and using that to make an inference about the whole population is a HUGE PROBLEM. Do not do that. Ever.

What's wrong with this conclusion?: "Since .625 is in the confidence interval (.458015, .772737) that tells us that..." OK, just stop right there. Anything that comes next is wrong. The fact that the sample proportion of successes is in the confidence interval for the population proportion of successes will never address a research question. We will never try to make an inference based on the fact that the sample statistic is one of the numbers in the estimate (confidence interval) for the analogous population parameter.

What's wrong with these interpretations of the confidence interval?

a) With 95% confidence we can say that the amount of all Statsville residents who are employed is between .458015 and .772737.

> The interval is not for the "amount" or number of employed residents—it's for the <u>proportion</u> of employed residents. Big difference.

b) With 95% confidence we can say that all Statsville residents who are employed is between .458015 and .772737.

> Huh?! What? That makes no sense.

c) With 95% confidence we can say that the proportion of Statsville residents in the sample who are employed is between .458015 and .772737.

It is true that the proportion for the sample (.625) is between .458015 and .772737, but that is not a useful observation. The manner in which confidence intervals are created ensures that the sample statistic will *always* be a number contained in the confidence interval. We know the exact sample proportion so the confidence interval is not estimating that. As usual, we estimate values we do not know—parameters for populations. Your confidence intervals are never estimating anything for samples. The whole purpose for making a confidence interval is to estimate a value for the population, never ever ever for the sample.

d) With 95% confidence we can say that the proportion of all Statsville residents who are employed is between the intervals of .458015 and .772737.

There is only one interval here, not two. The interval is between .458015 and .772737. Single numbers are not intervals.

Be sure to write thoughtfully when you provide interpretations. What you communicate needs to be accurate and make sense. Read what you write after you write it. If you know the statement you wrote makes no sense, like in part b), don't just move on to something else. Re-read the problem, think about what everything means, and write an interpretation that makes sense.

Exercises

1) Randomly selected diamonds from Sparkelz Gems and Javonn Jewelers were observed and their weight (in carats), size (in mm), color (letter rating), and clarity (letter/number rating) were recorded as shown below:

Sparkelz Gems

weight	size	color	clarity
.5	5.2	H	VVS1
1.5	7.4	N	VS1
2.25	8.6	H	VS1
2.5	9	N	VS1
0.75	5.9	N	VVS1
0.65	5.7	H	SI2
0.85	6.2	Z	SI2
1.75	7.8	Z	I1
0.75	5.8	N	IF
1.25	7	H	VVS1
2	8.2	D	I1
0.65	5.6	H	VS1
1	6.5	D	SI2
3	9.3	Z	VVS1
0.5	5	D	VS1

Javonn Jewelers

weight	size	color	clarity
1.25	7.0	G	VS1
0.50	5.2	D	SI1
0.75	6.0	K	VS1
2.25	8.5	H	VS1
2.00	8.2	F	VVS2
1.75	7.8	H	VS1
2.50	9.0	H	VVS1
1.50	7.5	H	VS1
1.25	6.9	N	VVS2
2.00	8.0	H	VVS1
2.25	8.4	N	I1
1.25	7.1	H	VS1
1.75	7.8	F	I2
1.50	7.4	G	I2
0.65	5.6	J	VS1
3.00	9.3	H	VVS1
1.50	7.5	H	VS1

Javonn Jewelers literature claimed that 43% of their diamonds are color H. Use a 93% confidence interval and address that claim.

2) Employees from two companies, Jerrell Enterprises and Jaden Industries, were surveyed. Of the 142 employees from Jerrell Enterprises, 64 were dissatisfied with their current position in the company, 46 were satisfied with their current position in the company, and the rest were neutral. 10 of the dissatisfied employees participate in professional development, 15 of the neutral employees participate in professional development, and 34 of the satisfied employees participate in professional development. Each employee is rated by a supervisor and then by a peer. The supervisor performance ratings for this sample had a mean of 157.3 with standard deviation 31. The peer performance ratings for these employees had a mean of 146.2 with standard deviation 46. These employees were absent an average of 3.71 days last year with standard deviation 3.42 days. The mean of the differences of the supervisor ratings and the peer ratings for this sample was 11.1 with standard deviation 51.8.

In the Jaden Industries sample, 58 employees were dissatisfied with their current position in their company (30 of them participate in professional development), 50 were satisfied with their current position in the company (24 of them participate in professional development), and 49 were neutral (25 of them participate in professional development). Each employee is rated by a supervisor and then by a peer. The supervisor performance ratings for this sample had a mean of 176.1 with standard deviation 42. The peer performance ratings for these employees had a mean of 180.4 with standard deviation 28. These employees were absent an average of 3.26 days last year with standard deviation 2.58 days. The mean of the differences of the supervisor ratings and the peer ratings for this sample was −4.3 with standard deviation 49.1.

Make a 99% confidence interval for the proportion of Jaden Industries employees who are satisfied with their current position in the company. Based on the interval, can we infer that less than half of Jaden Industries employees are satisfied with their current position in the company? Explain.

3) A manager at an air freshener company wants to estimate the percentage of the air fresheners that are still effective after 45 days. How many air fresheners must be tested so that the estimate is within four percentage points of the true percentage for all air fresheners with 97% confidence?

4) A Statistics instructor wants to get a sense of what proportion of students plan to earn graduate degrees. He plans to make a confidence interval using a process that has an 88% success rate. Before making the interval, though, he needs to figure out how many students to interview about their academic goals. How many students should be interviewed if the instructor wants to make sure his estimate is no more than .05 from the actual proportion of all students who plan to earn graduate degrees? Note that he will make use of a study he found in a published journal from a few years ago (in fact, this is what sparked his interest in the subject) in which it was found that 17% of those surveyed planned to earn graduate degrees.

5) Two drugs—Drug A and Drug B—are being tested in a clinical trial. The side effects experienced by each person in the trial are recorded as follows: nausea (n), itching (i), headaches (h), fatigue (f), dry mouth (d). Researchers believe that less than half of all Drug A users experience nausea. This is important to them as they start to think about the warning label for the drug. Assess that belief using the Minitab output below.

Test and CI for One Proportion: Drug A

```
Event = n
```

```
Variable    X   N  Sample p          95% CI
Drug A     13  50  0.260000  (0.146301, 0.403448)
```

6) Referring to the Minitab output in 5), what is WRONG with these conclusions?

a) The proportion of all Drug A users who experience nausea is .26 and, since that is less than .5, the researchers' belief appears to be correct.

b) Only 13 Drug A users experience nausea, and that is less than half of 50. The researchers' belief appears to be correct.

c) .26 is in the confidence interval (.146301, .403448) so the researchers' belief appears to be correct.

d) This research did not include all Drug A users but only a sample so we cannot test a belief about all Drug A users.

e) The number of Drug A users who experience nausea is between .146301 and .403448, so less than one person will experience nausea. The researchers' belief appears to be correct.

f) The proportion of all Drug A users who experience fatigue is between .146301 and .403448. Since the whole interval is less than .5, the researchers' belief appears to be correct.

7.3 Confidence Intervals for Means

Conditions: The data is from a simple random sample. The population is normally distributed and/or the sample size is bigger than 30. The standard deviation for the population is unknown.
(Assume all conditions are met for each example and exercise.)

Use the methods of this section when...

creating a confidence interval for a mean

Important formulas: $E = t_{\alpha/2}\dfrac{s}{\sqrt{n}}$ $\quad\quad \bar{x} - E < \mu < \bar{x} + E$

[Note: In this text we only consider estimating means when the population standard deviation is unknown. That is because, in reality, whenever the population mean is unknown (which is why we are estimating it) the population standard deviation is also unknown. In a contrived case in which the population standard deviation was known and a confidence interval for the population mean was being created, the method would be different from that described in this text.]

Before we consider confidence intervals for means, we need to become familiar with the relevant probability distribution: the t distribution. The t distribution also goes by the name "Student's t distribution" because of the manner in which it was originally published. The t distribution is actually a family of infinitely many distributions, just like the normal distribution is actually a family of infinitely many different specific distributions, distinguished by their mean and standard deviation. What distinguishes one t distribution from another is the number of degrees of freedom (df). Just like the normal distribution has a pdf (you saw it in Chapter 4), so does the t distribution. [The pdf for this distribution and all the other distributions we will see in the rest of the text are too obscure to be appreciated by students with only an intermediate algebra background, so there is no benefit to providing those functions in this text.] One of the pieces in the pdf for the t distribution is the number of degrees of freedom so, when using a particular t distribution, it is important to know the applicable number of degrees of freedom. When using a t distribution, the number of degrees of freedom is simply one less than the sample size: $df = n - 1$.

[Note regarding degrees of freedom: Degrees of freedom can be discussed in different ways, some of which can become rather mathematically complex, far beyond the level of math experience expected of students using this textbook. Trying to get into an in-depth discussion of the topic here is unproductive and unnecessary given students' backgrounds and the goals of a basic Statistics course. At this level, it is easiest to think of degrees of freedom as how many independent pieces of data there are. In a population, the sum of the deviations, $\Sigma(x-\mu)$, is zero. As you'll recall from the conceptual formula, when we use the sample variance we use the sample deviations $(x-\bar{x})$ to estimate the deviations in the population. In order for this to be a good estimate, $\Sigma(x-\bar{x})$ must also equal zero. That being the case, all but one (i.e., $n-1$) of the deviations are free to be whatever, but the last one is fixed (in order for the sum to be zero). Simple statistical processes involving one sample variance have $n-1$ degrees of freedom. More complicated procedures, and procedures with different estimate-based restrictions, have other numbers of degrees of freedom.]

No matter how many degrees of freedom there are, a t distribution has a symmetric bell curve like a normal distribution. The graphical difference from one t distribution to the next is the height and dispersion of the curve. In the display that follows you see graphs of t distributions with 3, 7, and 20 degrees of freedom. You can observe the differences in the heights and spread of the curves.

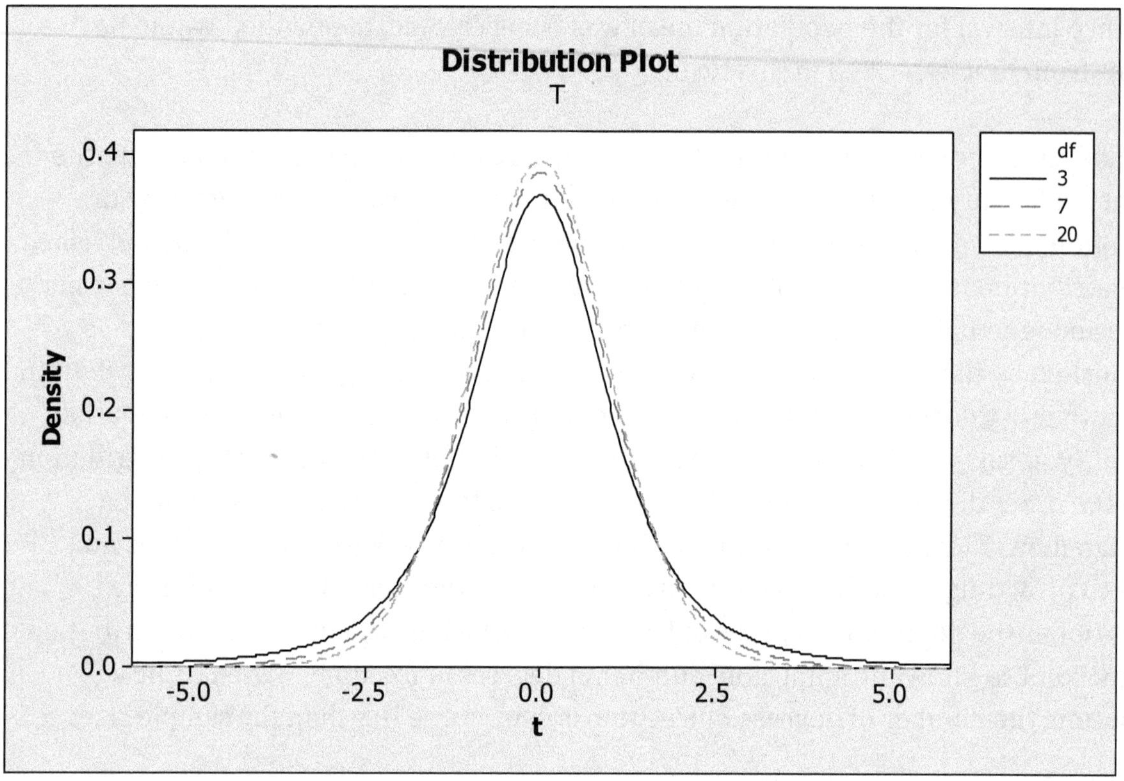

Notice that a smaller number of degrees of freedom results in the distribution having more spread. Extreme values (both large and small) are more likely in a distribution with fewer degrees of freedom. As the number of degrees of freedom increases, the curve gets closer to a standard normal distribution. Next is the graph of a t distribution with 20 degrees of freedom and the graph of the standard normal distribution. As you can see, the curves are already pretty close.

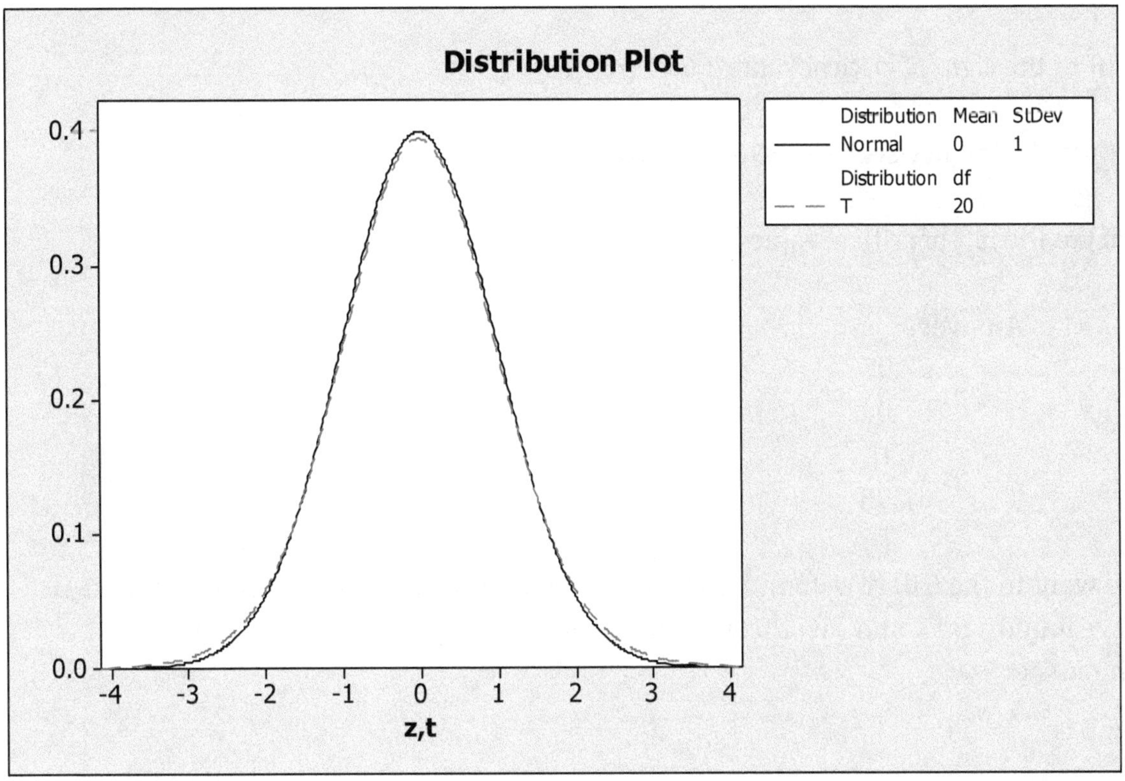

On the graph of a t distribution, the t-scores (analogous to z-scores in a standard normal distribution) are along the horizontal axis with zero in the middle because it is the mean. t-scores are also called critical values of the t distribution. As in the standard normal distribution, there are negative t-scores to the left of zero and positive critical values to the right. As with all probability distribution graphs, the total area under the curve is one.

Again, we will use a table to get areas under the curve (i.e., probabilities) for t distributions instead of doing the complicated calculus by hand. Table B (page 333) has areas and t-scores for many different specific t distributions so we will need to consider the degrees of freedom for the specific t distribution applicable to the problem. Each row of Table B represents a different t distribution, as indicated by the degrees of freedom listed in the first column of each row. For a particular row of the table, each column to the right of the degrees of freedom shows t-scores having the area to the right indicated by the label at the top of the page for that column. The table gives areas to the right of positive t-scores. If we need areas to the left of negative t-scores, we have to use the symmetry of the distribution and the area to the right of the opposite t-score.

Example: In the t-distribution, find the area...

a) to the right of t-score 1.943 (df = 6)

b) to the right of critical value 2.624 (n = 15)

c) to the left of t-score −1.313 (n = 29)

d) to the left of critical value 1.943 (n = 7)

Answers:

a) .05
b) .01
c) .10
d) .95

In a) we went to the 6 df row, found 1.943 in that row, looked at the top of the column where we found 1.943, and saw the area to the right labeled as .05. Graphically, that corresponds to:

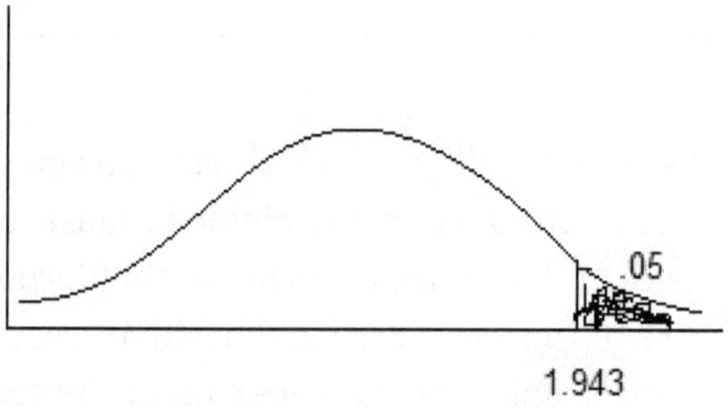

In b) since the degrees of freedom were not given, we had to do that easy calculation, df = 15 − 1 =14. [In most problems, you are given the sample size, not the df, so this quick calculation will be performed.] In df row 14 we found 2.624 and looked at the very top to see the area to the right, .01.

In c) since the degrees of freedom were not given, we had to again find the value using df = 29 − 1 = 28. We also had to remember that the table always gives the area to the right of positive t-scores. Using symmetry and the area to the right of the opposite t-score, we found 1.313 in df row 28, looked at the top of its row, and saw area .10. So t-score 1.313 has area .10 to its right, meaning that t-score −1.313 will also have area .10 to its left. Graphically, this looks like:

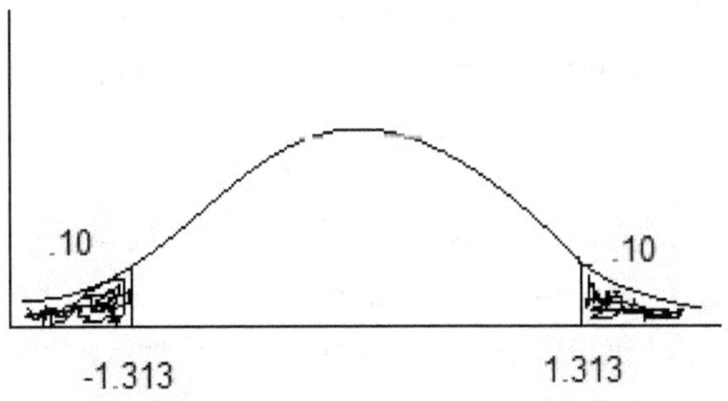

For d) notice that we use the procedure in a) to find that the area to the right of the given critical value is .05. Since the total area under the curve is one, the area to the left of the given critical value is 1−.05 = .95.

Note that you will never look for anything negative in Table B. It does not contain negative values.

The notation t_A stands for the t-score that has area A to its right. This is analogous to the meaning of z_A in the standard normal distribution.

Example: Find the indicated critical value.

a) $t_{.005}$ (n = 17)

b) $t_{.025}$ (n = 36)

c) $t_{.10}$ (n = 42)

d) $t_{.05}$ (n = 56)

Answers:

a) 2.921
b) 2.030
c) 1.303
d) 1.6735

In a) we need to find the t-score with area .005 to its right, with df = 17 – 1 = 16. So, in df row 16, we trace over to the column labeled .005 at the top of the page, and there we see the corresponding critical value: 2.921.

In b) we need to find the t-score with area .025 to its right, with df = 36 – 1 = 35. So, in df row 35, we trace over to the column labeled .025 at the top of the page, and there we see the corresponding critical value: 2.030.

In c) we need to find the t-score with area .10 to its right, with df = 42 – 1 = 41. Since the table does not have 41 degrees of freedom, we'll use the closest number of degrees of freedom that the table does have, which is 40. So, in df row 40, we trace over to the column labeled .10 at the top of the page, and there we see the corresponding critical value: 1.303.

In d) we need to find the t-score with area .05 to its right, with df = 56 – 1 = 55. Since the table does not have 55 degrees of freedom, we look for the closest number of degrees of freedom the table does show. Actually there are two that are equally close: 50 and 60. So in this case, only because there are two df's tied for equally close, we will get the mean of the respective critical values shown. For 50 df, the critical value in column .05 is 1.676. For 60 df, the critical value in column .05 is 1.671. What we will use for our critical value is $\frac{1.676+1.671}{2} = 1.6735$.

With that background, we can now talk about confidence intervals for means, beginning with the rationale.

<u>Rationale for creating a (1−α)100% confidence interval for μ</u>

As you can imagine, whenever you take a different sample from a population, and calculate the relevant mean for that sample, $\bar{x}$, there is a different $\bar{x}$ that comes from each sample. When the stated conditions for this section are met, those x-bars are approximated by a t-distribution with mean μ and standard deviation $\frac{s}{\sqrt{n}}$. The corresponding t-score for a particular $\bar{x}$ is given by $t = \frac{\bar{x}-\mu}{\frac{s}{\sqrt{n}}}$.

To create a (1−α)100% confidence interval for μ, we start by visualizing the distribution of x-bars as follows:

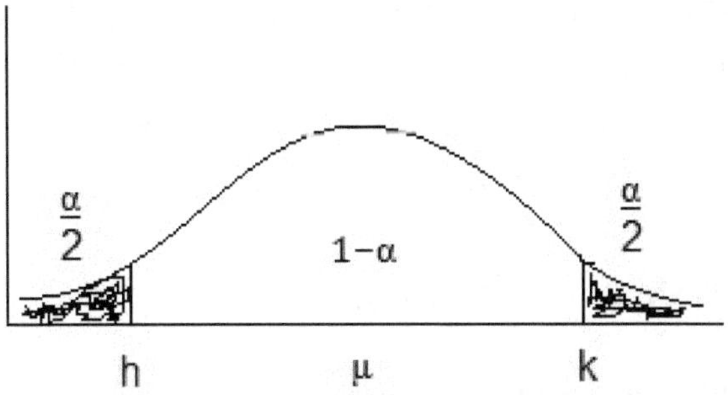

We focus our attention on the middle (1−α) proportion of the x-bar distribution, where α is the small proportion of unusually small and unusually large x-bars. μ is the mean of the distribution with h and k being the x-bars separating the middle (1−α) proportion of x-bars from the rest. The t-score that corresponds to h has the symbol $-t_{\alpha/2}$ and the t-score for k would be $t_{\alpha/2}$, similar to the z-score notation you learned previously (like, if $t_{\alpha/2}$ is t-score 1.313, then $-t_{\alpha/2}$ would be t-score −1.313). Using the t-score formula we have:

$$-t_{\alpha/2} = \frac{h-\mu}{\frac{s}{\sqrt{n}}} \qquad \text{and} \qquad t_{\alpha/2} = \frac{k-\mu}{\frac{s}{\sqrt{n}}}$$

solving for h and k gives:

$$h = \mu - t_{\alpha/2}\frac{s}{\sqrt{n}} \qquad \text{and} \qquad k = \mu + t_{\alpha/2}\frac{s}{\sqrt{n}}$$

So, any x-bar in the middle (1−α) proportion of all the x-bars is between h and k which also means for any such x-bar:

$$\mu - t_{\alpha/2}\frac{s}{\sqrt{n}} < \bar{x} < \mu + t_{\alpha/2}\frac{s}{\sqrt{n}}$$

Use the algebra that you learned in previous classes to get μ in the middle of the compound inequality by itself. After those steps, you get:

$$\bar{x} - t_{\alpha/2}\frac{s}{\sqrt{n}} < \mu < \bar{x} + t_{\alpha/2}\frac{s}{\sqrt{n}}$$

To make our confidence interval for μ, we will get a sample and use the indicated information from it to fill in that formula. That's our confidence interval!

[Again, we are hoping that we select a sample having a mean that comes from the middle (1−α) proportion of all the x-bars. That will usually be the case because we use pretty large (1−α) proportions. When we happen to get a sample with a $\bar{x}$ not in the middle (1−α) proportion of all the x-bars, the confidence interval fails to contain the true value of μ. (1−α)100% confidence intervals contain the true value of μ (1−α)100% of the time. If (1−α) = .95 then (1−α)100% = 95%, so 95% confidence intervals for μ contain the true value of μ 95% of the time.]

The process outlined above is simplified by using these steps:

Steps for creating a (1−α)100% confidence interval for μ

1) Find $t_{\alpha/2}$ for the desired confidence level and appropriate degrees of freedom ($t_{\alpha/2}$ is a positive value).

2) Calculate the margin of error, $E = t_{\alpha/2}\frac{s}{\sqrt{n}}$

3) Form the interval $\bar{x} - E < \mu < \bar{x} + E$

4) Interpret the interval in real world terms.

5) Respond to any given research question.

Note: The rationale for creating confidence intervals is the same for all parameters. The rationale was described in general in Section 7.1, and has been demonstrated for the specific parameters in this and the previous section. In future sections, we will omit the rationale and will just go directly to the steps. In the back of your mind, know that the steps are coming from the same rationale.

Example: Randomly selected current Statsville College students were asked how many units they have completed so far. Their responses were:

32 38 44 51 59 38 54 32 47 38 61 54

Of the students surveyed, 42% intend to major in Statistics and the standard deviation of the GPAs for everyone in the sample is .8017 with mean 2.4583. At Wudmoor College, 65% of the students have completed less than 50 units. 10 years ago, Statsville College students completed an average of 55 units with standard deviation 15.5 units. Make a 95% confidence interval to see if the mean number of units completed by current Statsville College students is the same as the mean number of units completed by Statsville College students from 10 years ago.

Answer:
1) First, we need $t_{\alpha/2}$ and we get that as follows:

$\alpha = 1 - .95 = .05 \quad \frac{\alpha}{2} = \frac{.05}{2} = .025 \qquad df = 12 - 1 = 11$

from Table B, $t_{\alpha/2} = t_{.025} = 2.201$

2) Next, we get the margin of error:

$$E = t_{\alpha/2}\frac{s}{\sqrt{n}} = 2.201\frac{10.1564}{\sqrt{12}} = 6.4531$$

(we need to calculate s using a formula from Section 3.2)

3) Now, we form the interval: $\bar{x} - E < \mu < \bar{x} + E$

$45.6667 - 6.4531 < \mu < 45.6667 + 6.4531$

$39.2136 < \mu < 52.1198$

(we need to calculate $\bar{x}$ using the formula from Section 3.1)

4) Interpretation: With 95% confidence we can say that the mean number of units completed by all current Statsville College students is between 39.2136 and 52.1198.

5) Respond to research question: Since the entire interval is less than 55 (the mean number of units completed by Statsville College students 10 years ago), it appears that the mean number of units completed by current Statsville College students is not the same as the mean number of units completed by Statsville College students from 10 years ago. Specifically, the mean number of units completed by current Statsville College students is smaller than the mean from 10 years ago.

$\bar{x}$ is the mean number of units completed by current Statsville College students in the sample. s is the standard deviation of the number of units completed by current Statsville College students in the sample. μ is the mean number of units completed by all current Statsville College students. We include any applicable units (ft., days, dollars, etc.) and, for this problem, the units are "units completed".

This interval is telling us that, if we got the number of units completed by every current Statsville College student, added all those numbers, and then divided by the total amount of current Statsville College students there are, the answer to that division problem would be a number between 39.2136 and 52.1198.

In some cases, the mean and standard deviation of the sample are given explicitly in the problem. When we are given the raw data, we have to calculate the mean and standard deviation as we did in this problem. We had to read carefully because the mean and standard deviation given in the problem had nothing to do with the number of units completed so they were irrelevant to our specific research interest (though they would certainly be useful for other research interests relevant to the population of Statsville College students).

If the confidence interval included 55, we would conclude that there is no difference in the current mean and the mean from 10 years ago. If the whole interval were greater than 55, we would conclude that the current mean is larger than the mean from 10 years ago.

What is wrong with this interpretation? "With 95% confidence we can say that the mean number of units completed by current Statsville College students in the sample is between 39.2136 and 52.1198." We are not estimating the mean for the sample. We know the mean for the sample is 45.6667 so there would never be a need to estimate that. With confidence intervals we are always estimating something we do not know. We do not know the population mean and that's what the confidence interval estimates. Of course, the sample mean is one of the numbers in the interval for the population mean because of the procedure used to create the interval. The interval itself, however, is an estimate for the mean of the whole population.

What is wrong with this interpretation? "With 95% confidence we can say that current Statsville College students have completed between 39.2136 and 52.1198 units." Depending on how you understand that sentence, you get the impression that 1) each student has completed between 39.2136 and 52.1198 units, or 2) the total number of units that all students have completed is between 39.2136 and 52.1198. The interval is not communicating either of those ideas. The interval is estimating the mean of a population, not each individual value in the population or the total of all the values in the population.

Be sure to avoid a couple common mistakes when making an inference for this type of problem. Notice that we did not compare 45.6667 to the interval for μ to get our response to the research question. The sample mean is always in the interval for the population mean (because of how the interval is made) so that observation has no usefulness when trying to make an inference. Also, we did not compare 45.6667 to 55. 45.6667 is a sample number but 55 is a population number so they cannot be compared in order to make an inference.

Example: The ages (in years) of randomly selected Statsville residents were used in the creation of the Minitab output below. Based on these results, what can we infer about the ages of all Statsville residents?

One-Sample T: SVAge

```
Variable   N   Mean  StDev  SE Mean       92% CI
SVAge     40  30.55  14.79     2.34  (26.35, 34.75)
```

Answer:
With 92% confidence we can say that the mean age of all Statsville residents is between 26.35 years and 34.75 years.

The column where the ages were stored in Minitab is labeled " **SVAge** ". The sample size is 40, mean age for the sample is 30.55 years, standard deviation of the sample is 14.79 years, the standard error of the mean (which we won't use explicitly) is 2.34 years, and the 92% confidence interval is (26.35, 34.75) years. Notice the label " **One-Sample T** ". In Minitab, any procedure that involves " **T** " relates to a mean (because we use a t distribution).

Example: Refer to the Minitab age output. What is wrong with these interpretations?

a) With 92% confidence we can say that the mean age of the 40 Statsville residents in the sample is between 26.35 years and 34.75 years.

b) With 92% confidence we can say that all Statsville residents have ages between 26.35 years and 34.75 years.

c) With 92% confidence we can say that the mean average age of all Statsville residents is between 26.35 years and 34.75 years.

d) With 92% confidence we can say that between 26.35% and 34.75% of Statsville residents have ages.

Answer:

a) The confidence interval is estimating the mean age for all Statsville residents (the population) not the mean for the sample. We know the exact mean age for the sample: 30.55. Why would we estimate something we know?! What we do not know is the mean age for all Statsville residents and that is what we are estimating.

b) The confidence interval does not tell us that all Statsville residents have ages in the interval. It is quite likely that many residents are younger than 26.35 years old or older than 34.75 years old. The interval is estimating the mean of the population values, it is not estimating each individual population value.

c) "Mean average"? Pick one. You can say "mean" or you can say "average", but not both.

d) What??????? First of all, the interval is for a mean, not a proportion (or percentage). Secondly, doesn't everyone have an age? When you write conclusions and interpretations, please make sure they make sense.

Example: Minitab gave us the interval $26.35 < \mu < 34.75$ for estimating the mean age of Statsville residents.

a) Just using that interval, what was the mean age of the sample of Statsville residents?

b) Just using that interval, what was the margin of error for the estimate of the mean age of all Statsville residents?

c) Write the confidence interval in the form $\bar{x}$ ± E.

Answer:

a) We know that the confidence interval was constructed by taking $\bar{x} - E < \mu < \bar{x} + E$, so $\bar{x}$ has to be in the middle of the interval. [For example, if you take 7−2 and 7+2 you get 5 and 9. What number is right in the middle of 5 and 9? 7] So we just need to find the number that is right in the middle of 26.35 and 34.75. As you learned in a previous class, to get the number right in the middle of two other numbers, we just add the numbers and divide by two. So, $\bar{x}$, the sample mean, was $\frac{26.35+34.75}{2} = 30.55$ years.

b) Again referring to the formula used to create the confidence interval, we realize that the margin of error is just half the total difference between the endpoints of the interval, 26.35 and 34.75. So, E, the margin of error, was $\frac{34.75-26.35}{2} = 4.2$. [To verify this, notice that 30.55 – 4.2 = 26.35 and 30.55 + 4.2 = 34.75]

c) Using our previous work we see that the original confidence interval can be written as: 30.55 ± 4.2

Exercises

1) Refer to the Sparkelz Gems/Javonn Jewelers data from Exercise 1) in Section 7.2. Last year, the mean size of diamonds at Sparkelz Gems was 7 mm. Does it appear that Sparkelz' diamonds currently have a smaller mean size? Use a 90% confidence interval to answer.

2) Refer to the Jerrell Enterprises/Jaden Industries data from Exercise 2) in Section 7.2. Test the assertion that the average supervisor rating at Jerrell Enterprises is 170. Use a process that correctly estimates population means 95% of the time.

3) There are four furniture stores in a certain city. One is in the north, one is in the south, one is in the east, and one is in the west. We collected random samples of prices (in dollars) from each store and used the appropriate data for the Minitab output below. Suppose stores that sell items with an average price of $1500 are considered "high end". Based on the output, is the north store a high end store? Explain.

One-Sample T: North

Variable	N	Mean	StDev	SE Mean	99% CI
North	91	1508.94	46.15	4.84	(1496.21, 1521.68)

4) Consider the Minitab output in 3). What is WRONG with these interpretations/conclusions?

a) 99% of the items at the north store have prices between $1496.21 and $1521.68.

b) The mean price of items sold at the north store is \$1508.94, not exactly \$1500, so this store may be in a different category than "high end".

c) \$1508.94 is in the interval (\$1496.21, \$1521.68) so it appears that the north store is high end.

d) With 99% confidence we can say that each item at the north store is more than \$1496.21 but less than \$1521.68.

e) With 99% confidence we can say that the mean average price of items at the north store is between \$1496.21 and \$1521.68.

f) With 99% confidence we can say that the average price of items at the north store in the sample was between \$1496.21 and \$1521.68.

5) Consider the Minitab output in 3). What was the margin of error for that confidence interval?

6) The statistic $\hat{y}$ follows a t distribution with mean y_p and standard deviation $s_{\hat{y}}$. Using this information, show the steps for creating a $(1 - \alpha)100\%$ confidence interval for y_p.

7.4 Confidence Intervals for Standard Deviations and Variances

Conditions: The data is from a simple random sample. The population must be normally distributed. [The methods of this section rely heavily on this requirement.]
(Assume all conditions are met for each example and exercise.)

Use the methods of this section when...

creating a confidence interval for a standard deviation or a variance

Important formulas: $\dfrac{(n-1)s^2}{\chi^2{}_{\frac{\alpha}{2}}} < \sigma^2 < \dfrac{(n-1)s^2}{\chi^2{}_{1-\frac{\alpha}{2}}}$ $\qquad \sqrt{\dfrac{(n-1)s^2}{\chi^2{}_{\frac{\alpha}{2}}}} < \sigma < \sqrt{\dfrac{(n-1)s^2}{\chi^2{}_{1-\frac{\alpha}{2}}}}$

Before we explore these new confidence intervals, we again start with the introduction of the distribution that will be applicable to this section, the χ^2 distribution. χ (pronounced like "sky" but without the s) is the Greek letter chi (resist the urge to pronounce this like it is energy or a type of tea). The chi-square (χ^2) distribution is also a family of distributions distinguished by their degrees of freedom, again being n – 1. The χ^2-scores, or critical values, are all non-negative. While the total area under the curve is still one, the distribution is not symmetric. The shape of the curve is called *skewed right*, with the direction of the skew corresponding to the direction of the tail. The plot that follows shows graphs for chi-square distributions with 3, 7, and 10 degrees of freedom.

The notation $\chi^2_{1-\frac{\alpha}{2}}$ stands for the χ^2 critical value having area $(1-\frac{\alpha}{2})$ to its right and the notation $\chi^2_{\frac{\alpha}{2}}$ indicates the χ^2 critical value having area $\frac{\alpha}{2}$ to its right.

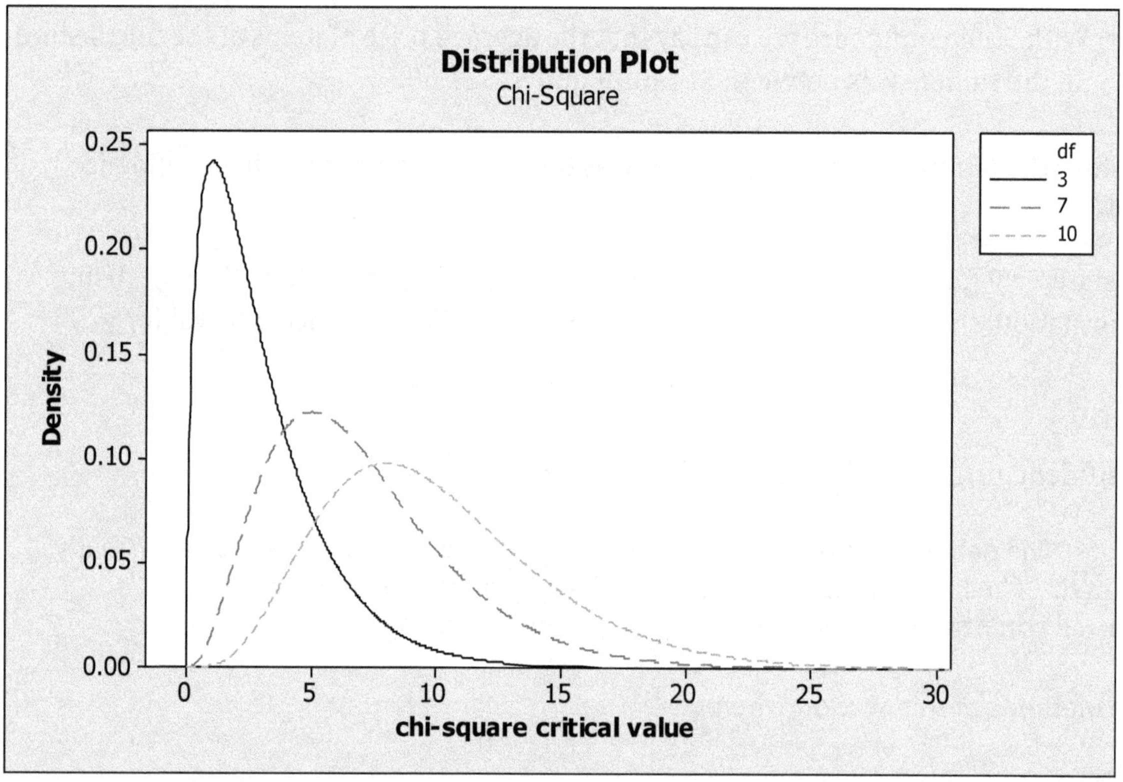

We will use Table C (page 334) to get needed probabilities and critical values for the chi-square distribution. Like the t table, the χ^2 table provides areas to the right of critical values.

Example: Find the critical values $\chi^2_{1-\frac{\alpha}{2}}$ and $\chi^2_{\frac{\alpha}{2}}$

a) $\alpha = .10$, n = 20

b) $\alpha = .02$, n = 13

c) α = .05 , n = 33

d) α = .01, n = 76

Answers:

a) $\chi^2_{1-\frac{\alpha}{2}} = 10.117, \chi^2_{\frac{\alpha}{2}} = 30.144$

b) $\chi^2_{1-\frac{\alpha}{2}} = 3.571,\ \chi^2_{\frac{\alpha}{2}} = 26.217$

c) $\chi^2_{1-\frac{\alpha}{2}} = 16.791, \chi^2_{\frac{\alpha}{2}} = 46.979$

d) $\chi^2_{1-\frac{\alpha}{2}} = 47.2235, \chi^2_{\frac{\alpha}{2}} = 110.268$

For a) $\frac{\alpha}{2} = \frac{.10}{2} = .05$ and $1 - \frac{\alpha}{2} = 1 - .05 = .95$. In a similar manner to how we read the t table, find the appropriate df row (19) and find the critical values shown in columns .95 and .05, which are 10.117 and 30.144, respectively. On a graph, this looks like:

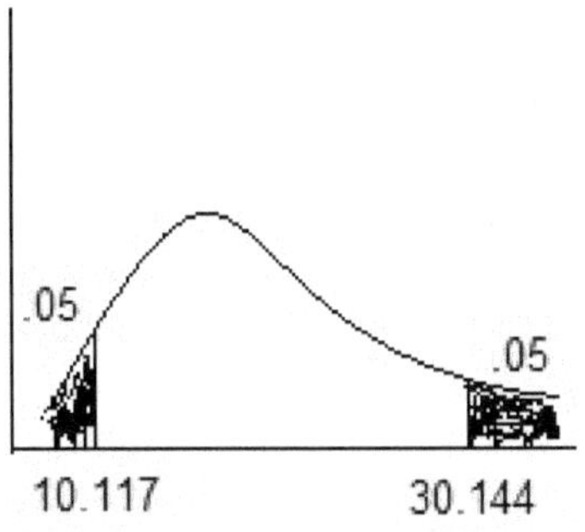

For b) $\frac{\alpha}{2} = \frac{.02}{2} = .01$ and $1 - \frac{\alpha}{2} = 1 - .01 = .99$. In a similar manner to how we read the t table, find the appropriate df row (12) and find the critical values shown in columns .99 and .01, which are 3.571 and 26.217, respectively.

For c) $\frac{\alpha}{2} = \frac{.05}{2} = .025$ and $1 - \frac{\alpha}{2} = 1 - .025 = .975$. In a similar manner to how we read the t table, find the appropriate df row (32 is our df, it is not in the table so use the closest: 30) and find the critical values shown in columns .975 and .025, which are 16.791 and 46.979, respectively.

For d) our exact degrees of freedom (75) do not appear and there is a tie for closest between 70 and 80. Just like we did with the t table, only when there is a tie, we find the mean of the critical values for the tied df's shown in the table. $\frac{\alpha}{2} = \frac{.01}{2} = .005$ and $1 - \frac{\alpha}{2} = 1 - .005 = .995$. In column .995 we use the mean of the critical values shown for

70 and 80 degrees of freedom ($\frac{43.275+51.172}{2} = 47.2235$) as our $\chi^2_{1-\frac{\alpha}{2}}$. In column .005 we use the mean of the critical values shown for 70 and 80 degrees of freedom ($\frac{104.215+116.321}{2} = 110.268$) as our $\chi^2_{\frac{\alpha}{2}}$.

When considering the distribution of sample variances (under the stated conditions for this section), the chi-square distribution is applicable. This fact leads to these steps for creating a confidence interval for standard deviations and variances:

Steps for creating a (1−α)100% confidence interval for σ^2 or σ

1) Find the critical values $\chi^2_{1-\frac{\alpha}{2}}$ and $\chi^2_{\frac{\alpha}{2}}$

2) Form the interval $\frac{(n-1)s^2}{\chi^2_{\frac{\alpha}{2}}} < \sigma^2 < \frac{(n-1)s^2}{\chi^2_{1-\frac{\alpha}{2}}}$ OR

$$\sqrt{\frac{(n-1)s^2}{\chi^2_{\frac{\alpha}{2}}}} < \sigma < \sqrt{\frac{(n-1)s^2}{\chi^2_{1-\frac{\alpha}{2}}}}$$

3) Interpret the interval in real world terms.

4) Respond to any given research question.

Example: Randomly selected current Statsville College students were asked how many units they have completed so far. Their responses were:

32	38	44	51	59	38	54	32	47	38	61	54

Of the students surveyed, 42% intend to major in Statistics and the standard deviation of the GPAs for everyone in the sample is .8017 with mean 2.4583. At Wudmoor College, 65% of the students have completed less than 50 units. 10 years ago, Statsville College students completed an average of 55 units with standard deviation 15.5 units. Make a 99% confidence interval that will help us assess if the numbers of units completed by current Statsville College students are less dispersed than for Statsville College students 10 years ago.

Answer:

(Note: Since the problem is asking about dispersion, we need to answer using a measure of dispersion—either the standard deviation or the variance.)

1) First, we need ${\chi^2}_{1-\frac{\alpha}{2}}$ and ${\chi^2}_{\frac{\alpha}{2}}$ and we get those as follows:

$$\alpha = 1-.99 = .01\ ;\ \frac{\alpha}{2} = \frac{.01}{2} = .005\ ;\ 1-\frac{\alpha}{2} = 1-.005 = .995\ ;\ df = 12-1 = 11$$

from Table C, ${\chi^2}_{1-\frac{\alpha}{2}} = \chi^2_{.995} = 2.603$, ${\chi^2}_{\frac{\alpha}{2}} = \chi^2_{.005} = 26.757$

2) Now, we form the interval: $\frac{(n-1)s^2}{{\chi^2}_{\frac{\alpha}{2}}} < \sigma^2 < \frac{(n-1)s^2}{{\chi^2}_{1-\frac{\alpha}{2}}}$

$$\frac{(12-1)10.1564^2}{26.757} < \sigma^2 < \frac{(12-1)10.1564^2}{2.603}$$

(we need to calculate s using a formula from Section 3.2)

$$42.4067 < \sigma^2 < 435.9113$$

taking the square root of the endpoints, we get:

$$6.5120 < \sigma < 20.8785$$

3) Interpretation: With 99% confidence we can say that the standard deviation of the numbers of units completed by all current Statsville College students is between 6.5120 and 20.8785 units.

4) Respond to research question: Since the interval contains 15.5 and is not completely below 15.5 (the standard deviation of the numbers of units completed by Statsville College students 10 years ago), there is no evidence that the numbers of units completed by current Statsville College students are less dispersed than for Statsville College students 10 years ago.

s is the standard deviation of the number of units completed by current Statsville College students in the sample. σ is the standard deviation of the number of units completed by all current Statsville College students. As always, squaring the standard deviation will give the variance of the respective sample or population.

This interval is telling us that, if we got the number of units completed by every current Statsville College student, and used a formula for σ to calculate the standard deviation of the population, the answer to that exercise would be some number between 6.5120 and 20.8785 units.

For this problem, we have the option of using the variance to address the question of dispersion. To do so, we would compare $15.5^2 = 240.25$ to the confidence interval we made for σ^2. Again, we would observe that the interval is not completely below 240.25 so there is no indication that current students have less dispersed numbers of units completed than students from 10 years ago.

Why does σ represent the standard deviation for all current students instead of the students from 10 years ago? Because we KNOW the standard deviation for all students from 10 years ago, so that is not being estimated. What we do not know is the standard deviation for all current students, so that must be estimated. We used a sample of current students to estimate the standard deviation for all current students. Remember that your sample will come from the population whose parameter you are estimating.

Be sure to avoid a few common mistakes when making an inference for this type of problem. Notice that we did not compare 10.1564 to the interval for σ to get our response to the research question. The sample standard deviation is always in the interval for the population standard deviation (because of how the interval is made) so that observation has no usefulness when trying to make an inference. Also, we did not compare 10.1564 to 15.5. 10.1564 is a sample number but 15.5 is a population number so they cannot be compared in order to make an inference. Finally, in the interpretation, we did not make reference to the sample. We know the exact standard deviation for the sample (10.1564) so that would never need to be estimated. As always, we are making an estimate of a population value!

Example: The ages (in years) of randomly selected Statsville residents were used in the creation of the Minitab output below. If someone said that the variance of the ages for all Statsville residents is more than 130 years^2, would that statement be supported by the results shown below?

Test and CI for One Variance: SVAge

```
Method

The chi-square method is only for the normal distribution.
The Bonett method is for any continuous distribution.


Statistics
```

```
Variable   N  StDev  Variance
SVAge     40   14.8       219


94% Confidence Intervals

                               CI for       CI for
Variable  Method               StDev        Variance
SVAge     Chi-Square   (12.2, 18.8)   (149, 353)
          Bonett       (12.4, 18.5)   (155, 341)
```

Answer:
With 94% confidence we can say that the variance of the ages for all Statsville residents is between 149 years^2 and 353 years^2. Since the whole interval is greater than 130 years^2, the statement is supported.

We will only concern ourselves with the chi-square method since the Bonett method is not discussed in this class. As usual, the bold label on the output reveals the type of procedure being performed (**Test and CI for One Variance**) as well as the data column being used (**SVAge**). The sample size is 40, the sample standard deviation is 14.8 years, and the variance of the ages in the sample is 219 years^2. The 94% confidence interval for the standard deviation of all Statsville residents' ages is (12.2, 18.8) years and for the variance of all Statsville residents' ages it is (149, 353) years^2.

What would be WRONG with this conclusion? "The variance of the ages of all Statsville residents is 219 years^2, and that is larger than 130 years^2, so the statement is supported." 219 is only the variance for the sample, the sample, the sample—NOT the population!! There's no way we can use that number by itself to make an inference.

Exercises

1) Refer to the Sparkelz Gems/Javonn Jewelers data from Exercise 1) in Section 7.2. Make a 98% confidence interval that would estimate how inconsistent (or, varied) the weights of Javonn Jewelers' diamonds are.

2) Refer to the Jerrell Enterprises/Jaden Industries data from Exercise 2) in Section 7.2. Suppose another company has peer performance ratings with a mean of 172.3 and standard deviation of 23.5. Are the peer performance ratings more dispersed at Jaden Industries? Use a process for creating an appropriate confidence interval that has an 80% success rate. (Note: Use the fact that, for 156 degrees of freedom, chi-square critical value 179 has area .10 to its right and chi-square critical value 133.8 has area .10 to its left.)

3) In a sample of 20 Tanayatown residents, the variance of the amount of time those residents spend acquiring news daily was 1513.74 minutes^2. Make a relevant 90% confidence interval for σ^2.

4) There are four furniture stores in a certain city. One is in the north, one is in the south, one is in the east, and one is in the west. We collected random samples of prices (in dollars) from each store and used the appropriate data for the Minitab output below. See if the standard deviation of prices at the south store is \$32, as was posited by a furniture magazine.

Test and CI for One Variance: South

```
Method

The chi-square method is only for the normal distribution.
The Bonett method is for any continuous distribution.


Statistics

Variable   N  StDev  Variance
South     82   48.4      2343


97% Confidence Intervals

                            CI for         CI for
Variable  Method            StDev         Variance
South     Chi-Square  (41.3, 58.3)  (1709, 3396)
          Bonett      (41.9, 57.4)  (1756, 3299)
```

7.5 Confidence Intervals for the Difference of Two Proportions

Conditions: The data is from two independent simple random samples (independent in that there is no pairing or relationship between values in the different samples). In each sample there are at least five successes and five failures.
(Assume all conditions are met for each example and exercise.)

Use the methods of this section when...

creating a confidence interval for the difference of two proportions

Important formulas: $E = z_{\alpha/2}\sqrt{\frac{\hat{p}_1\hat{q}_1}{n_1} + \frac{\hat{p}_2\hat{q}_2}{n_2}}$ $(\hat{p}_1 - \hat{p}_2) - E < p_1 - p_2 < (\hat{p}_1 - \hat{p}_2) + E$

When we consider differences of two sample proportions, under the stated conditions for this section, the normal distribution can be applied. This fact leads to these steps for creating a confidence interval for the difference of two proportions:

<u>Steps for creating a (1−α)100% confidence interval for $p_1 - p_2$</u>

1) Find the critical value, $z_{\alpha/2}$, that corresponds to the desired confidence level ($z_{\alpha/2}$ is a positive value).

2) Evaluate the margin of error, $E = z_{\alpha/2}\sqrt{\frac{\hat{p}_1\hat{q}_1}{n_1} + \frac{\hat{p}_2\hat{q}_2}{n_2}}$

3) Form the interval $(\hat{p}_1 - \hat{p}_2) - E < p_1 - p_2 < (\hat{p}_1 - \hat{p}_2) + E$

4) Interpret the interval in real world terms.

5) Respond to any given research question.

In this section, and in all sections dealing with differences, we'll take the first group mentioned as the first quantity in the difference and the second group as the second quantity.

Example: Randomly selected employed Americans were asked if they are employed by the government or a private employer. 62.2047% of the 127 people surveyed were employed by a private employer. The mean age of the sample was 37.5.

Employed non-Americans were asked if they are employed by their government or a private employer. Their responses are listed below (g: government, p: private—A couple respondents answered "other" with an o)

g g p g p g o g p g p p g p g p g g o p g

Create a 92% confidence interval to decide if a greater proportion of employed Americans are employed by private employers compared to the proportion of employed non-Americans who are employed by private employers.

Answer:

1) First, we need $z_{\alpha/2}$ and we get that as follows:

$\alpha = 1 - .92 = .08$ $\quad \frac{\alpha}{2} = \frac{.08}{2} = .04$ $\quad$ from Table A, $z_{\alpha/2} = z_{.04} = 1.75$

2) Next, we get the margin of error:

$$E = z_{\frac{\alpha}{2}}\sqrt{\frac{\hat{p}_1\hat{q}_1}{n_1} + \frac{\hat{p}_2\hat{q}_2}{n_2}} = 1.75\sqrt{\frac{(.622047)(.377953)}{127} + \frac{\left(\frac{8}{21}\right)\left(\frac{13}{21}\right)}{21}} = .2002$$

3) Now we form the interval: $(\hat{p}_1 - \hat{p}_2) - E < p_1 - p_2 < (\hat{p}_1 - \hat{p}_2) + E$

$$\left(.622047 - \frac{8}{21}\right) - .2002 < p_1 - p_2 < \left(.622047 - \frac{8}{21}\right) + .2002$$
$$.04089 < p_1 - p_2 < .4413$$

4) Interpretation: With 92% confidence we can say that the difference in the proportion of all employed Americans who are employed by private employers and the proportion of all employed non-Americans who are employed by private employers is between .04089 and .4413.

5) Respond to research question: Since the entire interval is positive, a greater proportion of employed Americans are employed by private employers compared to the proportion of employed non-Americans who are employed by private employers.

p_1 is the proportion of all employed Americans who are employed by private employers. p_2 is the proportion of all employed non-Americans who are employed by private employers. $\hat{p}_1$ is the proportion of employed Americans in the sample who are employed by private employers. $\hat{p}_2$ is the proportion of employed non-Americans in the sample who are employed by private employers.

To be clear, we have NO IDEA what is the proportion of all employed Americans who are employed by private employers. We have NO IDEA what is the proportion of all employed non-Americans who are employed by private employers. The interval we made does not give us this information. Is one of the population proportions equal to .04089 and the other equal to .4413? NO!!!!! The interval is telling us that if we took the proportion of all employed Americans who are employed by private employers (whatever that number is) and subtracted the proportion of all employed non-Americans who are employed by private employers (whatever that number is) the answer to that subtraction exercise would be a number between .04089 and .4413. So, the answer would be positive telling us that a greater proportion of employed Americans are employed by private employers compared to the proportion of employed non-Americans who are employed by private employers. Whenever an answer to a subtraction problem is positive, that is because we are taking a larger quantity and subtracting a smaller quantity (like, 7−3 = 4).

Suppose our interval was $-.3547 < p_1 - p_2 < -.1218$. Since the answer to the subtraction problem $p_1 - p_2$ is a negative number, that would tell us that p_2, the proportion of all employed non-Americans who are employed by private employers, is bigger than p_1, the proportion of all employed Americans who are employed by private employers. Whenever we perform subtraction and the answer is negative, that is because we took a smaller quantity and subtracted a bigger quantity (like, $3-7=-4$).

Suppose our interval was $-.1084 < p_1 - p_2 < .2035$. If the entire interval was positive, that would be a clear indication that p_1 is larger. If the entire interval was negative, that would be a clear indication that p_2 is larger. However, this interval contains zero so there is no clear indication that either population proportion is larger than the other. So our inference would be that there is no difference between the two population proportions.

Back to the original problem, what would happen if we switched the order of the groups and had p_1 stand for non-Americans and p_2 for Americans? We would, of course, need to switch the definitions of $\hat{p}_1$, $\hat{p}_2$, n_1, and n_2. The result would be that the interval would have the same endpoints in absolute value but the signs and therefore the locations of the endpoints would change. The new interval would be: $-.4413 < p_1 - p_2 < -.04089$. The real world conclusion would still be the same—a greater proportion of employed Americans are employed by private employers compared to the proportion of employed non-Americans who are employed by private employers. As usual, it is very important to have clear in mind what everything represents when working on problems so you can know exactly what the results indicate in real world terms. When stating the interpretation, the order in which you state the groups will indicate to your reader the order in which you setup the groups (first group mentioned is group 1, second group mentioned is group 2).

As you certainly know by now, the interval is estimating the difference in population proportions (not sample proportions). We can calculate the exact difference in the sample proportions because those are known. We do not know the exact population proportions so their difference is what we are estimating with the confidence interval. We always estimate something we do not already know when we use confidence intervals.

Example: Refer to the previous example. Suppose we performed that process 150 times with different samples of employed Americans and employed non-Americans each time. How many of those 150 92% confidence intervals would contain the actual difference in the proportion of all employed Americans who are employed by private employers and the proportion of all employed non-Americans who are employed by private employers?

Answer: About 138. [We get this from .92 times 150.]

Remember the confidence level gives use the success rate of the procedure used to create the interval. 92% confidence intervals contain the true population value 92% of the time.

Example: Suppose the first example of this section were changed as shown below. How would this problem be handled differently?

62.2047% of employed Americans are employed by a private employer. Employed non-Americans were asked if they are employed by their government or a private employer. Their responses are listed below (g: government, p: private—A couple respondents answered "other" with an o)

g g p g p g o g p g p p g p g p g g o p g

Create a 92% confidence interval to decide if a greater proportion of employed Americans are employed by private employers compared to the proportion of employed non-Americans who are employed by private employers.

> **Answer**: In this problem, we only have one sample—non-Americans. So, we would use the method in Section 7.2 to make a confidence interval for p (the proportion of all employed non-Americans who are employed by private employers). We then see if .622047 (the proportion of all employed Americans who are employed by private employers) is in the interval. If so, there is no difference in the population proportions. If the whole interval is greater than .622047 that tells us that the proportion of non-Americans is larger than the proportion of Americans. If the whole interval is less than .622047 that tells us that the proportion of non-Americans is smaller than the proportion of Americans.

Example: Some of the residents of Tanayatown and Statsville were researched in order to see if there is a difference in the proportion of residents from each area who are employed (i.e., the employment rate). Address the research question using the provided Minitab output.

Test and CI for Two Proportions: TTEmployed?, SVEmployed?

```
Event = y


Variable       X   N  Sample p
TTEmployed?   27  40  0.675000
SVEmployed?   25  40  0.625000


Difference = p (TTEmployed?) - p (SVEmployed?)
Estimate for difference:  0.05
```

```
99% CI for difference:   (-0.224344, 0.324344)
Test for difference = 0 (vs not = 0):   Z = 0.47   P-Value = 0.639
```

Answer:
With 99% confidence we can say that the difference in the proportion of Tanayatown residents who are employed and Statsville residents who are employed is between –.224344 and .324344. Since the interval contains zero, there is no difference in the proportion of employed residents (the employment rate) for these two areas.

In the sample of Tanayatown residents, 27 out of the 40 (67.5%) researched were employed. The proportion of employed Statsville residents observed was .625 (25 out of 40). We know that Minitab is viewing being employed as a success because "Event = y". We know that Tanayatown was the first population and Statsville is the second because of "`Difference = p (TTEmployed?) - p (SVEmployed?)`". The difference of the sample proportions was .05. The last row of output is not relevant now but will be very important in later sections. Review the output to make sure you see how the output displays the information given in this paragraph.

Side Note: The method in this section is for estimating the difference in independent (unrelated) population proportions using independent samples. Suppose we were dealing with matched pairs (dependent samples) proportion data. For example, the faculty in a Statistics department is considering a certain curriculum adjustment. For each faculty member we record whether the person is in favor of the adjustment or opposed to the adjustment. These faculty members then participate in a conference in which they are able to discuss the issue and exchange ideas. After the conference we again record for each faculty member whether the person is in favor of the adjustment or opposed to the adjustment. We are interested in whether or not this sort of conference is associated with changing the proportion of faculty in favor of the adjustment. The samples would not be independent because each of the before-conference responses (y or n) would be matched with an after-conference response because of belonging to the same individual. To make a confidence interval for the change in proportion of faculty in favor of the adjustment, we could recode the data. All of the yes votes could be coded "1" and all the no votes could be coded "0". We could then treat this data like a μ_d problem from section 7.7 and make a confidence interval using the method discussed in that section.

Exercises

1) Refer to the Sparkelz Gems/Javonn Jewelers data from Exercise 1) in Section 7.2. Use a 94% confidence interval to address the inquiry, "Which source has a larger percentage of VS1 diamonds—Sparkelz Gems or Javonn Jewelers?"

2) Refer to the Jerrell Enterprises/Jaden Industries data from Exercise 2) in Section 7.2. Use a process that has an 11% probability of not containing the true value of the population

quantity being estimated to see if there is a difference in the proportion of neutral employees who participate in professional development at Jerrell Enterprises compared to Jaden Industries.

3) Two drugs—Drug A and Drug B—are being tested in a clinical trial. The side effects experienced by each person in the trial are recorded as follows: nausea (n), itching (i), headaches (h), fatigue (f), dry mouth (d). What inference can you draw about all users of Drug A and Drug B using the output below?

Test and CI for Two Proportions: Drug A, Drug B

```
Event = n


Variable   X   N  Sample p
Drug A    13  50  0.260000
Drug B    23  46  0.500000


Difference = p (Drug A) - p (Drug B)
Estimate for difference:  -0.24
92% CI for difference:  (-0.408674, -0.0713263)
Test for difference = 0 (vs not = 0):  Z = -2.49  P-Value = 0.013
```

4) Refer to the confidence interval in 3). What is the margin of error for the estimate?

5) Refer to the confidence interval in 3). Suppose the interval is too wide for your research purposes. What is the best thing you could do to get an interval that is not so wide?

7.6 Confidence Intervals for the Difference of Two Means

Conditions: The data is from two independent simple random samples (independent in that there is no pairing or relationship between values in the different samples). We do not know the standard deviations of the two populations and do not assume they are equal. Both sample sizes are larger than 30 and/or their populations are normally distributed. (Assume all conditions are met for each example and exercise.)

Use the methods of this section when...

creating a confidence interval for the difference of two independent means

Important formulas: $E = t_{\alpha/2}\sqrt{\frac{s_1^2}{n_1} + \frac{s_2^2}{n_2}}$ $(\bar{x}_1 - \bar{x}_2) - E < \mu_1 - \mu_2 < (\bar{x}_1 - \bar{x}_2) + E$

df = smaller of $(n_1 - 1)$ and $(n_2 - 1)$

When we consider differences of two sample means, under the stated conditions of this section, the t distribution is applicable. This fact leads to the following steps for creating a confidence interval for the difference of two means:

Steps for creating a $(1-\alpha)100\%$ confidence interval for $\mu_1-\mu_2$

1) Find $t_{\alpha/2}$ for the desired confidence level ($t_{\alpha/2}$ is a positive value). For the degrees of freedom, use the smaller of (n_1-1) and (n_2-1).

2) Evaluate the margin of error, $E = t_{\alpha/2}\sqrt{\frac{s_1^2}{n_1}+\frac{s_2^2}{n_2}}$

3) Form the interval $(\bar{x}_1 - \bar{x}_2) - E < \mu_1 - \mu_2 < (\bar{x}_1 - \bar{x}_2) + E$

4) Interpret the interval in real world terms.

5) Respond to any given research question.

[Note: When creating these intervals by hand, the convention is to use for the degrees of freedom one less than the smallest sample size as described in Step 1. There is a more tedious formula, used by statistical software, to get the exact degrees of freedom. Using one less than the smallest sample size creates a more conservative estimate while avoiding the tedium of using that formula. Also, if the population standard deviations are known or assumed to be equal, the formula for the margin of error will be different. Those cases are rare.]

Example: Randomly selected Statsville College students were asked how many units they have completed so far. Their responses were:

32 38 44 51 59 38 54 32 47 38 61 54

Of the students surveyed, 42% intend to major in Statistics and the standard deviation of the GPAs for everyone in the sample is .8017 with mean 2.4583. In a sample of 31 Wudmoor College students, 20 completed less than 50 units. The mean number of units they completed was 37.1 with standard deviation 10.7. Create a 90% confidence interval for the difference in the mean number of units completed by Statsville College students and Wudmoor College students. Which college's students have completed more units, on average? Explain.

Answer:

1) First, we need $t_{\alpha/2}$ and we get that as follows:

$\alpha = 1-.90 = .10 \quad \frac{\alpha}{2} = \frac{.10}{2} = .05 \qquad \text{df} = 12-1 = 11$

from Table B, $t_{\alpha/2} = t_{.05} = 1.796$

2) Next, we get the margin of error:

$$E = t_{\alpha/2}\sqrt{\frac{s_1^2}{n_1} + \frac{s_2^2}{n_2}} = 1.796\sqrt{\frac{10.1564^2}{12} + \frac{10.7^2}{31}} = 6.2961$$

3) Now, we form the interval:

$$(\bar{x}_1 - \bar{x}_2) - E < \mu_1 - \mu_2 < (\bar{x}_1 - \bar{x}_2) + E$$
$$(45.6667 - 37.1) - 6.2961 < \mu_1 - \mu_2 < (45.6667 - 37.1) + 6.2961$$
$$2.2706 < \mu_1 - \mu_2 < 14.8628$$

4) Interpretation: With 90% confidence we can say that the difference in the mean number of units completed by all Statsville College students and all Wudmoor college students is between 2.2706 and 14.8628.

5) Respond to research question: Since the whole interval is positive, Statsville College students have completed more units, on average.

$\bar{x}_1$ is the mean number of units completed by Statsville College students in the sample. $\bar{x}_2$ is the mean number of units completed by Wudmoor College students in the sample. s_1 is the standard deviation of the number of units completed by Statsville students in the sample. s_2 is the standard deviation of the number of units completed by Wudmoor students in the sample. n_1 and n_2 are the Statsville and Wudmoor sample sizes, respectively. As we have done before, we had to calculate the sample statistics for Statsville, while they were given in the narrative for Wudmoor. μ_1 is the mean number of units completed by all Statsville College students. μ_2 is the mean number of units completed by all Wudmoor College students.

Remember that we have NO IDEA what is the mean number of units completed by all Statsville College students. We have NO IDEA what is the mean number of units completed by all Wudmoor College students. The interval is telling us that, if we took the mean number of units completed by all Statsville College students (whatever that number is) and subtracted the mean number of units completed by all Wudmoor College students (whatever that number is) the answer to that subtraction exercise would be a number between 2.2706 and 14.8628. So, the answer is positive. Whenever you subtract and you get a positive answer that is because the first number is bigger than the second (like, 10–6

= 4). That's how we can say, without knowing either population mean, that the first population mean is bigger than the second—the answer to the subtraction problem $\mu_1 - \mu_2$ is positive. We could go further and say that the mean number of units completed by all Statsville College students is between 2.2706 and 14.8628 units more than the mean for Wudmoor College students.

Suppose the confidence interval was $-12.546 < \mu_1 - \mu_2 < -6.524$. The whole interval is negative so μ_2, the mean number of units completed by all Wudmoor College students, is bigger than μ_1, the mean number of units completed by all Statsville College students. We know this because the answer to the subtraction problem $\mu_1 - \mu_2$ is negative so the second number must be bigger than the first number (like, 6−10 = −4). Again, we have no clue about the specific values of μ_1 or μ_2 but we don't need to know that to infer which one is larger because we have an inference about the answer to the subtraction problem $\mu_1 - \mu_2$.

Suppose the interval contained zero, like $-5.804 < \mu_1 - \mu_2 < 3.124$. If the interval were all positive, we would be confident that the first population mean is larger. If the interval consisted of only negative numbers we would conclude that the second population mean, μ_2, is larger. The interval gives us no clear indication that either population mean is larger than the other so we conclude that there is no difference in the population means.

As you can verify, if we approached the original problem making Wudmoor our first group and Statsville our second group, the interval would have the same numbers in absolute value but their signs, and therefore their relationship, would switch. The interval would be $-14.8628 < \mu_1 - \mu_2 < -2.2706$. Since the whole interval is negative that would tell us that μ_2, the mean number of units completed by all Statsville College students is larger. This is the same real world conclusion we obtained before! You decide, and inform your reader by your order of presentation in your interpretation, the assignment of the groups but, as long as you keep everything straight, you'll get the same real world conclusion either way.

Actually, did we really need to do all this work? Can't we just see that 45.6667 is bigger than 37.1 and move on? NO!!! The response to the research question was about the populations but the only means we have are for the samples. We can never just compare sample means and have enough information to make claims about entire populations. This is why we need inferential statistical methods, like confidence intervals.

Example: The following confidence interval was constructed for the difference in the average household size of Wudmoor residents (Group 1) and Chef County residents (Group 2).

$$-2.7891 < \mu_1 - \mu_2 < 2.6557$$

a) What is the margin of error for the estimate?

b) What was the difference in the sample mean household size for Wudmoor and Chef County?

c) Which location has a larger mean household size, Wudmoor or Chef County? How much larger is that location's mean household size?

d) What is the mean household size of Wudmoor residents?

Answer:

a) The margin of error is $\frac{2.6557-(-2.7891)}{2} = 2.7224$

b) The difference in the sample mean household size is $\frac{-2.7891+2.6557}{2} = -.0667$

c) The confidence interval contains zero so neither location has a larger mean household size.

d) We don't know the mean household size of Wudmoor residents. The confidence interval is not giving us any indication of the individual population means, but is estimating the *difference* in the population means.

If you want to break your Statistics teacher's heart, give −2.7891 as the answer for d). First of all, that shows that you do not understand how this confidence interval works. Even worse, you are giving an impossible answer to the question because the mean household size can't be negative!!!

Unless a problem specifically asks you for a value from a sample, questions are referring to the population. In c) you are being asked about the population. In b) you were asked specifically for a sample number so that is what you would provide. Remember that the whole purpose for confidence intervals is to make inferences about populations so most questions (in this class and, more importantly, in practice) are about populations. Anyone can answer a question about a sample without even taking a Statistics class!

For a) remember that the margin of error is just half the distance between the endpoints of the confidence interval. This is a direct result of the formula used to create the interval.

For b) we know the interval is created by calculating

$$(\bar{x}_1 - \bar{x}_2) - E < \mu_1 - \mu_2 < (\bar{x}_1 - \bar{x}_2) + E$$

so the difference of the sample means would be right in the middle of the endpoints of the interval.

Example: Selected residents from Statsville and Tanayatown had their ages (in years) recorded and entered into Minitab. Using the output below, which town's residents have a smaller mean age? How much smaller?

Two-Sample T-Test and CI: SVAge, TTAge

```
Two-sample T for SVAge vs TTAge

        N  Mean  StDev  SE Mean
SVAge  40  30.6   14.8      2.3
TTAge  40  62.0   45.6      7.2


Difference = mu (SVAge) - mu (TTAge)
Estimate for difference:  -31.50
97% CI for difference:  (-48.46, -14.54)
T-Test of difference = 0 (vs not =): T-Value = -4.16  P-Value = 0.000  DF = 47
```

Answer: With 97% confidence we can say that the difference in the mean age of all Statsville and Tanayatown residents is between −48.46 and −14.54 years. Since the whole interval is negative, Statsville residents have a smaller mean age. The mean age for Statsville residents is between 14.54 to 48.46 years smaller than for Tanayatown.

The "`SVAge`" row gives the sample age results for Statsville and the "`TTAge`" row gives the sample age results for Tanayatown. In the statement of the interpretation, Statsville was mentioned first which tells the reader the order of the subtraction. The output told us this was the order on the "`Difference = mu (SVAge) - mu (TTAge)`" line. The difference in the sample mean ages was −31.50. Notice that in the answer there was no mention of these sample results because these cannot be used to make a conclusion. The last row of output is not relevant now, but will be later. Stay tuned.

Exercises

1) Refer to the Sparkelz Gems/Javonn Jewelers data from Exercise 1) in Section 7.2. Is it true that Sparkelz diamonds typically weigh less than Javonn diamonds? Respond to this question using an appropriate 99% confidence interval.

2) Refer to the Jerrell Enterprises/Jaden Industries data from Exercise 2) in Section 7.2. Which company has a larger mean supervisor rating? Use a relevant 98% confidence interval to answer.

3) The following confidence interval estimates the difference in the mean tax return amount (in dollars) received last year by residents of Statsville (first population) and Wudmoor (second population):

$$-984.35 < \mu_1 - \mu_2 < -203.97$$

a) What was the mean tax return amount received by Statsville residents last year?

b) Which location's residents had the larger mean tax return amount last year? How much larger?

4) There are four furniture stores in a certain city. One is in the north, one is in the south, one is in the east, and one is in the west. We collected random samples of prices (in dollars) from each store and used the appropriate data for the Minitab output below. It is commonly believed that the south store sells more expensive products, on average, than does the north store. Test that belief using the output below.

Two-Sample T-Test and CI: North, South

```
Two-sample T for North vs South

        N    Mean  StDev  SE Mean
North  91  1508.9   46.2      4.8
South  82  1514.6   48.4      5.3


Difference = mu (North) - mu (South)
Estimate for difference:  -5.66
95% CI for difference:  (-19.89, 8.57)
T-Test of difference = 0 (vs not =): T-Value = -0.79  P-Value = 0.434  DF = 167
```

7.7 Confidence Intervals for the Mean of Differences

Conditions: The data is from two dependent simple random samples (dependent in that there is pairing due to a unique relationship between each value in one sample and a value in the other sample). We do not know the standard deviations of the two populations or their paired differences and make no assumptions about them. The number of sample pairs is larger than 30 and/or the population of paired differences is normally distributed. (Assume all conditions are met for each example and exercise.)

Use the methods of this section when...

creating a confidence interval for the mean of differences

Important formulas: $E = t_{\alpha/2}\frac{s_d}{\sqrt{n}}$ $\bar{x}_d - E < \mu_d < \bar{x}_d + E$

n is the number of pairs of data

This section deals with cases in which there are dependent samples. With dependent samples (also called paired data, or matched pairs data) there is a unique relationship between pairs of data across the two samples. For example, for randomly selected people we record the height of their parents. In the end, we would have lists of fathers' heights and mothers' heights. This would be paired data because each of the numbers in the fathers' list would have a match to one of the numbers in the mothers' list due to belonging to the same pair of parents. As another example, we collect CPE scores from students before and after attending a CPE preparation course. Each of the scores in the before list would have a unique relationship with a number in the after list, due to belonging to the same person.

When dealing with matched pairs data, the data that we care about is not the individual values, but the differences in the pairs. Since there will be variation from one pair to the next, working with differences is a more accurate way of dealing with the data than approaching the problem as though the variation observed is due solely to differences between the groups (as we do in Section 7.6, where the samples and their populations are independent). For example, in the CPE problem, there will be variation in before performance and after performance, but there is also natural variation from one student to the next. We would calculate the differences in each pair of before and after scores, and those differences would become our data. Any notation in formulas with the subscript "d" is referring to such differences. The sample size, n, is the number of differences in the sample.

Means of samples of paired differences, under the stated conditions for this section, can be modeled using a t distribution. This fact leads to the following steps for creating a confidence interval for the population mean of differences:

<u>Steps for creating a (1 – α)100% confidence interval for μ_d</u>

1) Find $t_{\alpha/2}$ for the desired confidence level and appropriate degrees of freedom.

2) Calculate the margin of error, $E = t_{\alpha/2}\frac{s_d}{\sqrt{n}}$

3) Form the interval $\bar{x}_d - E < \mu_d < \bar{x}_d + E$

4) Interpret the interval in real world terms.

5) Respond to any given research question.

Example: Randomly selected people were studied and had their hand length, foot length, and ear height (all measured in cm) measured. The measurements for each person in the sample are shown below:

Ear	6.4	6.1	5.9	5.7	6.3	5.8	6.3	6.0
Hand	20.9	18.2	15.6	17.3	19.3	17.0	19.7	18.4
Foot	29.5	27.4	21.6	26.0	25.9	22.8	28.8	24.0

Create a 99% confidence interval that will help us figure out if a person's foot is generally longer than the person's hand.

Answer:
Notice that this is a matched pairs problem because the hand and foot measurements are matched to the same people. If we had foot measurements from one sample of people representing one population and hand measurements from another sample of people from an unrelated population this would be a μ_1, μ_2 problem like those in the previous section (and it would be an odd way to setup the research!). The data that we work with will be the differences in the matched pairs, calculated in the "d" row of the following table. We will calculate the mean and standard deviation of those differences. Our work is made easier by organizing the values in a table:

									Totals
Hand	20.9	18.2	15.6	17.3	19.3	17.0	19.7	18.4	
Foot	29.5	27.4	21.6	26.0	25.9	22.8	28.8	24.0	
d	-8.6	-9.2	-6.0	-8.7	-6.6	-5.8	-9.1	-5.6	-59.6
d^2	73.96	84.64	36	75.69	43.56	33.64	82.81	31.36	461.66

(The d^2 row is used when finding the standard deviation of the differences.)

Using the formulas learned in previous sections, the mean of the differences in the sample is −7.45 cm, and their standard deviation is 1.5875 cm.

1) First, we need $t_{\alpha/2}$ and we get that as follows:

$\alpha = 1 - .99 = .01 \quad \frac{\alpha}{2} = \frac{.01}{2} = .005 \qquad df = 8 - 1 = 7$

from Table B $t_{\alpha/2} = t_{.005} = 3.499$

2) Next, we get the margin of error:

$$E = t_{\alpha/2}\frac{s_d}{\sqrt{n}} = 3.499\frac{1.5875}{\sqrt{8}} = 1.9639$$

3) Now we form the interval: $\bar{x}_d - E < \mu_d < \bar{x}_d + E$

$$-7.45 - 1.9639 < \mu_d < -7.45 + 1.9639$$

$$-9.4139 < \mu_d < -5.4861$$

4) Interpretation: With 99% confidence we can say that the mean of the differences in all people's hand length and foot length is between −9.4139 and −5.4861 cm.

5) Respond to the research question: Since the whole interval is negative, a person's foot is generally longer than the person's hand

To get the real world meaning of the interval, we had to remember how we calculated the differences: hand length minus foot length. We communicated this order of subtraction by the order the groups were mentioned in the interpretation. If the whole interval for μ_d is positive that indicates that the first quantity (in this case, hand length) is generally bigger. If the whole interval is negative that indicates that the second quantity (in this case, foot length) is generally bigger. If the interval contains zero that indicates that there is no difference, on average, in the two quantities.

If we got our same interval but it resulted from differences calculated as foot length minus hand length, the all-negative interval would indicate that hands are generally longer.

For this problem, μ_d represents the mean of all the differences in people's hand length and foot length. $\bar{x}_d$ is the mean of the differences in people's hand length and foot length in the sample. s_d is the standard deviation of those sample differences.

As usual, the interval is not estimating a value for a sample because we know the exact value for the sample (in this case, −7.45). What we do not know is the mean of the differences for ALL people, so that is what is being estimated. What this interval is telling us is that, if we got the hand length for all people, and we got the foot length for all people, and for each person we subtracted their foot length from their hand length, and then we added all those differences, and then divided that total by the number of people in the whole world, that final answer would be somewhere between −9.4139 and −5.4861 cm.

The ear data is obviously not needed to make the specific interval requested so we ignore it.

If the hand measurements came from mothers and the feet measurements came from the respective daughters, or if the hand measurements came from older siblings and the feet measurements came from the respective younger siblings, or if each hand measurement was taken in a different year and there was one foot measurement taken in each respective year, or if each hand measurement came from a different country and there was one foot measurement taken in each respective country, this would still be a μ_d problem. The point is that there needs to be some sort of matching. Each hand measurement would have had a special connection to one of the foot measurements.

Can't we just kinda glance at the measurements and see that the foot measurements are generally bigger than the hand measurements? NO! As always, we can't just look at the sample numbers and make a final interpretation about the entire population from which the sample was taken. Are the foot measurements in the sample so much bigger than the hand measurements in the sample that we can conclude that in the whole population the foot measurements are generally larger? Yes, but we only know that because the whole confidence interval is negative. We would not know that merely from glancing at the sample measurements.

Example: 40 homes from Statsville were investigated and the value of the homes (in dollars) before the construction of the new local sports arena were recorded. After the construction of the arena, the value of these same homes was recorded. The real estate agent collecting the data wanted to see if home values changed (on average) after the construction of the arena. What conclusion can you draw based on the Minitab output below?

Paired T-Test and CI: ValueNow, ValueBefore

```
Paired T for ValueNow - ValueBefore

              N    Mean  StDev  SE Mean
ValueNow     40  203729  41787     6607
ValueBefore  40  187924  37566     5940
Difference   40   15805  29318     4636


96% CI for mean difference: (5956, 25654)
T-Test of mean difference = 0 (vs not = 0): T-Value = 3.41  P-Value = 0.002
```

Answer: With 96% confidence we can say that the mean of the differences in the value of all Statsville homes after the construction of the sports arena and before the construction of the sports arena is between \$5956 and \$25654. Since the whole interval is positive, home values are generally higher after the construction of the arena than they were before the construction of the arena.

Again, the label on this output tells everything you need to know to identify what is being estimated. The "`ValueNow`" column of data in Minitab had the current (after arena) values and the "`ValueBefore`" column had the values before, as the variable names indicate. As usual, the first chunk of output is giving you the sample results, none of which can be used to make the final inference about the whole population. It's of great importance to observe how the differences are being considered: "`Paired T for ValueNow - ValueBefore`". That makes a huge difference in how to understand the results. As usual, the order in which the groups are mentioned in the interpretation communicates the order of the differences, that is why after was mentioned first and then before. Again, the last line of output can be ignored for now.

This interval is telling us that, if we considered each home in Statsville, took its current value minus its value before the arena, added each of those differences for each home, and then divided by the total number of homes in Statsville, the answer would be some number between 5956 and 25654. The interval is positive telling us that the home prices (in the whole population!) after the arena are generally larger than they were before. There could certainly be exceptions but, on average, this is the case. If the whole interval were negative, that would tell us that the before values are generally larger, and if the interval contained zero that would mean that there was no change (on average) in home values .

This is a μ_d problem because every "now" home value has a unique match with a "before" home value—they belong to the same house. So we have pairs of now and before values that are matched. If, on the other hand, we randomly selected some homes and got their "now" values, but randomly selected some other homes when we got the "before" values we would use the method from Section 7.6 to analyze the data from those independent samples representing independent populations.

Example: Consider the previous example. Suppose our interval was too wide to get useful information. How could we rectify that situation?

> **Answer**: The best thing to do is to get a bigger sample. That would mean we would need more pairs of after and before values for Statsville homes. We then would make a new confidence interval using the bigger sample. If that was impossible (say, because now that the arena is built we have no way of going back and capturing the "before" values) we could settle for reducing the confidence level.

Exercises

1) Curious shoppers recorded the prices (in dollars) for identical items from Statsville Market and from Statsville Grocers. Using the price data in the following table, determine if there is a difference (on average) between the price of items at one store versus the price of the same items at the other store, using a 90% confidence interval.

item	a	b	c	d	e	f	g	h
Market	6.78	12.54	9.99	13.65	12.09	5.99	8.99	9.27
Grocers	6.99	12.53	9.98	13.89	12.09	6.15	8.98	9.26

2) Refer to the Jerrell Enterprises/Jaden Industries data from Exercise 2) in Section 7.2. An employee at Jerrell Enterprises has surmised that supervisor performance ratings are generally smaller than peer performance ratings (for respective employees) at his company. Use an 80% confidence interval to assess the employee's conjecture.

3) Randomly selected siblings from Tanayatown were contacted. Researchers studied pairs of siblings in which one sibling had no college degree and the other sibling had a college degree. The researchers wanted to see if there is a difference in income for siblings with no college degree (Group 1) and siblings with a college degree (Group 2). With income being measured in dollars, the following confidence interval was constructed:

$$-11925 < \mu_d < -5961$$

a) Interpret the interval and make a response to the research question.

b) What was the relevant sample mean for this problem? What was the value of that statistic?

4) Consider the interval in 3). What is WRONG with these statements?

a) The difference in the mean incomes of all Tanayatown siblings who do not have a college degree and those who do have a college degree is between −11925 and −5961 dollars.

b) The mean of the differences in incomes of all Tanayatown siblings who have a college degree and those who do not have a college degree is between −11925 and −5961 dollars.

c) The mean income for all Tanayatown siblings with no college degree is −11925 dollars.

d) The mean income for all Tanayatown siblings without and with college degrees is between −11925 and −5961 dollars.

e) The mean of the differences in incomes for Tanayatown siblings in the sample who do not have a college degree and those who do have a college degree is between −11925 and −5961 dollars.

f) The mean of college in Tanayatown is −11925 degrees for siblings with no siblings and the difference with college dollars is −5961.

5) There are four furniture stores in a certain city. One is in the north, one is in the south, one is in the east, and one is in the west. We collected random samples of prices (in dollars) from each store and used the appropriate data for the Minitab output below. Make an inference about the relevant populations using the output.

Paired T-Test and CI: North, East

```
Paired T for North - East

             N     Mean   StDev  SE Mean
North       91  1508.94   46.15     4.84
East        91  2005.90   37.18     3.90
Difference  91  -496.95   62.49     6.55


93% CI for mean difference: (-508.97, -484.94)
T-Test of mean difference = 0 (vs not = 0): T-Value = -75.86  P-Value = 0.000
```

7.8 Confidence Intervals for the Ratio of Two Variances or Standard Deviations

Conditions: The data comes from two independent simple random samples. Each population must be normally distributed. [The normality requirement is very strict for this test.]
(Assume all conditions are met for each example and exercise.)

Use the methods of this section when...

creating a confidence interval for the ratio of two variances or standard deviations

Important formulas: $\frac{s_1^2/s_2^2}{F_{\alpha/2}} < \frac{\sigma_1^2}{\sigma_2^2} < \frac{s_1^2/s_2^2}{F_{1-(\alpha/2)}}$ $\sqrt{\frac{s_1^2/s_2^2}{F_{\alpha/2}}} < \frac{\sigma_1}{\sigma_2} < \sqrt{\frac{s_1^2/s_2^2}{F_{1-(\alpha/2)}}}$

We begin by introducing a new family of distributions: the F distribution. This distribution is applicable to ratios of sample variances. The number of degrees of freedom associated with the sample variance in the numerator of the ratio is called the numerator degrees of freedom, or df_1. The number of degrees of freedom associated with the sample variance in the denominator is called the denominator degrees of freedom, or df_2. This distribution is skewed right and has no negative F-scores, or critical values. The diagram that follows depicts an F distribution with different pairs of degrees of freedom values, as indicated in the key for the graph.

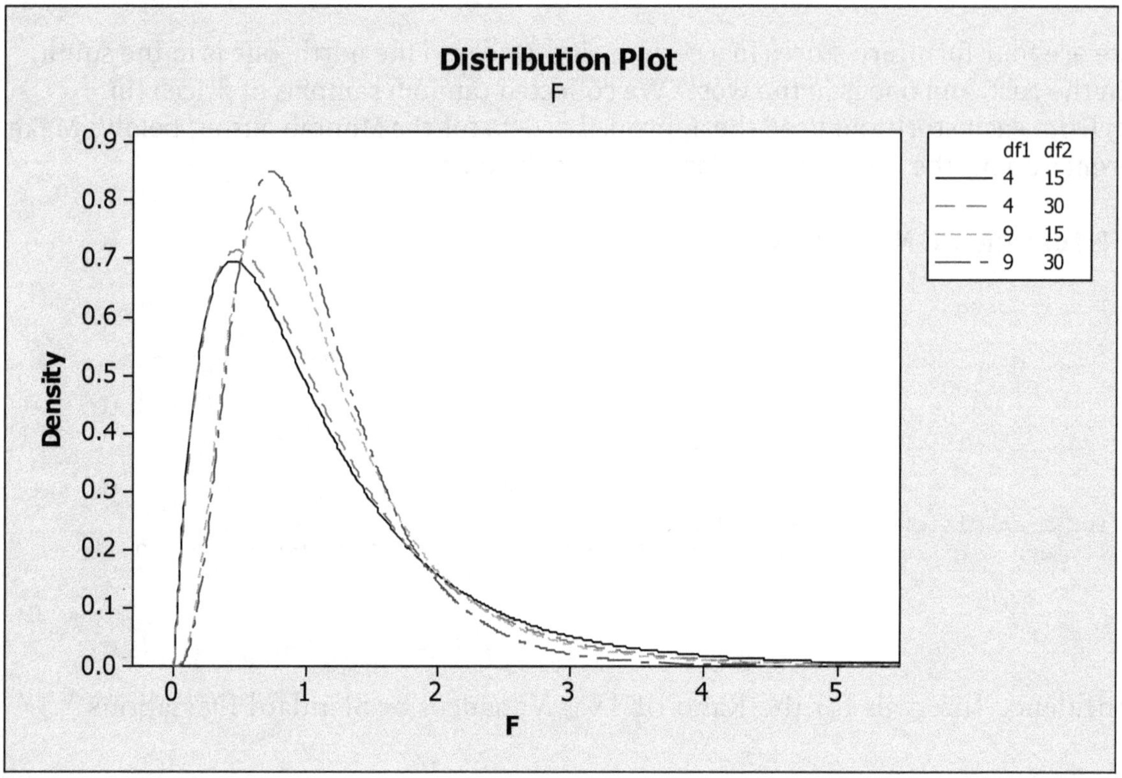

Table D (pages 335—338) gives the area to the right of F critical values. We will need two critical values, $F_{1-(\alpha/2)}$ and $F_{\alpha/2}$, for our confidence intervals. The table provides the right critical value, $F_{\alpha/2}$, directly when the appropriate degrees of freedom are used on the page of the table with area = $\frac{\alpha}{2}$. To get the left critical value, $F_{1-(\alpha/2)}$, swap the degrees of freedom and then take the reciprocal of the critical value shown on the page of the table with area = $\frac{\alpha}{2}$. Across the four pages of the F table (Table D) there are only two right-tail areas represented: .025 and .05. Two pages are for .025 in the right tail and two pages are for .05 in the right tail (as labeled). On each page of the table, df_1 is shown along the top and df_2 is shown down the left side.

Example: Find $F_{1-(\alpha/2)}$ and $F_{\alpha/2}$

a) $n_1 = 7, n_2 = 21, \alpha = .10$

b) $n_1 = 25, n_2 = 15, \alpha = .05$

c) $n_1 = 28$, $n_2 = 32$, $\alpha = .10$

Answers:

a) $F_{1-(\alpha/2)} = .2581$ $F_{\alpha/2} = 2.5990$
b) $F_{1-(\alpha/2)} = .4103$ $F_{\alpha/2} = 2.7888$
c) $F_{1-(\alpha/2)} = .5307$ $F_{\alpha/2} = 1.86415$

In a) $\frac{\alpha}{2} = \frac{.10}{2} = .05$, so we will use a page of the table for area .05 in the right tail when finding each critical value. To find $F_{\alpha/2}$ we look for the intersection of the column where $df_1 = 6$ and the row where $df_2 = 20$, to find critical value 2.5990. To find $F_{1-(\alpha/2)}$ we swap the degrees of freedom so that $df_1 = 20$ and $df_2 = 6$. Still on a .05 area page, we look for the intersection of the column where $df_1 = 20$ and the row where $df_2 = 6$ to find 3.8742. Our left critical value is the reciprocal of that number, $\frac{1}{3.8742} = .2581$

The graph looks like this:

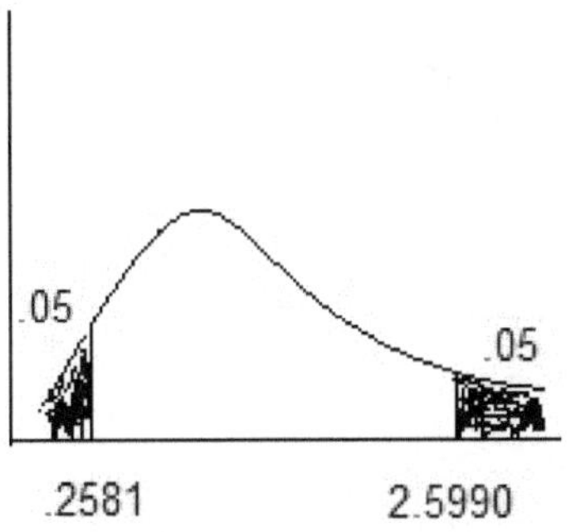

In b) $\frac{\alpha}{2} = \frac{.05}{2} = .025$, so we will use a page of the table for area .025 in the right tail when finding each critical value. To find $F_{\alpha/2}$ we look for the intersection of the column where $df_1 = 24$ and the row where $df_2 = 14$, to find critical value 2.7888. To find $F_{1-(\alpha/2)}$ we swap the degrees of freedom so that $df_1 = 14$ and $df_2 = 24$. Still on a .025 area page, we look for the intersection of the column where $df_1 = 14$ (it's not there so we use the closest: 15) and the row where $df_2 = 24$ to find 2.4374. Our left critical value is the reciprocal of that number, $\frac{1}{2.4374} = .4103$

In c) $\frac{\alpha}{2} = \frac{.10}{2} = .05$, so we will use a page of the table for area .05 in the right tail when finding each critical value. To find $F_{\alpha/2}$ we look for the intersection of the column where

df_1 = 27 and the row where df_2 = 31 (it's not there so we use the closest: 30). For df_1, 27 is not in the table and there are two df_1's that are equally close: 24 and 30. For $F_{\alpha/2}$ we'll use the mean of the critical value for df_1 = 24,df_2 = 30 and df_1 = 30,df_2 = 30 which is $\frac{1.8874+1.8409}{2} = 1.86415$. To find $F_{1-(\alpha/2)}$ we swap the degrees of freedom so that df_1 = 31 and df_2 = 27. Still on a .05 area page, we look for the intersection of the column where df_1 = 31 (it's not there so we use the closest: 30) and the row where df_2 = 27 to find 1.8842. Our left critical value is the reciprocal of that number, $\frac{1}{1.8842} = .5307$

When the stated conditions for this section are met, the F distribution can be used to describe the distribution of ratios (or quotients) of two sample variances. This fact leads to the following steps for creating a confidence interval for the ratio of variances or standard deviations:

<u>Steps for creating a (1 – α)100% confidence interval for</u> $\frac{\sigma_1^2}{\sigma_2^2}$ or $\frac{\sigma_1}{\sigma_2}$

1) Find the critical values $F_{1-(\alpha/2)}$ and $F_{\alpha/2}$

2) Form the interval $\frac{s_1^2/s_2^2}{F_{\alpha/2}} < \frac{\sigma_1^2}{\sigma_2^2} < \frac{s_1^2/s_2^2}{F_{1-(\alpha/2)}}$ OR

$$\sqrt{\frac{s_1^2/s_2^2}{F_{\alpha/2}}} < \frac{\sigma_1}{\sigma_2} < \sqrt{\frac{s_1^2/s_2^2}{F_{1-(\alpha/2)}}}$$

3) Interpret the interval in real world terms.

4) Respond to any given research question.

[Technical note: As you know, whenever we deal with quotients in the whole world of mathematics the denominator cannot be zero. That would mean that, when we deal with the ratio $\frac{\sigma_1^2}{\sigma_2^2}$, σ_2^2 cannot be zero. If we don't know the value of σ_2^2 how can we be sure that it is not zero? Remember that, in order for the population variance to be zero, all the numbers in the population would be the same. If you get a sample from a population, and all of those numbers are not the same (i.e., the variance of the sample is not zero) then the population from which the sample was taken cannot have all identical numbers. If a sample has a variance of zero, the population does not necessarily have a variance of zero. However, if the sample variance is not zero, there is no way that population can have a variance of zero. So, as long as the sample variance is some positive number, we know the population variance is some positive number.]

Example: Randomly selected Statsville College students were asked how many units they have completed so far. Their responses were:

32 38 44 51 59 38 54 32 47 38 61 54

Of the students surveyed, 42% intend to major in Statistics and the standard deviation of the GPAs for everyone in the sample is .8017 with mean 2.4583. In a sample of 31 Wudmoor College students, 20 completed less than 50 units. The mean number of units they completed was 37.1 with standard deviation 10.7. Create a 95% confidence interval to see if there is a difference in the variance of the number of units completed by Statsville College students and the variance of the number of units completed by Wudmoor College students.

Answer:
Note that while we are looking to see if there is a "difference" there is no distribution we can use for the difference of two variances $(\sigma_1^2 - \sigma_2^2)$. We do have a distribution we can use for the ratio of two variances. So, we will see if the variances are different by estimating their ratio.

1) First, we need $F_{\alpha/2}$ and $F_{1-(\alpha/2)}$ and we get them as follows:

$\alpha = 1 - .95 = .05$ $\quad \frac{\alpha}{2} = \frac{.05}{2} = .025$ $\quad df_1 = 12-1 = 11,\ df_2 = 31 - 1 = 30$

from Table C, $F_{\alpha/2} = F_{.025} = 2.4616$ (since 11 is not in the table for df_1, but 11 is halfway between 10 and 12, we will use the F critical value that is halfway between 2.5112 and 2.4120—the F critical values for df_1=10 df_2=30, and df_1 = 12 df_2 = 30)

$F_{1-(\alpha/2)}$ is obtained by switching the degrees of freedom (df_1 = 30, df_2 = 11), finding that critical value, and then taking the reciprocal of that F critical value. That would be $\frac{1}{3.1176} = .3208$

2) Next we form the interval: $\frac{s_1^2/s_2^2}{F_{\alpha/2}} < \frac{\sigma_1^2}{\sigma_2^2} < \frac{s_1^2/s_2^2}{F_{1-(\alpha/2)}}$

$$\frac{10.1564^2/10.7^2}{2.4616} < \frac{\sigma_1^2}{\sigma_2^2} < \frac{10.1564^2/10.7^2}{.3208}$$

$$.3660 < \frac{\sigma_1^2}{\sigma_2^2} < 2.8085$$

3) Interpretation: With 95% confidence we can say that the ratio of the variance of the number of units completed by all Statsville College students and the variance of the number of units completed by all Wudmoor College students is between .3660 and 2.8085.

4) Respond to research question: Since the interval contains one, there appears to be no difference in the variance of the number of units completed by Statsville College students and the variance of the number of units completed by Wudmoor College Students

σ_1^2represents the variance of the number of units completed by all Statsville College students. σ_2^2represents the variance of the number of units completed by all Wudmoor College students. If the whole interval was less than one that would indicate that the variance of the number of units completed by all Wudmoor College students is larger than the variance of the number of units completed by all Statsville College students. If the whole interval was bigger than one that would indicate that the variance of the number of units completed by all Statsville College students is larger than the variance of the number of units completed by all Wudmoor College students. This comes from the simple arithmetic facts that (for fractions that are all positive, which we know must be true for fractions involving two variances) if the fraction is bigger than one the numerator is bigger than the denominator (like, $\frac{6}{3} = 2$), if the fraction is smaller than one the numerator is smaller than the denominator (like, $\frac{3}{6} = .5$), and if the fraction is equal to one the numerator and denominator are the same (like, $\frac{6}{6} = 1$). Since our interval is not completely larger than one or smaller than one, there is no clear indication that one college has a larger variance than the other. So, we conclude that there is no difference in the two variances.

As was the case with intervals for differences, the order in which we mention the groups in the interpretation tells the reader the order of the ratio (numerator is mentioned first, denominator is mentioned second).

As always, our interval and our conclusion refer to the population variances, not the sample variances!! Also, we have NO IDEA what is the value of either population variance. The interval is just for the ratio of the variances, not the individual variances. If we were to evaluate the variance of the number of units completed by all Statsville College students divided by the variance of the number of units completed by all Wudmoor College students the answer to that division problem would be a number between .3660 and 2.8085. Notice that, even though the individual variances would have units "units^2", the ratio has no units because the original units cancel out.

If we wanted an interval for the ratio of the standard deviations, we could just take the square root of the endpoints of the confidence interval we made to get (.6050, 1.6759).

Example: Suppose the previous example was changed as shown below. How would we approach this problem?

Randomly selected Statsville College students were asked how many units they have completed so far. Their responses were:

32 38 44 51 59 38 54 32 47 38 61 54

Of the students surveyed, 42% intend to major in Statistics and the standard deviation of the GPAs for everyone in the sample is .8017 with mean 2.4583. The mean number of units completed by Wudmoor College students is 37.1 with standard deviation 10.7. Create a 95% confidence interval to see if there is a difference in the variance of the number of units completed by Statsville College students and the variance of the number of units completed by Wudmoor College students.

Answer: This time we have the population variance ($10.7^2 = 114.49$) for the number of units completed by all Wudmoor students. We only have one sample and it is from Statsville. We would use the method in Section 7.4 to make a confidence interval for σ^2 (the variance of the number of units completed by all Statsville college students) and compare it to 114.49. If the interval contains 114.49, there is no difference in the variance for the two schools. If the whole interval is greater than 114.49, Statsville has the larger variance. If the whole interval is smaller than 114.49, Statsville has the smaller variance.

Example: 11 randomly selected meal combinations from one restaurant chain had fat contents with a standard deviation of 5.8g. Four randomly selected meal combinations from another restaurant chain had fat contents with a standard deviation of 3.4g. Make a 95% confidence interval for a relevant ratio of two standard deviations.

Answer:

1) First, we need $F_{\alpha/2}$ and $F_{1-(\alpha/2)}$ and we get them as follows:

$\alpha = 1-.95 = .05$ $\frac{\alpha}{2} = \frac{.05}{2} = .025$ $df_1 = 11-1 = 10$, $df_2 = 4-1 = 3$

from table C, $F_{\alpha/2} = F_{.025} = 14.419$. $F_{1-(\alpha/2)}$ is obtained by switching the degrees of freedom ($df_1 = 3$, $df_2 = 10$), finding that critical value, and then taking the reciprocal of that F critical value. That would be $\frac{1}{4.8256} = .2072$

2) Next we form the interval:

$$\frac{s_1^2/s_2^2}{F_{\alpha/2}} < \frac{\sigma_1^2}{\sigma_2^2} < \frac{s_1^2/s_2^2}{F_{1-(\alpha/2)}}$$

$$\frac{5.8^2/3.4^2}{14.419} < \frac{\sigma_1^2}{\sigma_2^2} < \frac{5.8^2/3.4^2}{.2072}$$

$$.2018 < \frac{\sigma_1^2}{\sigma_2^2} < 14.04457$$

Taking the square root of the endpoints we get the confidence interval for the ratio of the standard deviations: (.4492, 3.7476).

3) Interpretation: With 95% confidence we can say that the ratio of the standard deviation of the fat content of all meal combinations at the first restaurant and the standard deviation of the fat content of all meal combinations at the second restaurant is between .4492 and 3.7476.

(no research question was asked)

σ_1^2 is the variance of the fat content of all meal combinations at the first restaurant and σ_2^2 is the variance of the fat content of all meal combinations at the second restaurant. Each of these variances would have units g^2 but in the ratio the units cancel. Once we take the square root of the endpoints of the interval for the ratio of the variances, we get the interval for the ratio of the standard deviations (which also has no units).

If the entire interval was smaller than one, that would indicate that the numerator (the standard deviation of the fat content of all meal combinations at the first restaurant) is smaller than the denominator (the standard deviation of the fat content of all meal combinations at the second restaurant). If the entire interval was bigger than one, that would indicate that the numerator (the standard deviation of the fat content of all meal combinations at the first restaurant) is bigger than the denominator (the standard deviation of the fat content of all meal combinations at the second restaurant). Since our interval contains one, there is no indication that either standard deviation is bigger than the other so we would conclude there is no difference in the standard deviation of the fat content of meal combinations at these two restaurants.

Is the standard deviation of the fat content of all meal combinations at the first restaurant .4492g and the standard deviation of the fat content of all meal combinations at the second restaurant 3.7476g? NO!!!!! This confidence interval tells us nothing about the value of the individual population standard deviations. It only gives us information about the ratio of the standard deviations. We have no idea what the individual population standard deviations are.

Notice that the two examples considered above explicitly mentioned variance or standard deviation in the directions. Of course, the problems could be presented in terms of "variability", "dispersion", "spread", etc. because that's what the standard deviation and the variance measure.

Example: Refer to the previous example. State why this conclusion would make no sense: "Since the whole interval is positive, the standard deviation of the fat content of all meal combinations at the first restaurant is larger than the standard deviation of the fat content of all meal combinations at the second restaurant."

> **Answer**: We know that standard deviations are always non-negative and, since our sample standard deviations were not zero, neither are the population standard deviations. Each of the population standard deviations are positive so the ratio of the two standard deviations would have to be positive. The observation that "the whole interval is positive" tells nothing about which one is bigger. This is an interval for a quotient (or ratio) of two positive numbers, not the difference (subtraction) of two numbers.

As usual, we must clearly understand exactly what everything stands for when we make and use confidence intervals.

Example: Selected residents from Statsville and Tanayatown had their ages (in years) recorded and entered into Minitab. Which location has residents with less dispersed ages?

Test and CI for Two Variances: SVAge, TTAge

```
Method

Null hypothesis          Sigma(SVAge) / Sigma(TTAge) = 1
Alternative hypothesis   Sigma(SVAge) / Sigma(TTAge) not = 1
Significance level       Alpha = 0.07
```

```
Statistics

Variable   N   StDev  Variance
SVAge     40  14.791   218.767
TTAge     40  45.599  2079.279

Ratio of standard deviations = 0.324
Ratio of variances = 0.105


93% Confidence Intervals

                                     CI for
Distribution    CI for StDev        Variance
of Data             Ratio             Ratio
Normal         (0.242, 0.435)  (0.058, 0.189)
Continuous     (0.236, 0.548)  (0.056, 0.300)
```

Answer: With 93% confidence we can say that the ratio of the standard deviation of the ages of all Statsville residents and the standard deviation of the ages of all Tanayatown residents is between .242 and .435. Since the entire interval is less than one, the ages of all Statsville residents are less dispersed than the ages of all Tanayatown residents.

For now we can ignore the "Method" section of output. Under the "Statistics" section we find the size, standard deviation, and variance for each sample (Statsville first and Tanaytown second). The ratio of the sample standard deviations and variances are shown as labeled. "CI for StDev Ratio" and "CI for Variance Ratio" show the confidence intervals for the ratio of the population standard deviations and variances, respectively. We'll use the "Normal" results because this course assumes normality of the populations. The "Continuous" row is beyond the level of this course.

If the whole interval were greater than one, we would conclude that the ages of all Statsville residents are more dispersed than the ages of all Tanayatown residents. If the interval contained one we would conclude there is no difference in the dispersion of ages for the two places.

We could use the variance confidence interval to answer the question since that would also address the question of dispersion. Our conclusion would be the same.

Why can't we just compare 14.791 to 45.599 and make the conclusion based on that? I know you know the answer. Right? All together now...BECAUSE THOSE ARE THE SAMPLE STANDARD DEVIATIONS AND CANNOT BE USED TO MAKE A CONCLUSION ABOUT THE POPULATIONS!!!!!

Excellent.

Exercises

1) Refer to the Sparkelz Gems/Javonn Jewelers data from Exercise 1) in Section 7.2. With a confidence level of 95%, see if there is less spread in the sizes for Sparkelz diamonds compared to Javonn diamonds.

2) Refer to the Jerrell Enterprises/Jaden Industries data from Exercise 2) in Section 7.2. An industrial journal reported that peer ratings at Jerrell Enterprises have more variability than the peer ratings at Jaden Industries. Respond to that claim using a 95% confidence interval. (Use these facts for an F distribution: df_1=141, df_2=156 area in right tail = .025, critical value = 1.380; df_1=156, df_2=141, area in right tail = .025, critical value = 1.384)

3) There are four furniture stores in a certain city. One is in the north, one is in the south, one is in the east, and one is in the west. We collected random samples of prices (in dollars) from each store and used the appropriate data for the Minitab output below. Which store has more inconsistent pricing among its items?

Test and CI for Two Variances: West, South

```
Method

Null hypothesis          Sigma(West) / Sigma(South) = 1
Alternative hypothesis   Sigma(West) / Sigma(South) not = 1
Significance level       Alpha = 0.1


Statistics

Variable   N   StDev  Variance
West      73  79.113  6258.907
South     82  48.404  2342.953

Ratio of standard deviations = 1.634
Ratio of variances = 2.671


90% Confidence Intervals

                                      CI for
Distribution    CI for StDev        Variance
of Data            Ratio              Ratio
Normal          (1.354, 1.978)  (1.833, 3.913)
Continuous      (1.252, 1.941)  (1.567, 3.767)
```

4) Refer to 3). Suppose we generated output like that based on 300 samples each from the West and South stores. How many of the confidence intervals would contain the true value of the ratio of the standard deviations of the prices at those stores?

7.9 Confidence Intervals: A Mixture

We have now considered all of the confidence intervals that we will cover in this course. As is the case with most math and Statistics books, each section of this text contains similar problems. So, if you're not really paying attention, you could breeze through the exercises for each section because once you figure out how to answer the first one, the others are all pretty similar. Unfortunately, that is not enough to show that you actually know the material. The use and understanding of Statistics is based on knowing what kind of procedure is appropriate for the data you have and the research question at hand. You have to be able to look at the situation, look at the data, and do the right thing with it. To that end, this section includes problems from the previous confidence interval sections. There is no order to them. Read the situation, figure out the correct type of confidence interval to create, and then do it!

One tip to keep in mind is that the number of samples will match the number of populations involved in your confidence interval. For example, if the problem involves proportion and there is one sample, the interval will be for p. If there are two separate samples, the interval will be for p_1-p_2. Before trying the exercises in this section, you may want to go back and take note of how the intervals for a single parameter from one population (p, μ, σ, σ^2) used a single sample, whereas the intervals for differences or ratios from two populations (p_1-p_2, $\mu_1-\mu_2$, μ_d, $\frac{\sigma_1^2}{\sigma_2^2}$, $\frac{\sigma_1}{\sigma_2}$) used two samples.

As always, it is important to read problems carefully to know what information is relevant and what is not relevant to the specific research question at hand. When using statistical methods, it is quite common to do several different things with one set of data to answer several different research questions. As you go from one question to the next, you have to know what information is relevant to each inquiry. The pieces of data that are not relevant to one inquiry do not disappear, and then reappear when needed for another question. All the data stays where it is but you have to decipher what is needed and when.

There are no trick questions. Read carefully and answer what is asked. There's nothing tricky about that.

As you know by now, there are no "key words" in Statistics problems because the English language affords us the ability to express the same research questions using a wide variety of words. It's all about understanding what's happening. If research questions are about the mean, the average, one group being generally larger than another, one group tending to have smaller values than the other, etc. this is addressed using the mean. If a problem is about a proportion, a percentage, a rate, etc. this is addressed using the proportion. If a problem is about variation, dispersion, spread, inconsistency, variability, etc. we can use the standard deviation or the variance to address that. Notice the repeated use of "etc." in this paragraph. Once again, you cannot be fixated on certain words but think about the context of what is being researched—this is not just for "passing this class" but for being able to use Statistics far beyond the class you're currently taking.

For any confidence interval question, the first step is figuring out which type of confidence interval will be used and why. Once that is clear to you, then you can proceed to determine the correct formula to use and the relevant information that will be needed for the formula. After that, choose the correct critical value(s) from the applicable distribution table, and carefully perform the calculations.

If all else fails, and you cannot figure out what kind of confidence interval to make from reading the problem, you may need to try process of elimination. "Well, I know it's not a [__] confidence interval because of [__], and I know it's not [__] because [__]..." There are only a handful of confidence intervals you learned to make so, if you can rule out some, that can help you pick the correct type for the problem.

In case you did not notice this pattern yet, certain distributions are matched with certain parameters. If you're working with the mean(s), the t distribution will be used. This is for μ, $\mu_1-\mu_2$, μ_d. If you are working with the proportion(s), the z distribution will be used. This is for p, p_1-p_2. If you are estimating a single measure of variation (σ or σ^2), χ^2 is your distribution. If you are working with the ratio of measures of variation ($\frac{\sigma_1^2}{\sigma_2^2}, \frac{\sigma_1}{\sigma_2}$), the F distribution is what you use.

In addition to the exercises in this section, you may also find it helpful to randomly pick exercises from the earlier sections in this chapter and try the exercises again without looking at work you already did or similar examples from the section. Approach each problem like you are seeing it for the first time and see if you can figure out what to do. Hopefully, the only help you will need from your notes or the text is the applicable formula.

Speaking of formulas, you should be able to recognize what formula(s) go with each type of confidence interval. You probably don't need to have the formulas memorized (verify this with your instructor) but you should be able to recognize them. For example, when I see

$$E = z_{\alpha/2}\sqrt{\frac{\hat{p}_1\hat{q}_1}{n_1} + \frac{\hat{p}_2\hat{q}_2}{n_2}}$$

I immediately know that is the margin of error formula when creating a confidence interval for $p_1 - p_2$. How? Well, I know it is a margin of error formula of some kind because of the E. I know it applies to two proportions because I see the two $\hat{p}$ symbols in the radicals. Review the other formulas and look for the indications that tell you which formula is used for which type of interval.

Finally, remember that confidence intervals always always always estimate something for population(s), never sample(s)!!! That's kind of a big deal.

Exercises

1) Randomly selected students from Chef County College were surveyed. Nine of them reported liking lacrosse. Six of them were first-generation college students. Their mean household income was $64397.12 with standard deviation $1923.44. For 85% of them both of their parents are still alive. The variance of their GPAs was .6714. These students were also asked their typical commute time to school, typical commute time from school, typical amount of time grooming on a school day, and whether they typically walk to and from school or ride to and from school. Those responses are recorded below:

Student	A	B	C	D	E	F	G	H	I	J	K	L	M
To	30	10	25	15	45	60	5	25	30	20	50	15	12
From	60	10	60	40	60	60	5	20	35	20	60	15	12
Groom	20	40	30	15	90	35	40	20	30	15	10	25	30
Walk/Ride	r	w	r	r	r	r	w	r	r	w	r	w	w

(times are in minutes)

Randomly selected students from Twin Streams College were surveyed. Of the 19 students surveyed, their mean typical commute time to school was 26.27 minutes with variance 265.69 minutes^2, mean typical amount of time spent grooming on a school day was 41.32 minutes with standard deviation 20.1 minutes, average typical commute time from school was 32.89 minutes with variance 338.56 minutes^2, and 57.9% of them typically walk to and from school. Eight of them were first-generation college students. Their mean household income was $73219.34 with standard deviation $1899.37. 31.58% typically spend less than 20 minutes commuting to school. Their mothers averaged 3.84 years of higher education with a median of three years.

a) Create an appropriate 95% confidence interval to see if Chef County College students generally have a longer commute to school than from school.

b) Is there a difference in the percentage of Twin Streams College students who generally walk to and from school and the percentage of Chef County College students who generally walk to and from school? Create a 93% confidence interval that would address that question and respond to the question.

c) Respond to the claim that 47% of Chef County College students are first generation college students. Use a 96% confidence interval.

d) Students at Turner College spend an average of 45.3 minutes grooming on a typical school day, with standard deviation 32.3 minutes. Use a 90% confidence interval to test the claim that there is less dispersion in grooming times among Turner College students compared to Twin Streams College students.

e) 13 selected students at Turner College spend an average of 45.3 minutes grooming on a typical school day, with standard deviation 32.3 minutes. Use a 90% confidence interval to test the claim that there is less dispersion in grooming times among Turner College students compared to Twin Streams College students.

f) Suppose we asked all Chef County College students their typical commute time to school and we asked all Chef County College students their typical commute time from school. And then, for each student, we subtract their typical commute time from school from their typical commute time to school. Finally, we add all those differences and divide by the total number of Chef County College students. Give an estimate for what that final number would be.

2) Residents from Statsville (Group 1) and Tanayatown (Group 2) were surveyed and asked how many hours they worked last week. Based on the results, the following confidence interval was created:

$$.4987 < \frac{\sigma_1}{\sigma_2} < .8402$$

Does this support the claim that Tanayatown residents have greater variability in the amount of time they spend working? Explain.

3) In a study of the use of free time by Americans, the following confidence interval was made for the mean amount of time (hours) Americans spend watching TV each week:

$$18 < \mu < 22$$

The *Journal of Free Time* reports that, "Americans watch TV an average of 19.7 hours per week." Evaluate that claim using the confidence interval.

4) For Exercises 2) and 3) why would it be incorrect to have an inference that starts with, "Since the interval is positive..."?

5) Some Chef County (Group 1) and Wudmoor (Group 2) residents were asked if they would continue working if they suddenly (from a lottery, inheritance, etc.) got enough money to live as comfortably as they would like for the rest of their lives. Based on "yes" responses, the following confidence interval was created:

$$.1568 < p_1 - p_2 < .2128$$

a) Draw a conclusion about the populations based on the interval.

b) What is the margin of error for this estimate?

c) What was the difference in the sample proportion of "yes" responses?

d) What proportion of all Chef County residents would continue working if they suddenly (from a lottery, inheritance, etc.) got enough money to live as comfortably as they would like for the rest of their lives?

6) Moviegoers were asked their favorite soft drink. 14 chose Popsee and 58 chose something else. Create a 99% confidence interval for the proportion of moviegoers who prefer Popsee.

7) The same moviegoers from 6) who preferred Popsee were observed during their stay in the movie theater on the day of the interview. They stayed at the theater (from the time they entered until the time they left) an average of 145.6 minutes with variance 600.3 minutes^2. Use this information to make a 90% confidence interval for μ.

8) A movie theater manager wants to estimate the proportion of moviegoers who generally buy popcorn when visiting the movie theatre. What is the smallest number of moviegoers he needs to survey to make sure the estimate is within .07 of the true proportion with 87% confidence?

9) Residents from Chef County and residents from Twin Streams were asked if they **r**ent or **o**wn their home (variable: **status**), how long (years) they have lived at their current residence (variable: **residence**), if children live in the household (yes or no, variable:

children), how long (years) they have worked at their current job (variable: **job**). Renters were also categorized based on how long they have rented their current home as short term (less than a year), moderate term (from one to five years), and long term (over fivc years) using the variable **rent-type,** with responses coded as s, m, or l. Each respondent's city of residence was recorded in the **location** variable (c: Chef County, t: Twin Streams). The following Minitab output was created based in that data:

One-Sample T: residence

```
Variable     N   Mean  StDev  SE Mean       95% CI
residence   67  5.263  4.800    0.586  (4.092, 6.433)
```

Test and CI for One Proportion: status

```
Event = r


Variable    X   N  Sample p          98% CI
status     31  67  0.462687  (0.319539, 0.610391)
```

Test and CI for Two Variances: job vs location

```
Method

Null hypothesis         Sigma(c) / Sigma(t) = 0.5
Alternative hypothesis  Sigma(c) / Sigma(t) not = 0.5
Significance level      Alpha = 0.1


Statistics

location   N  StDev  Variance
c         33  3.150     9.925
t         34  4.958    24.582

Ratio of standard deviations = 0.635
Ratio of variances = 0.404


90% Confidence Intervals

                                        CI for
Distribution   CI for StDev           Variance
of Data            Ratio                Ratio
Normal         (0.474, 0.852)   (0.225, 0.726)
Continuous     (0.464, 0.897)   (0.215, 0.805)


Tests

                                                  Test
Method                           DF1  DF2    Statistic  P-Value
F Test (normal)                   32   33         1.62    0.176
Levene's Test (any continuous)     1   65         1.67    0.201
```

Two-Sample T-Test and CI: residence, location

```
Two-sample T for residence

location   N  Mean  StDev  SE Mean
c         33  5.60   4.82     0.84
t         34  4.94   4.83     0.83


Difference = mu (c) - mu (t)
Estimate for difference:  0.66
99% CI for difference:  (-2.47, 3.79)
T-Test of difference = 0 (vs not =): T-Value = 0.56  P-Value = 0.575  DF = 64
```

a) Is there a difference in the mean amount of time Chef County residents have been at their current residence compared to Twin Streams residents?

b) A local paper claims that the proportion of residents in Chef County and Twin Streams who rent their homes is .84. Respond to that claim.

c) What percentage of the intervals that are created using the process that was used to create the first interval in the Minitab output contain the true value of the population mean?

8 Basic Hypothesis Testing

8.1 Hypothesis Testing Foundations

We have considered the estimation aspect of inferential statistics—confidence intervals. Now we consider formal tests of parameters—hypothesis testing.

When we say, "hypothesis" we are referring to a statement about one or more parameters. There are two types of hypotheses: null and alternative. The null hypothesis (symbolized as H_0) is a statement of equality regarding a parameter or parameters. The alternative hypothesis (symbolized as H_1 in this text, H_a or H_A in some other sources) is a transformation of the null hypothesis into a statement of inequality. The alternative to the null hypothesis' statement of equality can be "less than", "greater than", or "not equal to". The alternative hypothesis is sometimes called the *research hypothesis* because it is often the motivation of the research.

Hypothesis testing essentially investigates mathematically which of the two hypotheses is more likely based on the observations from a sample. If the sample results are "pretty close" to what the null hypothesis states, then we do not reject that hypothesis. If the sample results are "pretty far" from what the null hypothesis states, then we reject the null hypothesis in favor of the alternative hypothesis.

The null hypothesis is the direct object of the test. Technically, we never "accept" the null hypothesis (although you sometimes hear people say this). We either reject the null hypothesis or we fail to reject the null hypothesis. We will never prove the null hypothesis to be true but we will obtain enough evidence from the sample to say that the null hypothesis is wrong, or we will not obtain enough evidence to say the null hypothesis is wrong.

Suppose a classmate tells you he had cereal for breakfast this morning. OK. Great. You weren't there and you don't have any reason to reject that statement so you do not. This is not the same thing, though, as proving that the person did have cereal for breakfast; you simply do not have evidence that the person did not have cereal for breakfast. This is like failing to reject a claim.

Suppose that same classmate tells you he is 11 feet tall. Well, without even getting out a measuring stick, you can look at the person and safely conclude that his claim is false. In that case, you do have evidence to reject that claim, so you would. You don't have to know exactly how tall he is to safely conclude that, whatever his height is, it is not 11 feet.

When we write our hypotheses, we generally use parameter symbols like we used earlier in the course (p, μ, etc.). These will always stand for quantities we do not already know. We will be testing hypotheses about those unknown quantities.

Each hypothesis test has a significance level, which has the symbol α. The significance level is the probability of rejecting the null hypothesis when it is actually true. If α is not stated it is understood that $\alpha = .05$.

Each parameter or set of parameters has its own test statistic that follows (or is closely approximated by) a known probability distribution, provided certain conditions are met. [These conditions will be mentioned when each process is discussed, but you can assume that all problems in this book meet the necessary conditions. In practice, the processes we will discuss generally provide useful results as long as all necessary conditions are close to being met, if not perfectly. When doing legitimate research the conditions should be verified and, when data grossly fails to meet necessary conditions, there is a whole body of other methods—bootstrapping and nonparametric statistics—that should be used. There are upper division and graduate level classes and textbooks for these special topics.] Each test statistic will have a different formula based on the distribution of the sample statistic (p-hat, x-bar, etc.). We will continue to use the z distribution (standard normal) for proportions, the t distribution for means, the χ^2 distribution for one measure of variation, and the F distribution for two measures of variation. If the test statistic falls in the critical region (or rejection region) of the distribution of the sample statistic, we will reject the null hypothesis in favor of the alternative hypothesis. This is traditional hypothesis testing. The idea here is that we construct the distribution of the statistic that is indicated by the null hypothesis. Based on the alternative hypothesis, we notice the extremely small and/or large values of the hypothetical distribution in the left and/or right tails, the total proportion of which is α. If our sample has a statistic that falls in the "extremes" of the distribution then we will reject the null hypothesis' claim about the distribution, in favor of the alternative hypothesis' claim. Otherwise, we fail to reject the null hypothesis.

These are the steps for traditional hypothesis testing:

Step 1: State the hypotheses

Step 2: Calculate the test statistic

Step 3: Setup a diagram of the relevant distribution with the appropriate tail(s) shaded to indicate the rejection region

> Details: If the alternative hypothesis has "<" we shade the left tail, if it has ">" we shade the right tail, if it has "≠" we shade both tails. We indicate a total area of α in the tail(s). Using the indicated area we find the score(s) from the applicable distribution to serve as our cutoff(s), called critical value(s), separating the rejection region from the rest of the distribution.

Step 4: Plot the test statistic in the diagram from Step 3. If the test statistic falls in the rejection region, reject H_0; otherwise fail to reject H_0

Step 5: State the real world conclusion using the Guide for Stating Real World Conclusions (located later in this section)

Another method for hypothesis testing involves p-values. You can think of the p-value as the probability of getting a sample like we got, assuming the null hypothesis is true. [Technically, "a sample like we got" is "a sample that produces a test statistic at least as extreme as the one in our sample."] If this probability is “small” (less than α) that suggests that H_0 is unlikely so we reject H_0. Otherwise, we fail to reject H_0. As a formally stated rule this is:

p-value < α reject H_0
otherwise fail to reject H_0

The idea here is that we again construct the distribution of the statistic that is indicated by the null hypothesis. We consider the probability of a sample having a test statistic at least as extreme as the test statistic our sample had [where "extreme" is determined by the alternative hypothesis]. If this probability is small, i.e. less than α, we reject the null hypothesis because the claim in the null hypothesis appears to be quite unlikely.

These are the steps for p-value hypothesis testing:

Step 1: State the hypotheses

Step 2: Calculate the test statistic

Step 3: Setup a diagram of the relevant distribution with the appropriate tail(s) shaded and find the area in the tail(s); the total shaded area is the p-value

> Details: If the alternative hypothesis has "<" we shade the left tail, if it has ">" we shade the right tail, if it has "≠" we shade both tails. We place the test statistic (and its opposite, when both tails are shaded) at the border(s) for the tail(s) and then find the total area in the tail(s).

Step 4: If the p-value is less than α, reject H_0; otherwise fail to reject H_0

Step 5: State the real world conclusion using the Guide for Stating Real World Conclusions (located later in this section)

We'll use both methods when completing examples and you should do the same when completing the exercises! Both methods are used in practice, so you should understand both.

Some software packages label the p-value as "significance" or abbreviate that as "sig." The reason is because the p-value is considered the significance of the test (not to be confused with the *significance level* of the test, which is α).

The following is a guide for stating real world conclusions. It works every time (not just in this text and your introductory Statistics course, but in the whole universe of Statistics). This takes all the mystery out of how to state the conclusion to any hypothesis test (my apologies to those who enjoy mysteries).

Guide for stating real world conclusions:

Decision	Real World Conclusion
Reject H_0	H_1 (using words from the problem)
Fail to reject H_0	We did not find that H_1 (using words from the problem)

When you reject the null hypothesis, your conclusion is whatever the alternative hypothesis says, using words from the original problem.

When you fail to reject the null hypothesis, your conclusion is "We did not find that" followed by whatever the alternative hypothesis says, using words from the original problem.

Works every time.

When you reject H_0, it is said that there is a "significant" result.

When we use samples to test hypotheses about populations sometimes, despite our best efforts to use good data collection methods (mentioned in Chapter 6), the test procedures lead us to conclusions that do not match the truth about the populations. When this happens, the methods of hypothesis testing have committed an error. These errors come in two types: Type I and Type II. A Type I error results when we reject H_0 but the reality of the population dictates that was the wrong conclusion. A Type II error results when we fail to reject H_0 but the reality of the population dictates that was the wrong conclusion. The probability of a Type I error is α, but the probability of Type II error is more complicated to figure. That is tackled in upper division courses. Depending on the situation, one of the types of errors could be more severe than the other so the research needs to be designed with that in mind. Again, the details of that await you in your next-level Statistics course.

We are still working with inferential statistics which means that all of our hypothesizing and our conclusions are about populations—NEVER samples. We use samples, yes, but for the purpose of drawing conclusions about the entire populations from which the samples are drawn.

You might notice that some questions in this Chapter could be informally addressed using the confidence intervals we studied in Chapter 7. We are now looking at formal methods of testing hypotheses. All exercises in this chapter should use these formal hypothesis testing methods, not confidence intervals.

OK, much of what you just read went completely over your head. That is unavoidable. This is all very abstract at this point and is difficult to grasp without examples. The main purpose of this opening section was to give you something to refer back to later. The steps, the guide, etc. will be helpful references once you start reading examples and trying exercises. Speaking of reading examples and trying exercises...

8.2 Hypothesis Testing for a Proportion

Conditions: The data is from a simple random sample. The requirements from the binomial distribution (given in Chapter 4) are met. $np \geq 5$ and $nq \geq 5$.
(Assume all conditions are met for each example and exercise.)

Use the methods of this section when...

testing a claim about a proportion

Important formula: $z = \frac{\hat{p}-p}{\sqrt{\frac{pq}{n}}}$

We will use the methods and steps described in Section 8.1, so refer back to them as needed for the description of each step.

Example: Randomly selected Americans were asked if they are employed by the government or a private employer. 62.2047% of the 127 people surveyed were employed by a private employer. The mean age of the sample was 37.5. 13.4% of respondents reported being "very dissatisfied" with their current employment and 20 respondents reported having searched for a job in the past year. Using a .05 significance level, test the claim that the majority of Americans are employed by private employers.

Answer:
Traditional method

Step 1: H_0: p = .5
H_1: p > .5 ("majority" = p > .5)

Step 2: Test statistic: $z = \frac{\hat{p}-p}{\sqrt{\frac{pq}{n}}} = \frac{.622047-.5}{\sqrt{\frac{(.5)(.5)}{127}}} = 2.7508$

Step 3:

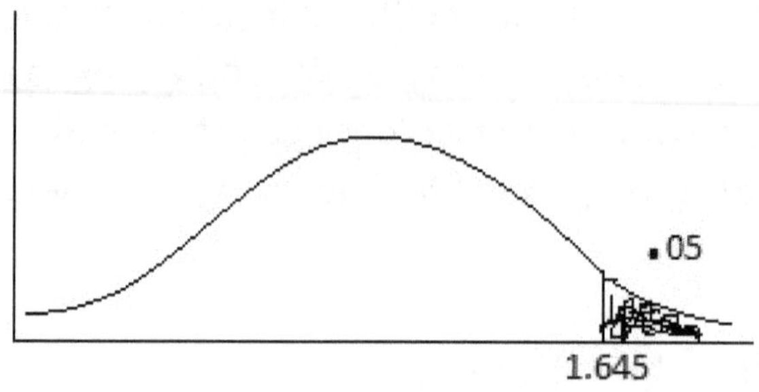

Step 4: The test statistic falls in the rejection region; reject H_0

Step 5: Conclusion: The majority of Americans are employed by private employers.

P-value method

Step 1: H_0: p = .5
H_1: p > .5

Step 2: Test statistic: $z = \frac{\hat{p}-p}{\sqrt{\frac{pq}{n}}} = \frac{.622047-.5}{\sqrt{\frac{(.5)(.5)}{127}}} = 2.7508$

Step 3:

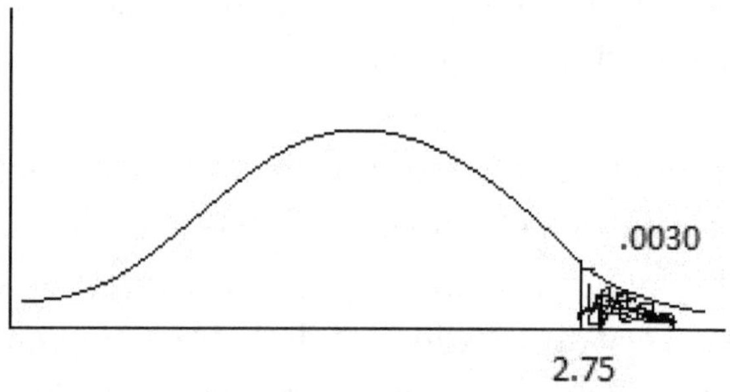

Step 4: total shaded area = p-value = .0030; given α = .05; p-value < α; reject H_0

Step 5: Conclusion: The majority of Americans are employed by private employers.

Notice that we got the same conclusion using both methods. That will always happen.

p stands for "the proportion of all Americans who are employed by private employers." The direct translation of H_1 is "the proportion of all Americans who are employed by private employers is greater than .5" Using words from the original problem, H_1 says, "the majority of Americans are employed by private employers." Using the guide for stating real world conclusions, since we rejected H_0 our conclusion was H_1 using words from the original problem.

For Step 3 of the traditional method, we shaded the right tail because ">" was the symbol used in H_1. The total shaded area is α, which is .05. Since we are in the z distribution, we found the z-score that has area .05 to its right. That value, 1.645, is our critical value. When we plot our test statistic (Step 4), we see that it is in the rejection region, so we reject H_0. (For this problem, any test statistic bigger than 1.645 would land in the rejection region; all other test statistics would not. The shaded portion is the rejection region.)

For Step 3 of the p-value method, we again shaded the right tail because ">" was the symbol used in H_1. We placed our test statistic at the border of the shaded tail, and then found the area in the tail, .0030. That area is the p-value. The p-value is less than α, .05, so we reject H_0 (Step 4).

Please refer back to the explanation of the steps in Section 8.1 so you can see how we followed them. We will follow those steps for all hypothesis tests, so these explanations for each specific test will not be provided for each example.

By the way, can't we just look at 62.2047%, see that is more than 50%, and then determine that the majority of Americans are employed by private employers? NO!!!!! The conclusions that we state at the end of hypothesis tests are always inferences about POPULATIONS, not samples. We know exactly what happened in the sample so there is nothing to test as it relates to drawing a conclusion about a sample. Furthermore, we don't really care about this particular sample we happened to get. What we care about is what we can infer about the entire population from which this sample was taken. Suppose the sample proportion was 51%. You can verify that the test statistic would have been equal to .2254, which is not in the rejection region so we would have failed to reject H_0. The p-value would have been .4090 which is greater than α, also causing us to fail to reject H_0. The conclusion in that case? "We did not find that the majority of Americans are employed by private employers." Sure, in the sample there was a majority but not a strong enough majority for us to make the inference that, not only was there a majority in our sample, but such a majority exists in the whole population. To see if the observed majority is strong enough to make that inference about the entire population, the hypothesis test is needed.

We can NEVER, EVER take a sample statistic (proportion, mean, etc.) and compare that to a claimed value about a population parameter to make an inference. That will be wrong...every time.

Remember that conclusions from hypothesis tests, like interpretations of confidence intervals, are always referring to ALL of something (the whole population) and never the sample.

The formula used for the test statistic is always the one that will be used when performing a test for a proportion. Each different type of hypothesis test requires a different test statistic. The same notation we learned before tells us where to plug in the correct numbers.

Example: Randomly selected current Statsville College students were asked how many units they have completed so far. Their responses were:

32	38	44	51	59	38	54	32	47	38	61	54

Of the students surveyed, 42% intend to major in Statistics and the standard deviation of the GPAs for everyone in the sample is .8017 with mean 2.4583. At Wudmoor College, 65% of the students have completed less than 50 units. 10 years ago, Statsville College students completed an average of 55 units with standard deviation 15.5 units. Using $\alpha = .09$, see if there is a difference between the percentage of current Statsville College students who have completed less than 50 units and the percentage of Wudmoor College students who have completed less than 50 units.

Answer:
Traditional method

Step 1: H_0: p = .65
H_1: p ≠ .65 ("≠" because of "a difference")

Step 2: Test statistic: $z = \frac{\hat{p}-p}{\sqrt{\frac{pq}{n}}} = \frac{\frac{7}{12}-.65}{\sqrt{\frac{(.65)(.35)}{12}}} = -.4842$

Step 3:

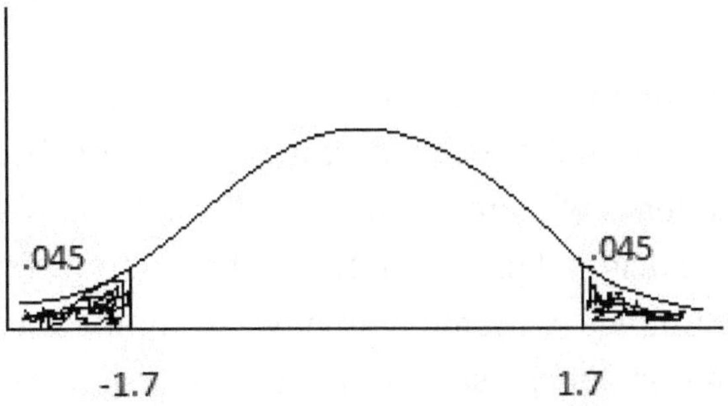

Step 4: The test statistic does not fall in the rejection region; fail to reject H_0

Step 5: Conclusion: We did not find that there is a difference between the percentage of current Statsville College students who have completed less than 50 units and the percentage of Wudmoor College students who have completed less than 50 units.

P-value method

Step 1: H_0: p = .65
H_1: p ≠ .65

Step 2: Test statistic: $z = \frac{\hat{p}-p}{\sqrt{\frac{pq}{n}}} = \frac{\frac{7}{12}-.65}{\sqrt{\frac{(.65)(.35)}{12}}} = -.4842$

Step 3:

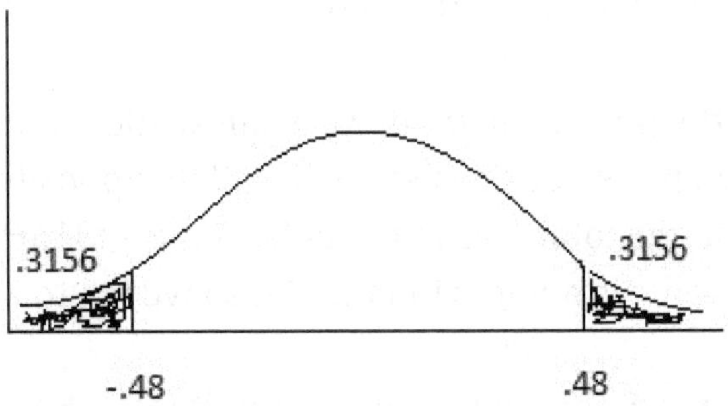

Step 4: p-value = .3156 + .3156 = .6312; given α = .09; p-value>α; fail to reject H_0

Step 5: Conclusion: We did not find that there is a difference between the percentage of current Statsville College students who have completed less than 50 units and the percentage of Wudmoor College students who have completed less than 50 units.

p stands for "the proportion of all current Statsville College students who have completed less than 50 units" which makes the direct translation of H_1, "the proportion of all current Statsville College students who have completed less than 50 units is not equal to .65." Using words from the original problem, H_1 is "there is a difference between the percentage of current Statsville College students who have completed less than 50 units and the percentage of Wudmoor College students who have completed less than 50 units." If we preface that with "We did not find that" as directed in the guide for stating real world conclusions, we easily have our conclusion.

How do we know that p refers to Statsville and not Wudmoor? Remember that the whole purpose of hypothesis testing is to make an inference about an unknown parameter. We know the proportion for the Wudmoor population (.65), so there is nothing to wonder about or test. We do not know the population proportion for Statsville so that is the subject of the hypothesis test. As mentioned before, the parameter symbol (in this case p) always stands for some population value we do not know—that is true in confidence intervals and hypothesis tests.

For Step 3 of the traditional method, we shaded both tails because "≠" was the symbol used in H_1. The total shaded area is α, which is .09 (half of that is in the left tail and half of that is in the right tail). Since we are in the z distribution, we found the z-score that has area .045 to its left (−1.7) and the z-score that has area .045 to its right (1.7). Those values are our critical values. When we plot our test statistic (Step 4), we see that it is not in the rejection region, so we fail to reject H_0. (For this problem, any test statistic smaller than −1.7 or bigger than 1.7 would land in the rejection region; all other test statistics would not. The shaded portion is the rejection region.)

For Step 3 of the p-value method, we again shaded both tails because "≠" was the symbol used in H_1. We placed our test statistic and its opposite at the borders of the shaded tails, and then found the total area in the tails, .3156 + .3156 = .6312. That total area is the p-value. The p-value is not less than α, .09, so we fail to reject H_0 (Step 4).

Please refer back to the explanation of the steps in Section 8.1 so you can see how we followed them. We will follow those steps for all hypothesis tests, so these explanations for each specific test will not be provided for each example.

Wait a minute...the percentage for Statsville is 58% ($\frac{7}{12} \times 100$) and the percentage for Wudmoor is 65%, so why did we say that we did not find a difference? Aren't those percentages different? As always, the conclusion is an inference about the whole population! Yes, in the Statsville sample we had a percentage of 58% but that is not different enough (as determined by our hypothesis testing) from 65% to allow us to make the inference that these two populations have different percentages.

Isn't failing to reject H_0 really just the same as accepting H_0? Isn't this just semantics? Consider the following.

Using this same sample, we could have performed the following tests:

H_0: p = .65	H_0: p = .60	H_0: p = .70	H_0: p = .7135
H_1: p ≠ .65	H_1: p ≠ .60	H_1: p ≠ .70	H_1: p ≠ .7135

For each of these tests, and for tests of infinitely many other hypothesized values, we would fail to reject H_0. So, does this mean that the percentage of Statsville students who have completed less than 50 units is equal to 65% and is equal to 60% and is equal to 70% and is equal to 71.35% and is equal to infinitely many other numbers? Obviously not because the population proportion is equal to one number! All of these fail to reject results do not say that the population percentage equals each of those values. In all those cases, the sample percentage is "close enough" to the hypothesized population percentage so that we would not find that the population percentage is not equal to (or, replace "is not equal to" with "is different from") the hypothesized value. Failing to reject H_0 is not the same as accepting H_0! We will never accept H_0. [In later classes, you may hear an instructor who is loose with the terminology and may say "accept H_0" but technically you are failing to reject H_0, not accepting it.]

Example: Refer to the "units" example just completed.

a) Describe the meaning and probability of a Type I error for that problem.

b) Describe the meaning of a Type II error for that problem.

c) Describe the meaning of the p-value for that problem.

d) Is the result of the hypothesis test significant?

Answers:

a) A Type I error would be concluding that there is a difference between the percentage of current Statsville College students who have completed less than 50 units and the percentage of Wudmoor College students who have completed less than 50 units when, in fact, there is no difference. The probability of making that error is α which is .09.

b) A Type II error would be concluding that we did not find that there is a difference between the percentage of current Statsville College students who have completed less than 50 units and the percentage of Wudmoor College students who have completed less than 50 units when, in fact, there is a difference.

c) The p-value is the probability of getting a sample like we got if there is no difference between the percentage of current Statsville College students who have completed less than 50 units and the percentage of Wudmoor College students who have completed less than 50 units.

d) No. We failed to reject the null hypothesis.

A Type I error is rejecting the null hypothesis when it is, in fact, true. To figure out what that means for a particular problem, simply write what the conclusion would be if you reject H_0. Then add what would make that conclusion wrong (i.e., what would make that conclusion an error.) α is the probability of making a Type I error.

A Type II error is failing to reject the null hypothesis when it is, in fact, wrong. To figure out what that means for a particular problem, simply write what the conclusion would be if you fail to reject H_0. Then add what would make that conclusion wrong (i.e., what would make that conclusion an error.)

The p-value is the probability of getting a sample like we got assuming the condition stated in H_0. To figure out what that means for a particular problem, it will be necessary to translate the null hypothesis into words.

A significant result occurs when H_0 is rejected. It is "significant" because the difference observed in the sample is so huge that it allows us to infer that there is some difference in the population as well.

If necessary, you may find it helpful to review the descriptions of types of errors, p-value, and significant results initially provided in Section 8.1.

Behind the scenes

Now that you have seen a couple hypothesis tests, let's look at what is actually happening when we perform the steps for each method. For this explanation, we will use the "units" example completed earlier in this section. This explanation using a specific example should clarify the description of the process provided in Section 8.1.

In the "units" example, we got a sample. As always, the sample we got was one of the many possible samples that we could have randomly selected. With each different sample that could have been selected there is a different p-hat belonging to that sample. Under the conditions stated in this section, those p-hats have a distribution that is approximately normal with mean p and standard deviation $\sqrt{\frac{pq}{n}}$. Based on the specific distribution indicated by H_0, the population of p-hats from which our p-hat came has a mean of .65 and a standard deviation of $\sqrt{\frac{(.65)(.35)}{12}}$. The alternative hypothesis is that the true p is something either bigger than .65 or less than .65 so we will use both tails of the distribution to represent that. A p-hat that is far from .65 in either direction will alert us to reject H_0 in favor of H_1.

<u>Traditional method</u>

We setup the hypothesized distribution with the mean .65 and indicate a total area of α in the tails. Since the total area in both tails will be $\alpha = .09$ that means that half of that (.045) will be in each tail. There is some p-hat, we'll call it H, in the distribution of all p-hats such that .045 (4.5%) of the p-hats are less than it. There is another p-hat, we'll call it K, in the distribution of all p-hats such that .045 (4.5%) of the p-hats are greater than it. Here's a sketch of what we have so far with this p-hat distribution:

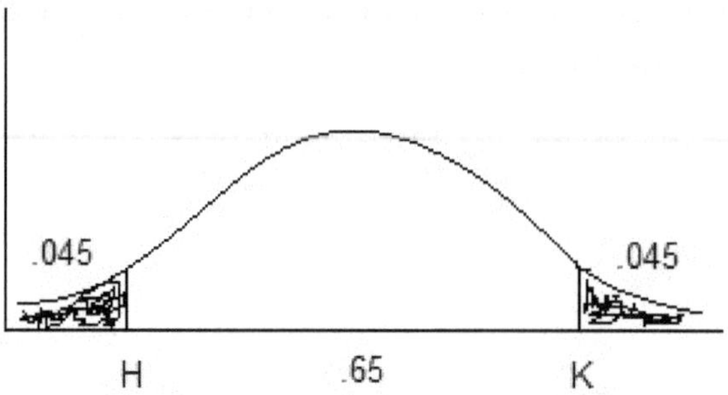

If the p-hat that we actually have from our sample is less than H or bigger than K, then we consider it "far away" from the hypothesized population proportion, .65. While it's certainly possible to randomly select a sample having a p-hat in one of those extremes of the distribution, it's improbable. Only 9% of all the p-hats are less than H or greater than K! Therefore, if our p-hat is less than H or greater than K, we will reject the null hypothesis with its claim that p is .65. However, if our p-hat is any one of the p-hats in the middle 91% of all the p-hats, it would be "close to" .65 so we have no reason to reject the null hypothesis. Notice that it takes a lot for us to reject the null hypothesis.

So, how do we know if our p-hat is smaller than H or bigger than K? As it turns out, it is more efficient to transform everything to z-scores. While we don't know H and we don't know K, we do know their corresponding z-scores. The z-score that corresponds to the p-hat H has area .045 to its left. That z-score is −1.7. The z-score that corresponds to the p-hat K has area .045 to its right. That z-score is 1.7. Using z-scores, our new diagram looks like this:

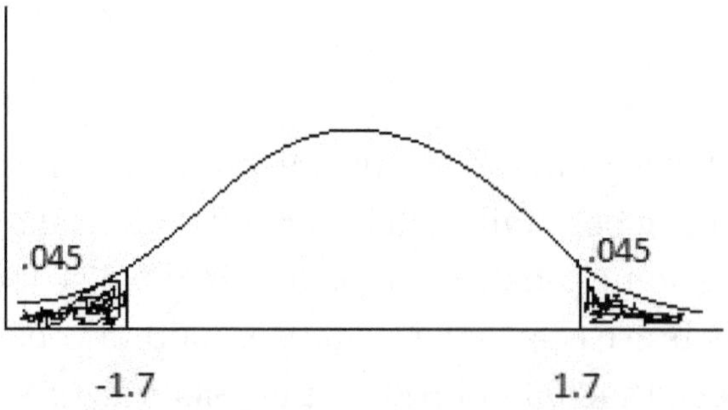

All we need to do now is convert the p-hat from our sample into a z-score. Using the z-score formula and the values for this hypothesized distribution, that is:

$$z = \frac{x-\mu}{\sigma} = \frac{\hat{p}-p}{\sqrt{\frac{pq}{n}}} = \frac{\frac{7}{12}-.65}{\sqrt{\frac{(.65)(.35)}{12}}} = -.4842$$

So, if that z-score is less than −1.7 that means that our p-hat $\left(\frac{7}{12}\right)$ is less than H, making our p-hat far away from .65. If that z-score is greater than 1.7 that means that our p-hat is greater than K, making our p-hat far away from .65. In either case, we would reject the null hypothesis. Otherwise, we fail to reject the null hypothesis. That z-score is not in the extremes of the distribution so we fail to reject the null hypothesis. Notice that we haven't proven the null hypothesis to be true, we just don't have a reason to say it is wrong.

Does the z-score that we just calculated look familiar? That's the test statistic!

When we perform the steps of the traditional hypothesis test, we are doing what we just described but in a more efficient manner.

<u>P-value Method</u>

For the p-value method, we again start with the hypothesized distribution and label the extremes (tails). We want to determine how likely it is that we would randomly select a sample having a p-hat at least as far away from .65 as ours was. We use probability to make that determination. Our p-hat was a certain distance below .65. There is another p-hat the same distance from .65 but above it. We'll call it M. We want to know the probability of randomly selecting a sample having a p-hat that is smaller than ours or bigger than M to address the matter of, "at least as far away from .65 as our p-hat was". Here's the sketch of what we have so far for this p-hat distribution:

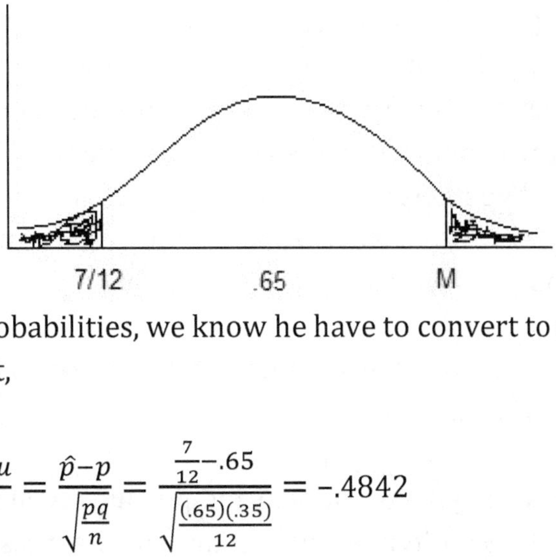

To find the desired probabilities, we know he have to convert to z-scores so we can use Table A. For our p-hat,

$$z = \frac{x-\mu}{\sigma} = \frac{\hat{p}-p}{\sqrt{\frac{pq}{n}}} = \frac{\frac{7}{12}-.65}{\sqrt{\frac{(.65)(.35)}{12}}} = -.4842$$

How can we get the z-score for M? Well, we could do a calculation to figure out M, but there's no need! Remember that M is as far above .65 as 7/12 is below .65. What does that tell you about the z-score? It's the opposite of the one we calculated. So, with that, we are looking for the probability to the left of z-score –.4842 (rounded to –.48 for the table lookup) and the probability to the right of z-score .4842 (rounded to .48 for the table lookup). Here's a new diagram of our activities using z-scores:

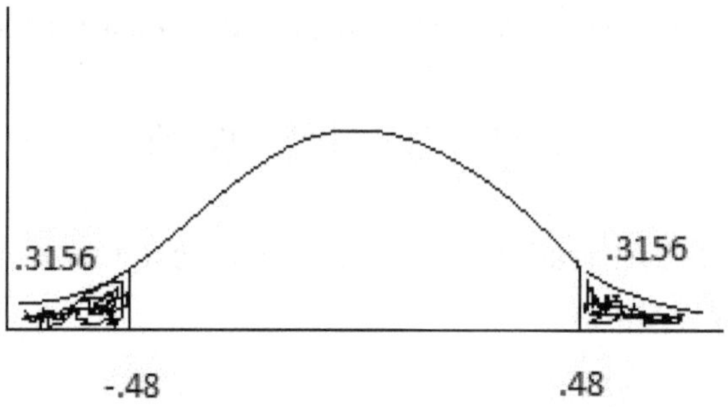

The probability of randomly selecting a sample with a p-hat less than 7/12 is .3156 and the probability of randomly selecting a sample with a p-hat greater than M is .3156. So, the probability of randomly selecting a sample having a p-hat at least as far away from .65 as our p-hat was is .3156+.3156 = .6312. [Stated in terms of the z-score test statistic we could say that that probability of getting a sample that produces a test statistic at least as extreme as the one in our sample is .6312, assuming the null hypothesis to be true.] Said another way, 63.12% of the time a sample having a p-hat at least as far away from .65 as our p-hat was would be selected. We could also say that 63.12% of the p-hats in this hypothesized distribution are at least as far away from .65 as our p-hat was. The probability of getting sample results like we got, assuming the null hypothesis to be true, is .6312.

If that probability is "small" (less than α) we consider the sample results to be unlikely based on the hypothesized distribution. So, we would reject the null hypothesis which gave us that hypothesized distribution. Otherwise, we would consider the sample results reasonably likely based on the hypothesized distribution so we would have no grounds upon which to reject the null hypothesis. Our p-value (.6312) is not less than α (.09) so we do not reject the null hypothesis. Notice that we haven't proven the null hypothesis to be true, we just don't have a compelling reason to say it is wrong.

When we perform the steps of the p-value hypothesis test, we are doing what we just described but in a more efficient manner.

What we just considered regarding the traditional and p-value methods is ALWAYS the background motivating the steps for hypothesis testing. For that reason, this explanation will not be provided with future problems. We will simply use the steps. As we're doing so, though, remember that there is a reason behind the steps and it is all based on testing how realistic the null hypothesis seems to be based on the sample results we obtained.

Comparing the methods

When using the steps for hypothesis testing, the primary difference in the two methods occurs at Step 3. In the traditional method we start with area (alpha) and then find score(s) from a table (critical value(s)). In the p-value method we start with a score (test statistic) and then use a table to find area (p-value). The traditional method has area first, then score. The p-value method has score first, then area.

Example: A sociologist interviewed some Statsville residents and asked, among other things, if each respondent was currently employed. She believes that the employment rate in Statsville is less than the 70% employment rate in Chef County. Assess her belief using the Minitab output below and a .05 significance level.

Test and CI for One Proportion: Employed?

```
Test of p = 0.7 vs p < 0.7

Event = y


                                    95% Upper     Exact
Variable     X   N  Sample p          Bound   P-Value
Employed?  25  40  0.625000     0.752705     0.193
```

Answer: The p-value (.193) is greater than α (.05) so fail to reject H_0. We did not find that the employment rate in Statsville is less than the 70% employment rate in Chef County.

As you know by now, Minitab output is always labeled. In addition to the main label for the section of output, "**Test and CI for One Proportion: Employed?**", Minitab also shows the hypotheses being tested: "`Test of p = 0.7 vs p < 0.7`". For all different types of hypothesis tests, Minitab does not always show the hypotheses in the same manner but, one way or another, the hypotheses are displayed. As we know, the statement with the "=" is the null hypothesis and the other is the alternative hypothesis. As you saw before, Minitab gives the sample results: 25 successes ("Event = y" telling you that a yes response is a success), sample size 40, $\hat{p}$ = .625. You can ignore any references to "upper bound" or "lower bound" in any output in this course. This refers to one-sided confidence intervals which are not considered in introductory Statistics courses. The p-value is shown as .193. As with the hypotheses, the p-value is not always shown in a consistent manner for all hypothesis tests but, one way or another, the p-value will be there. All we have to do is compare the p-value to alpha and then make the appropriate conclusion. Minitab has done all the tedious calculations for us and all we have to do is correctly use the results.

Exercises

For Exercises 1) and 2) do the following: a) Write the null and alternative hypotheses using symbols. b) Write the direct translation of H_1. c) Write the translation of H_1 using words from the original problem. d) Write what the conclusion would be if you rejected H_0. e) Write what the conclusion would be if you failed to reject H_0.

1) Selected Americans were asked to describe the place they lived when they were 16 years old. Test the claim that 30% of Americans lived in a large city when they were 16 years old.

2) "With whom were you living on your 16th birthday?" was a question asked of several Americans. See if the proportion of Americans who lived with their own father and a stepmother on their 16th birthday is less than .4.

3) Refer to the Sparkelz Gems/Javonn Jewelers data from Exercise 1) in Section 7.2. Test the claim that 71% of Sparkelz Gems diamonds are VS1 or VVS1. Use a significance level of .04.

4) Refer to the Jerrell Enterprises/Jaden Industries data from Exercise 2) in Section 7.2. *ProfDev* magazine says that, "8 out of 10 employees who are satisfied with their current position in their companies participate in professional development." See if the proportion of satisfied employees who participate in professional development is smaller at Jerrell Enterprises than what is reported by *ProfDev*. ($\alpha = .03$)

5) Two drugs—Drug A and Drug B—are being tested in a clinical trial. The side effects experienced by each person in the trial are recorded as follows: nausea (n), itching (i), headaches (h), fatigue (f), dry mouth (d). Using alpha = .025 and the Minitab output, test the assertion that less than half of all Drug A users experience nausea.

Test and CI for One Proportion: Drug A

```
Test of p = 0.5 vs p < 0.5

Event = n


                                  95% Upper    Exact
Variable    X   N  Sample p           Bound  P-Value
Drug A     13  50  0.260000        0.381264    0.000
```

6) Refer to 5). What is the probability of getting a sample like we got, assuming that the proportion of all Drug A users who experience nausea is .5? What is the probability that we would conclude that less than half of all Drug A users experience fatigue if, in fact, that is not true?

8.3 Hypothesis Testing for a Mean

Conditions: The data is from a simple random sample. The population is normally distributed and/or the sample size is bigger than 30. The standard deviation for the population is unknown.
(Assume all conditions are met for each example and exercise.)

Use the methods of this section when...

testing a claim about a mean

Important formula: $t = \frac{\bar{x}-\mu}{\frac{s}{\sqrt{n}}}$

[Note: In this text we only consider testing means when the population standard deviation is unknown. That is because, in reality, whenever the population mean is unknown (which is why we are performing a test) the population standard deviation is also unknown. In a contrived case in which the population standard deviation was known and hypothesis testing for the population mean was being performed, the method would be different from that described in this text.]

Review the steps for hypothesis testing in Section 8.1, as well as the examples in Section 8.2, before continuing. We will be using the same steps for hypothesis testing.

Example: Randomly selected current Statsville College students were asked how many units they have completed so far. Their responses were:

32	38	44	51	59	38	54	32	47	38	61	54

Of the students surveyed, 42% intend to major in Statistics and the standard deviation of the GPAs for everyone in the sample is .8017 with mean 2.4583. At Wudmoor College, 65% of the students have completed less than 50 units. 10 years ago, Statsville College students completed an average of 55 units with standard deviation 15.5 units. Use hypothesis testing to see if the mean number of units completed by current Statsville College students is the same as the mean number of units completed by Statsville College students from 10 years ago.

Answer:
Traditional method

Step 1: H_0: μ = 55
H_1: μ ≠55 ("≠" because of "the same")

Step 2: Test statistic: $t = \frac{\bar{x}-\mu}{\frac{s}{\sqrt{n}}} = \frac{45.6667-55}{\frac{10.1564}{\sqrt{12}}} = -3.1834$

Step 3:
(recall: since α is not given, we use α = .05)

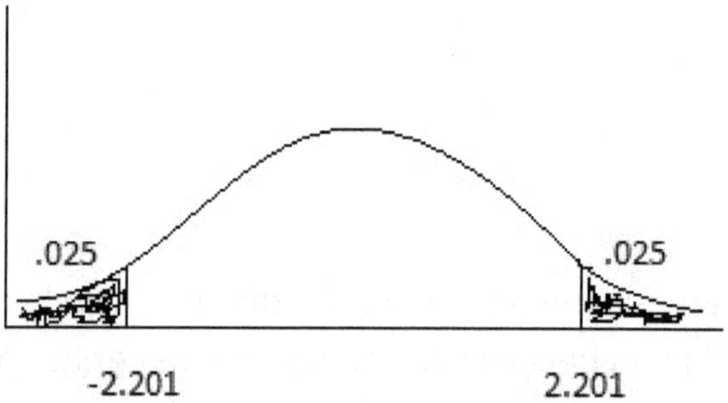

Step 4: The test statistic is in the rejection region; reject H_0

Step 5: Conclusion: The mean number of units completed by current Statsville College students is not the same as the mean number of units completed by Statsville College students from 10 years ago.

P-value method

Step 1: H_0: μ = 55
H_1: μ ≠55

Step 2: Test statistic: $t = \frac{\bar{x}-\mu}{\frac{s}{\sqrt{n}}} = \frac{45.6667-55}{\frac{10.1564}{\sqrt{12}}} = -3.1834$

Step 3:

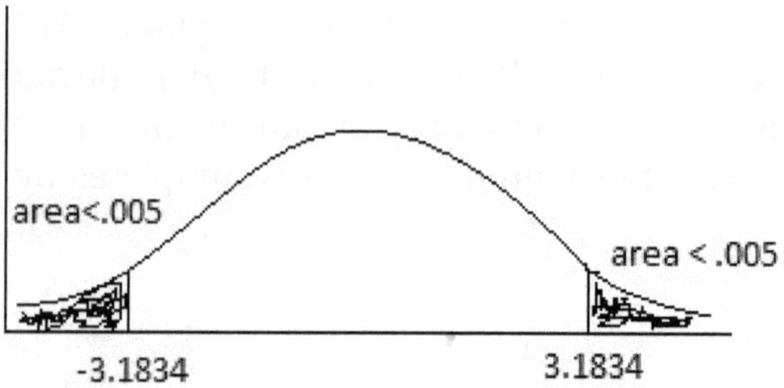

Step 4: area < .005 + area < .005 says that p-value < .01; α not given so use .05; p-value < α; reject H_0

Step 5: Conclusion: The mean number of units completed by current Statsville College students is not the same as the mean number of units completed by Statsville College students from 10 years ago.

To clarify the p-value, start by noticing that, for 11 degrees of freedom, the t-score 3.1834 is off the table (Table B) to the left:

	Area to the RIGHT				
Degrees of Freedom	.005	.01	.025	.05	.10
1	63.657	31.821	12.706	6.314	3.078
2	9.925	6.965	4.303	2.920	1.886
3	5.841	4.541	3.182	2.353	1.638
4	4.604	3.747	2.776	2.132	1.533
5	4.032	3.365	2.571	2.015	1.476
6	3.707	3.143	2.447	1.943	1.440
7	3.499	2.998	2.365	1.895	1.415
8	3.355	2.896	2.306	1.860	1.397
9	3.250	2.821	2.262	1.833	1.383
10	3.169	2.764	2.228	1.812	1.372
11	3.106	2.718	2.201	1.796	1.363

So, the area to the right of t-score 3.1834 is some number smaller than .005 (areas in the table get smaller as you move left). Since the t-distribution is symmetric, the area to the left of t-score -3.1834 would be that same number smaller than .005. If you take any number less than .005 and double it, the answer is a number that is smaller than .01. So the p-value, the total area in the tails, is a number that is smaller than .01. Any number that is smaller than .01 is smaller than .05. Therefore, we know the p-value is less than α which is all we need to know to make a decision about H_0.

Suppose the test statistic had been −1.934. Since the t-table only gives areas to the right of positive t-scores, we would need to look for 1.934 in row 11 of the table. We observe that it falls between 2.201 and 1.796. That tells us that the area to the right of t-score 1.934 is between .025 and .05. Due to symmetry, the area to the left of t-score −1.934 would also be between .025 and .05. So the p-value, the total area in the tails, is a number between .05 and .10. (If you take any number between .025 and .05 and double it, the answer will be between .05 and .10. As a shortcut, you can just double the endpoints of the interval for one-tail area when you have a two-tailed test to get the interval for the p-value.) Any number strictly between .05 and .10 is bigger than .05, so the p-value would be bigger than α.

Suppose the test statistic had been −1.109. We look for 1.109 in row 11 and see it is off the table to the right. The area to the right of t-score 1.109 is the same as the area to the left of t-score −1.109, and that is a number greater than .10 (areas in the table get bigger as you move right) . Total area in the tails? Some number bigger than .20. That tells us p-value > .20. Any number bigger than .20 is bigger than .05 so the p-value would be bigger than α.

In the traditional method, Step 3 required us to get critical values. After we appropriately shaded both tails, because of the "≠" in H_1, we looked in the .025 column of the t table, df row 11, and found the critical value 2.201. We knew from the symmetry of the distribution that −2.201 has area .025 to its left, so that is our other critical value.

μ stands for the mean number of units completed by all current Statsville College students. That value is unknown and is the subject of the test. We already know the population mean from 10 years ago. The direct translation of H_1 is, "the mean number of units completed by all current Statsville College students is not equal to 55." Using words from the original problem H_1 is, "The mean number of units completed by current Statsville College students is not the same as the mean number of units completed by Statsville College students from 10 years ago." That statement was our conclusion since we rejected H_0.

Suppose the problem was presented as, "... to see if the mean number of units completed by current Statsville College students is different from the mean..." The problem would be completed exactly the same way as demonstrated above. If you are testing for "the same" or "different" it is a two-tailed test either way. Just be sure that H_1 is represented correctly in the conclusion using words from the problem. Notice how we had to add the word "not" to the chunk of words we took from the problem to help us write the conclusion.

Had we failed to reject H_0, we could conclude, "We did not find that the mean number of units completed by current Statsville College students is not the same as the mean number of units completed by Statsville College students from 10 years ago. There appears to be no difference between the current mean and the mean from 10 years ago."

Hold on here...the mean for current students is 45.6667 and the mean 10 years ago was 55 so can't we just look at those numbers and see that the means are obviously different without messing around with this hypothesis testing? No, no, no! The mean of the current students' sample was 45.6667, but the mean for the whole population of current students is unknown. As always, our conclusion is an inference about the whole population. What our conclusion is telling us is that the sample mean was so different (as determined by our hypothesis test) from the population mean from 10 years ago that we can safely make the inference that the current population mean (whatever it is) is different from the population mean from 10 years ago.

Example: Refer to the example we just completed. What if the mean number of units completed by current Statsville College students is actually the same as the mean number of units completed by Statsville College students from 10 years ago. What type of error occurred?

Answer: Type I error.

Remember, a Type I error is when you reject the null hypothesis (as we did) but the null hypothesis is actually true.

Example: The ages (in years) of randomly selected Statsville residents were used in the creation of the Minitab output below. Is the mean age of Statsville residents greater than 25 years? ($\alpha = .04$)

One-Sample T: SVAge

```
Test of mu = 25 vs > 25


                                          92% Lower
Variable   N   Mean  StDev  SE Mean        Bound     T      P
SVAge     40  30.55  14.79     2.34        27.20  2.37  0.011
```

Answer: p-value (.011) < α (.04) , reject H_0. The mean age of Statsville residents is greater than 25 years.

Again, the bold heading indicates what test is being performed (**One-Sample T**) and what variable is being used (**SVAge**). Whenever there is any T testing, you know a mean is involved, just like with the confidence intervals. The usual sample statistics are provided. After verifying that we are looking at the correct output, including the hypotheses (`Test of mu = 25 vs > 25`), all we really need for our conclusion is the p-value.

Example: If the mean age of all Statsville residents is really 25 years, what is the probability of getting a sample of Statsville residents like the one we got?

Answer: .011

Remember, the p-value is the probability of getting sample results like we got if whatever H_0 says is true.

Exercises

For Exercises 1) and 2) do the following: a) Write the null and alternative hypotheses using symbols. b) Write the direct translation of H_1. c) Write the translation of H_1 using words from the original problem. d) Write what the conclusion would be if you rejected H_0. e) Write what the conclusion would be if you failed to reject H_0.

1) Employed Tanayatown residents were randomly selected and asked, "About how many people work at the location where you work?" Use hypothesis testing to see if over 140 people, on average, work at the locations where employed Tanayatown residents work.

2) After completing their grocery shopping, randomly selected shoppers had their bags weighed. Is the average weight of filled shopping bags 10 pounds?

3) Refer to the Sparkelz Gems/Javonn Jewelers data from Exercise 1) in Section 7.2. Suppose a supplier is considered "mid-size" if the average size of its diamonds is less than 8

mm. See if Javonn Jewelers is a mid-size supplier. Perform a test that has probability .10 of concluding that the mean size is less than 8 mm when that is not actually true.

4) Refer to the Jerrell Enterprises and Jaden Industries data from Exercise 2) in Section 7.2. Suppose a company is considered "employee friendly" if average supervisor ratings are greater than 160. Use a test having a probability of .01 for making a Type I error to investigate whether or not Jaden Industries is employee friendly.

5) There are four furniture stores in a certain city. One is in the north, one is in the south, one is in the east, and one is in the west. We collected random samples of prices (in dollars) from each store and used the appropriate data for the Minitab output below. Using significance level .01, write a conclusion for the hypothesis test that was performed.

One-Sample T: North

```
Test of mu = 1500 vs not = 1500


Variable   N     Mean  StDev  SE Mean          99% CI              T      P
North     91  1508.94  46.15     4.84  (1496.21, 1521.68)   1.85  0.068
```

6) Refer to 5).

a) If it turns out that the true state of the population does not match the conclusion you wrote, what type of error would that be?

b) What is the p-value telling you?

8.4 Hypothesis Testing for a Standard Deviation or Variance

Conditions: The data is from a simple random sample. The population must be normally distributed. [The methods of this section rely heavily on this requirement.]
(Assume all conditions are met for each example and exercise.)

Use the methods of this section when...

testing a claim about a standard deviation or a variance

Important formula: $\chi^2 = \frac{(n-1)s^2}{\sigma^2}$

[Due to the limitations of the distribution table we use in this section, we will use only the traditional method when tests are two-tailed. For one-tailed tests, we will continue to use both methods. Of course, whenever Minitab performs any hypothesis test, we will use the p-value it provides.]

Example: Randomly selected current Statsville College students were asked how many units they have completed so far. Their responses were:

32	38	44	51	59	38	54	32	47	38	61	54

Of the students surveyed, 42% intend to major in Statistics and the standard deviation of the GPAs for everyone in the sample is .8017 with mean 2.4583. At Wudmoor College, 65% of the students have completed less than 50 units. 10 years ago, Statsville College students completed an average of 55 units with standard deviation 15.5 units. At the .005 significance level, see if there is less variation in the numbers of units completed by current Statsville college students compared to 10 years ago.

Answer:
(Note: Since we are being asked about variation, we need to address the problem using a measure of variation—standard deviation or variance)

Traditional method

Step 1: H_0: $\sigma = 15.5$
H_1: $\sigma < 15.5$ ("<" because of "less variation")

Step 2: Test statistic: $\chi^2 = \frac{(n-1)s^2}{\sigma^2} = \frac{(12-1)10.1564^2}{15.5^2} = 4.7229$

Step 3:

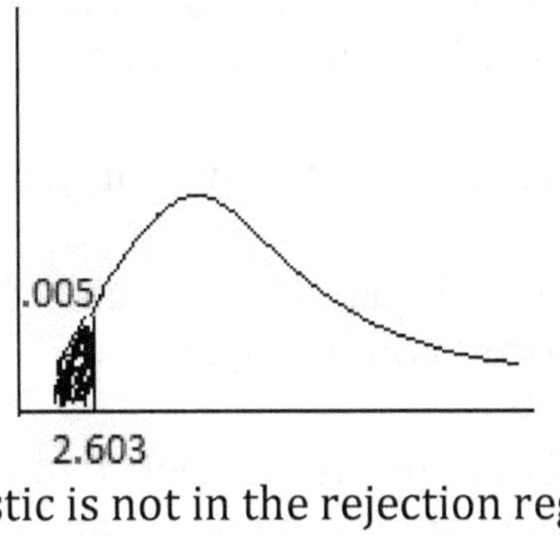

Step 4: The test statistic is not in the rejection region; fail to reject H_0

Step 5: Conclusion: We did not find that there is less variation in the number of units completed by current Statsville college students compared to 10 years ago.

P-value method

Step 1: H_0: $\sigma = 15.5$
H_1: $\sigma < 15.5$

Step 2: Test statistic: $\chi^2 = \frac{(n-1)s^2}{\sigma^2} = \frac{(12-1)10.1564^2}{15.5^2} = 4.7229$

Step 3:

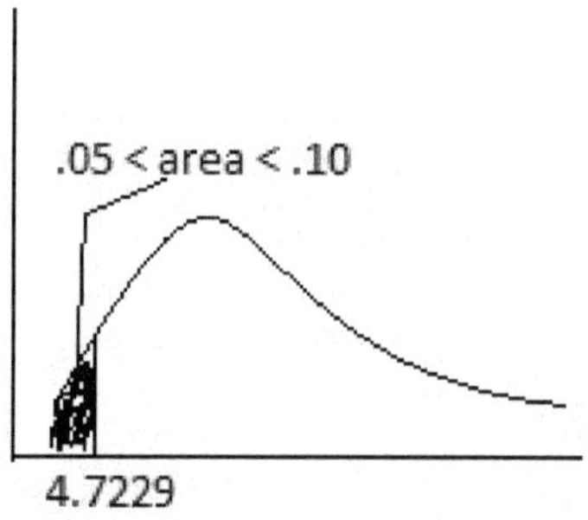

Step 4: total shaded area = p-value; $.05 <$ p-value $< .10$; given $\alpha = .005$; p-value $> \alpha$; fail to reject H_0

Step 5: Conclusion: We did not find that there is less variation in the number of units completed by current Statsville college students compared to 10 years ago.

Getting the p-value from Table C (chi-square table) is similar to how it's done when using Table B (t table). Looking in Table C, 11 degrees of freedom, we observe that the test statistic 4.7229 is between 4.575 and 5.578. So, the area to the right of chi-square score 4.7229 is between .90 and .95. Since the total area under a probability density curve is always 1, the area to the left of chi-square score 4.7729 is between 1−.95 = .05 and 1−.90 = .10. This is the reason for the statement .05 < p-value < .10. The p-value is some number between .05 and .10, any number between .05 and .10 is bigger than .005, so p-value > α.

Always remember: the chi-square table is providing area to the RIGHT of chi-square scores!

When we got the critical value, we had to be in column 1−.005 = .995 because, again, this table is organized according to area on the right!

σ stands for the standard deviation of the number of units completed by all current Statsville College students. That value is unknown and is the subject of the test. We know the population standard deviation from 10 years ago. If we wanted to address variation using the variance, the hypotheses would have been H_0: $\sigma^2 = 240.25$, H_1: $\sigma^2 < 240.25$, where 240.25 is 15.5^2. The test statistic formula would be the same as the one we used.

The direct translation of H_1 is, "the standard deviation of the number of units completed by all current Statsville College students is less than 15.5." Using words from the original problem H_1 is, "there is less variation in the number of units completed by current Statsville college students compared to 10 years ago." Had we rejected H_0, the conclusion would have been, "There is less variation in the number of units completed by current Statsville college students compared to 10 years ago." Since we failed to reject H_0, we prefaced that phrase with "We did not find that".

Wait a minute...we concluded that, "We did not find that there is less variation in the number of units completed by current Statsville college students compared to 10 years ago." But didn't we have a standard deviation of 10.1564 for current students and isn't the standard deviation for the students from 10 years ago 15.5? Doesn't that mean that there is less variation for the current students? Uggghhh...the 10.1564 number is for the sample. We can never ever ever take a sample statistic and use it to make an inference about the parameter for the whole population. The conclusion we stated is an inference about the whole population!! Yes, the sample had a standard deviation that was less than 15.5 but it was not so much less than 15.5 that we could infer that the population of current students has a standard deviation less than 15.5. Hence, the "we did not find".

Remember, hypothesis tests are always inferences about the entire population from which the sample was taken.

Example: Referring to the previous example, what is the probability that we would conclude that there is less variation in the number of units completed by current Statsville college students compared to 10 years ago if, really, that is not the case?

Answer: .005

The significance level, α, gives the probability of rejecting a true H_0.

Example: The ages (in years) of randomly selected Statsville residents were used in the creation of the Minitab output below. The standard deviation of the ages of Statsville residents six years ago was 14 years. Are Statsville ages more varied than they were back then?

Test and CI for One Variance: SVAge

```
Method

Null hypothesis         Sigma = 14
Alternative hypothesis  Sigma > 14

The chi-square method is only for the normal distribution.
The Bonett method is for any continuous distribution.


Statistics

Variable   N  StDev  Variance
SVAge     40   14.8       219
```

```
94% One-Sided Confidence Intervals

                              Lower
                              Bound
                               for    Lower Bound
Variable  Method              StDev  for Variance
SVAge     Chi-Square           12.6           159
          Bonett               12.8           165


Tests

                                  Test
Variable  Method              Statistic  DF  P-Value
SVAge     Chi-Square              43.53  39    0.285
          Bonett                      —   —    0.269
```

Answer: p-value (.285) > α (.05), fail to reject H_0. We did not find that Statsville ages more varied than they were back then (six years ago).

Notice here that, in the "Method" section, the hypotheses are spelled out more completely than in the tests for p and μ. Sample statistics are provided. Since we are dealing with hypothesis testing, all we need is the p-value. As mentioned before, we are not using the Bonett method in this course. Since α was not given, we use .05.

Example: If the methods of hypothesis testing caused us to make a wrong conclusion for the previous example, what type of error would that be?

Answer: Type II

A Type II error is failing to reject the null hypothesis, as we did here, when the null hypothesis actually turns out to be false.

Exercises

For Exercises 1) and 2) do the following: a) Write the null and alternative hypotheses using symbols. b) Write the direct translation of H_1. c) Write the translation of H_1 using words from the original problem. d) Write what the conclusion would be if you rejected H_0. e) Write what the conclusion would be if you failed to reject H_0.

1) Randomly selected married people were asked how many hours per week their spouses typically work. Historically, the standard deviation of the hours per week worked by spouses was 22.7 hours. See if there has been a reduction in the standard deviation of the hours per week typically worked by spouses.

2) Before an adjustment of overtime laws, the variance of the number of days per month people usually worked overtime was 27.9 days^2. We got data from a sample of current

workers to perform testing to see if there has been a change in the dispersion of the number of days per month people usually work overtime.

3) Refer to the Sparkelz Gems/Javonn Jewelers data from Exercise 1) in Section 7.2. In the past, the mean weight of Sparkelz diamonds was 1.33 carats with standard deviation .58 carat. Sparkelz has changed its operations over the past few years and now it's believed that the diamonds have more variation in their weights. Test this belief.

4) Refer to the Jerrell Enterprises and Jaden Industries data from Exercise 2) in Section 7.2. A new bonus structure has been implemented at Jerrell Enterprises and we now want to see if there has been a change in the variance of peer ratings, which used to be 2218.41. Perform an appropriate test ($\alpha = .01$).

5) There are four furniture stores in a certain city. One is in the north, one is in the south, one is in the east, and one is in the west. We collected random samples of prices (in dollars) from each store and used the appropriate data for the Minitab output below. Test at the .03 significance level the claim that the standard deviation of all prices from the south store is $32.

Test and CI for One Variance: South

```
Method

Null hypothesis         Sigma = 32
Alternative hypothesis  Sigma not = 32

The chi-square method is only for the normal distribution.
The Bonett method is for any continuous distribution.


Statistics

Variable   N  StDev  Variance
South     82   48.4      2343


97% Confidence Intervals

                            CI for         CI for
Variable  Method            StDev        Variance
South     Chi-Square  (41.3, 58.3)  (1709, 3396)
          Bonett      (41.9, 57.4)  (1756, 3299)


Tests

                         Test
Variable  Method      Statistic  DF  P-Value
South     Chi-Square     185.33  81    0.000
          Bonett              –   –    0.000
```

6) Is the result in 5) significant? Explain.

7) Suppose the standard deviation of all prices from the south store really is \$32. What type of error was made in your conclusion for 5) ?

8.5 Hypothesis Testing for Two Proportions

Conditions: The data is from two independent simple random samples (independent in that there is no pairing or relationship between values in the different samples). In each sample there are at least five successes and five failures.
(Assume all conditions are met for each example and exercise.)

Use the methods of this section when...

testing a claim about the relationship between two proportions, using one sample from each population

Important formulas: $z = \frac{(\hat{p}_1-\hat{p}_2)-(p_1-p_2)}{\sqrt{\frac{\bar{p}\bar{q}}{n_1}+\frac{\bar{p}\bar{q}}{n_2}}}$ $\bar{p} = \frac{x_1+x_2}{n_1+n_2}$

[In this section, we will focus on cases where the null hypothesis is $p_1 = p_2$ (or, equivalently, $p_1 - p_2 = 0$). To test for a non-zero difference, the formulas would be different from what are used in this section. Exercise 5 will give you an opportunity to practice with a case like that.]

Example: Randomly selected employed Americans were asked if they are employed by the government or a private employer. 62.2047% of the 127 people surveyed were employed by a private employer. The mean age of the sample was 37.5.

Employed non-Americans were asked if they are employed by their government or a private employer. Their responses are listed below (g: government, p: private—A couple respondents answered "other" with an o)

g g p g p g o g p g p p g p g p g g o p g

Use hypothesis testing (α = .04) to test the claim that a greater proportion of employed Americans are employed by private employers compared to the proportion of employed non-Americans who are employed by private employers.

Answer:

Traditional method

Step 1: H_0: $p_1 = p_2$

H_1: $p_1 > p_2$ (p_1 is for Americans; ">" because of "greater")

Step 2: Test statistic z = $\frac{(\hat{p}_1-\hat{p}_2)-(p_1-p_2)}{\sqrt{\frac{\bar{p}\bar{q}}{n_1}+\frac{\bar{p}\bar{q}}{n_2}}} = \frac{\left(.622047-\frac{8}{21}\right)-0}{\sqrt{\frac{(.5878)(.4122)}{127}+\frac{(.5878)(.4122)}{21}}}$

= 2.07921

$p_1 - p_2 = 0$ because H_0 claims $p_1 = p_2$

x_1 was calculated from (.622047)(127) = 79 Americans

$\bar{p} = \frac{x_1+x_2}{n_1+n_2} = \frac{79+8}{127+21} = .5878\,, \bar{q} = 1 - \bar{p} = .4122$

Step 3:

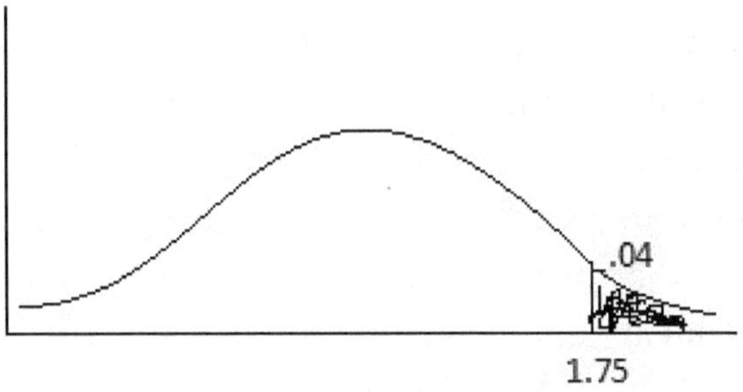

Step 4: The test statistic is in the rejection region; reject H_0

Step 5: Conclusion: A greater proportion of employed Americans are employed by private employers compared to the proportion of employed non-Americans who are employed by private employers.

P-value method

Step 1: H_0: $p_1 = p_2$

H_1: $p_1 > p_2$

Step 2: Test statistic z = $\frac{(\hat{p}_1-\hat{p}_2)-(p_1-p_2)}{\sqrt{\frac{\bar{p}\bar{q}}{n_1}+\frac{\bar{p}\bar{q}}{n_2}}} = \frac{\left(.622047-\frac{8}{21}\right)-0}{\sqrt{\frac{(.5878)(.4122)}{127}+\frac{(.5878)(.4122)}{21}}}$

= 2.07921

Step 3:

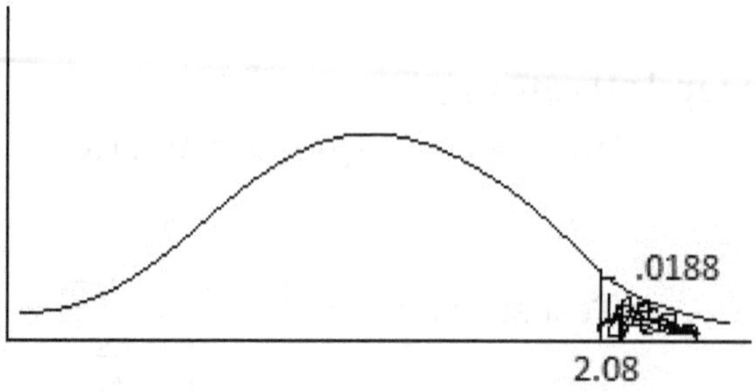

Step 4: Total shaded area = p-value = .0188; given $\alpha = .04$; p-value < α; reject H_0

Step 5: Conclusion: A greater proportion of employed Americans are employed by private employers compared to the proportion of employed non-Americans who are employed by private employers.

p_1 is the proportion of all employed Americans who are employed by private employers. p_2 is the proportion of all employed non-Americans who are employed by private employers. $\hat{p}_1$ is the proportion of employed Americans in the sample who are employed by private employers. $\hat{p}_2$ is the proportion of employed non-Americans in the sample who are employed by private employers. x_1 is the number of employed Americans in the sample who are employed by private employers and x_2 is the number of employed non-Americans in the sample who are employed by private employers. Clearly, x_1 and x_2 would both have to be whole numbers, so rounding is often needed when either of these needs to be calculated to find $\bar{p}$.

The direct translation of H_1 is, "the proportion of all employed Americans who are employed by private employers is greater than the proportion of all employed non-Americans who are employed by private employers." Using words from the original problem H_1 is, "a greater proportion of employed Americans are employed by private employers compared to the proportion of employed non-Americans who are employed by private employers" and that was our conclusion since we rejected H_0. If we failed to reject H_0, we would conclude, "We did not find that a greater proportion of employed Americans are employed by private employers compared to the proportion of employed non-Americans who are employed by private employers."

Why didn't we just compare .622047 to 8/21 and make our conclusion that way? That would have been so much simpler. Oh, yeah—those are the sample proportions! Our inference is about the entire populations of all employed Americans and all employed non-

Americans. .622047 is bigger than 8/21 but the hypothesis tests tells us that .622047 is so much bigger than 8/21 that we can infer that the population proportion for the first population is bigger than the population proportion for the second population. We got a significant result in this hypothesis test.

Example: Some of the residents of Tanayatown and Statsville were researched in order to see if there is a difference in the proportion of residents from each area who are employed (i.e., the employment rate). Use hypothesis testing with a probability of .03 of rejecting a true null hypothesis.

Test and CI for Two Proportions: TTEmployed?, SVEmployed?

```
Event = y


Variable      X   N  Sample p
TTEmployed?  27  40  0.675000
SVEmployed?  25  40  0.625000


Difference = p (TTEmployed?) - p (SVEmployed?)
Estimate for difference:  0.05
99% CI for difference:  (-0.224344, 0.324344)
Test for difference = 0 (vs not = 0):  Z = 0.47  P-Value = 0.639
```

Answer: p-value (.639) > α (.03), fail to reject H_0. We did not find that there is a difference in the proportion of residents from each area who are employed.

What would be wrong with this answer: "The proportion of Tanayatown residents who are employed is .675 and that is greater than the proportion of Statsville residents who are employed because that is .625."? Those are the SAMPLE proportions!! Remember Minitab always gives the sample results but we can never compare the sample statistics to get an inference about the populations.

The way Minitab shows the hypotheses is a little different than we have seen with previous tests. "`Test for difference = 0 (vs not = 0)`" is the statement of the hypotheses with the difference being shown as "`Difference = p (TTEmployed?) - p (SVEmployed?)`". Written more explicitly this is H_0: $p_1 - p_2 = 0$, H_1: $p_1 - p_2 \neq 0$, where Tanayatown is population 1. Using simple algebra you can see this is the same thing as H_0: $p_1 = p_2$, H_1: $p_1 \neq p_2$. Minitab and other sources write the hypotheses in terms of differences (because that is actually what we are testing) but the algebraically equivalent forms are used in this and many other books. As we have seen many times before, once we use the bold headings to verify that this output addresses our research question, all we need is the p-value to get an inference.

Example: Let's suppose that the employment rate is the same for Tanayatown and Statsville residents. What is the probability of getting sample results like we did?

Answer: .639

The p-value gives you the probability of getting samples like you did assuming H_0.

Side note: The method in this section is for hypothesis testing with independent (unrelated) population proportions using independent samples. Suppose we were dealing with matched pairs (dependent samples) proportion data. For example, the faculty in a Statistics department is considering a certain curriculum adjustment. For each faculty member we record whether the person is in favor of the adjustment or opposed to the adjustment. These faculty members then participate in a conference in which they are able to discuss the issue and exchange ideas. After the conference we again record for each faculty member whether the person is in favor of the adjustment or opposed to the adjustment. We are interested in whether or not the conference is associated with a change in the proportion of faculty in favor of the adjustment. The samples would not be independent because each of the before-conference responses (y or n) would be matched with an after-conference response because of belonging to the same individual. There is a special type of hypothesis test, the McNemar Test, used for this type of situation.

Exercises

For Exercises 1) and 2) do the following: a) Write the null and alternative hypotheses using symbols. b) Write the direct translation of H_1. c) Write the translation of H_1 using words from the original problem. d) Write what the conclusion would be if you rejected H_0. e) Write what the conclusion would be if you failed to reject H_0.

1) A defense attorney is considering requesting a change of venue for his client in what has been a highly publicized case. He surveys a sample of Wudmoor residents and a sample of Chef County residents to see if there is a difference in the proportion who believe his client is guilty.

2) Randomly selected customers from Deli 1 and Deli 2 are asked about their food choices. Test the claim that a smaller proportion of Deli 1 customers prefer turkey and Swiss sandwiches compared to Deli 2 customers.

3) Refer to the Sparkelz Gems/Javonn Jewelers data from Exercise 1) in Section 7.2. Perform an appropriate hypothesis test to see if a smaller percentage of Sparkelz Gems diamonds are color H compared to Javonn Jewelers diamonds. Use a .005 significance level.

4) Refer to the Jerrell Enterprises and Jaden Industries data from Exercise 2) in Section 7.2. A headhunter believes there is a higher proportion of Jerrell Enterprises employees dissatisfied with their current position at the company than there is at Jaden Industries. Does this appear to be a valid belief? With $\alpha = .10$, use an appropriate hypothesis test.

5) Refer to the example in this section dealing with employed Americans and non-Americans. Suppose we wanted to see if the proportion of employed Americans who are employed by private employers is .15 higher than the proportion of employed non-Americans who are employed by private employers. Since we are looking for a non-zero difference, we need the following test statistic formula:

$$z = \frac{(\hat{p}_1 - \hat{p}_2) - (p_1 - p_2)}{\sqrt{\frac{\hat{p}_1\hat{q}_1}{n_1} + \frac{\hat{p}_2\hat{q}_2}{n_2}}}$$

where $(p_1 - p_2)$ does not equal zero

Complete the test at the .05 significance level.

6) Two drugs—Drug A and Drug B—are being tested in a clinical trial. The side effects experienced by each person in the trial are recorded as follows: nausea (n), itching (i), headaches (h), fatigue (f), dry mouth (d). With alpha = .04, see if the percentage of users who experience nausea with Drug B is more than 20% higher than for Drug A.

Test and CI for Two Proportions: Drug B, Drug A

```
Event = n


Variable   X   N  Sample p
Drug B    23  46  0.500000
Drug A    13  50  0.260000


Difference = p (Drug B) - p (Drug A)
Estimate for difference:  0.24
92% lower bound for difference:  0.104625
Test for difference = 0.2 (vs > 0.2):  Z = 0.42  P-Value = 0.339
```

7) What is the meaning of the significance level in 6) ?

8.6 Hypothesis Testing for Two Means

Conditions: The data is from two independent simple random samples (independent in that there is no pairing or relationship between values in the different samples). We do not know the standard deviations of the two populations and do not assume they are equal. Both sample sizes are larger than 30 and/or their populations are normally distributed. (Assume all conditions are met for each example and exercise.)

Use the methods of this section when...

> testing a claim about the relationship between two means, using one sample from each population

Important formula: $t = \frac{(\bar{x}_1-\bar{x}_2)-(\mu_1-\mu_2)}{\sqrt{\frac{s_1^2}{n_1}+\frac{s_2^2}{n_2}}}$ df = smaller of $(n_1 - 1)$ and $(n_2 - 1)$

[Note: As we did with confidence intervals, the df we will use is one less than the smallest sample size. There is a more tedious formula, used by statistical software, to get the exact degrees of freedom. Using one less than the smallest sample size results in a test less prone to error while avoiding the tedium of using that formula. Also, if the population standard deviations are known or assumed to be equal, the formula for the test statistic will be different. Those cases are rare.]

Example: Randomly selected Statsville College students were asked how many units they have completed so far. Their responses were:

32 38 44 51 59 38 54 32 47 38 61 54

Of the students surveyed, 42% intend to major in Statistics and the standard deviation of the GPAs for everyone in the sample is .8017 with mean 2.4583. In a sample of 31 Wudmoor College students, 20 completed less than 50 units. The mean number of units they completed was 37.1 with standard deviation 10.7. Test the belief, at the .05 significance level, that Statsville College students have completed a higher mean number of units than Wudmoor College students.

Answer:

Traditional method

Step 1: H_0: $\mu_1 = \mu_2$

H_1: $\mu_1 > \mu_2$ (μ_1 is for Statsville; ">" for "higher")

Step 2: Test statistic: $t = \frac{(\bar{x}_1-\bar{x}_2)-(\mu_1-\mu_2)}{\sqrt{\frac{s_1^2}{n_1}+\frac{s_2^2}{n_2}}} = \frac{(45.6667-37.1)-(0)}{\sqrt{\frac{10.1564^2}{12}+\frac{10.7^2}{31}}} = 2.4437$

$\bar{x}_1$ and s_1^2 are calculated from the raw Statsville data

$\mu_1 - \mu_2 = 0$ because H_0 claims $\mu_1 = \mu_2$

Step 3:

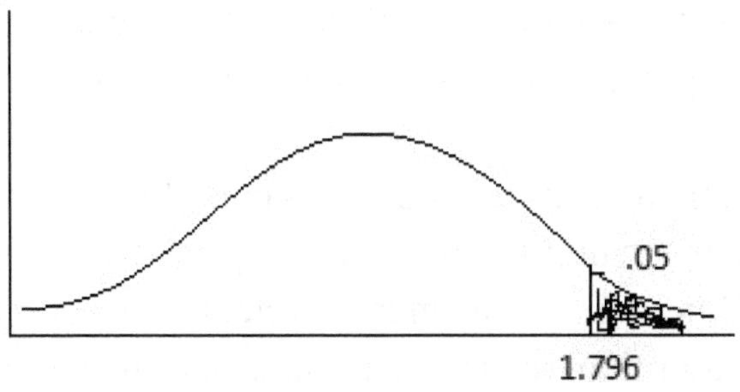

(df is the smaller of 12−1 and 31−1, which is 11)

Step 4: The test statistic is in the rejection region; reject H_0

Step 5: Conclusion: Statsville College students have completed a higher mean number of units than Wudmoor College students.

P-value method

Step 1: H_0: $\mu_1 = \mu_2$
H_1: $\mu_1 > \mu_2$

Step 2: Test statistic: $t = \frac{(\bar{x}_1 - \bar{x}_2) - (\mu_1 - \mu_2)}{\sqrt{\frac{s_1^2}{n_1} + \frac{s_2^2}{n_2}}} = \frac{(45.6667 - 37.1) - (0)}{\sqrt{\frac{10.1564^2}{12} + \frac{10.7^2}{31}}} = 2.4437$

Step 3:

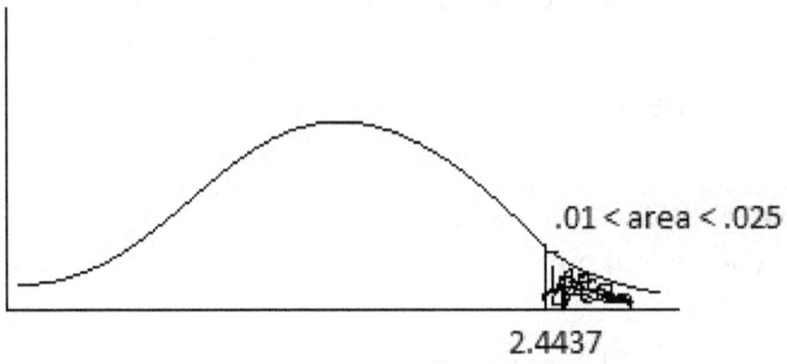

Step 4: total shaded area between .01 and .025 so .01 < p-value < .025; given $\alpha = .05$; p-value < α; reject H_0

Step 5: Conclusion: Statsville College students have completed a higher mean number of units than Wudmoor College students.

We get the p-value in the same manner described in section 8.3 (review that section to refresh your memory, if necessary). Since t-score 2.4437 is between 2.201 and 2.718 for 11 degrees of freedom, the area in the right tail is between .01 and .025.

μ_1 represents the mean number of units completed by all Statsville College students. μ_2 represents the mean number of units completed by all Wudmoor College students. The direct translation of H_1 is, "the mean number of units completed by all Statsville College students is greater than the mean number of units completed by all Wudmoor College students." Using words from the original problem H_1 says, "Statsville College students have completed a higher mean number of units than Wudmoor College students." Since we rejected H_0, that last phrase was our conclusion. Had we failed to reject H_0, our conclusion would have been, "We did not find that Statsville College students have completed a higher mean number of units than Wudmoor College students."

But wait...couldn't we just look at 45.6667 and see that it is higher than 37.1? NOOO!!!! Those are sample means! Our conclusion is an inference about the means for the populations. Yes, 45.6667 is bigger than 37.1 but is it so much bigger that we can conclude that the discrepancy is not simply due to random chance but is because the mean for the Statsville population is really bigger than the mean for the Wudmoor population? The answer is yes, and that is what the hypothesis test results tell us.

Example: In the previous example, we concluded that Statsville College students completed a higher mean number of units than Wudmoor college students. Suppose that Statsville simply having a larger mean is not that noteworthy. For your research, there is only a practical interest if Statsville has a larger mean by more than seven units. Perform hypothesis testing to determine if the mean number of units completed by Statsville college students is more than seven units higher than the mean number of units completed by Wudmoor College students.

Answer:

Traditional method

Step 1: H_0: $\mu_1 = \mu_2 + 7$

H_1: $\mu_1 > \mu_2 + 7$ (μ_1 is for Statsville; "more than seven units higher")

Step 2: Test statistic: $t = \dfrac{(\bar{x}_1 - \bar{x}_2) - (\mu_1 - \mu_2)}{\sqrt{\frac{s_1^2}{n_1} + \frac{s_2^2}{n_2}}} = \dfrac{(45.6667 - 37.1) - (7)}{\sqrt{\frac{10.1564^2}{12} + \frac{10.7^2}{31}}} = .4469$

$\mu_1 - \mu_2 = 7$ because H_0 says $\mu_1 = \mu_2 + 7$

Step 3:

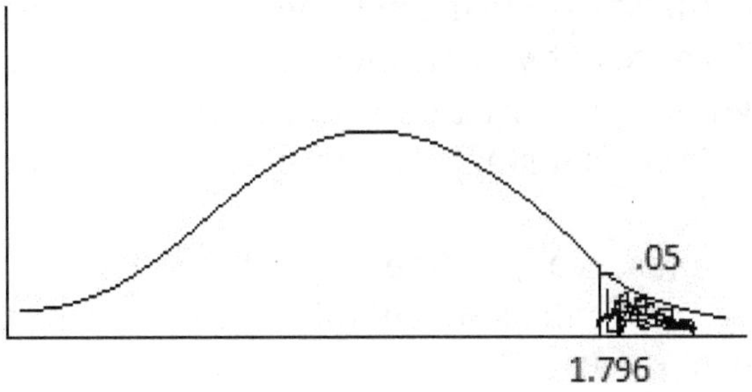

(df is the smaller of 12−1 and 31−1, which is 11)

Step 4: The test statistic is not in the rejection region; fail to reject H_0

Step 5: Conclusion: We did not find that the mean number of units completed by Statsville College students is more than seven units higher than the mean number of units completed by Wudmoor College students.

P-value method

Step 1: H_0: $\mu_1 = \mu_2 + 7$
H_1: $\mu_1 > \mu_2 + 7$

Step 2: Test statistic: $t = \frac{(\bar{x}_1-\bar{x}_2)-(\mu_1-\mu_2)}{\sqrt{\frac{s_1^2}{n_1}+\frac{s_2^2}{n_2}}} = \frac{(45.6667-37.1)-(7)}{\sqrt{\frac{10.1564^2}{12}+\frac{10.7^2}{31}}} = .4469$

Step 3:

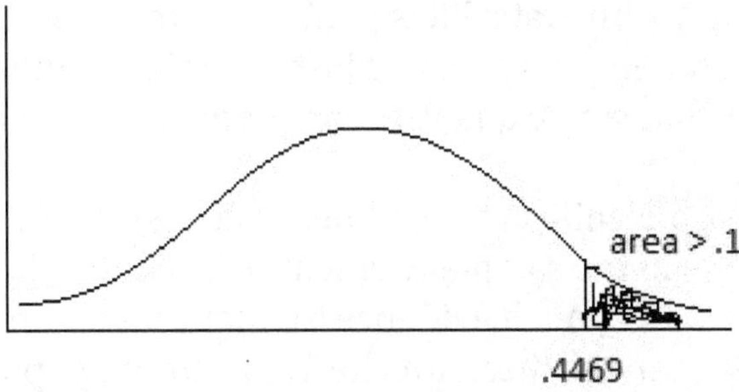

Step 4: Total shaded area is greater than .10 so p-value > .10; α not given so use α = .05; p-value > α; fail to reject H_0.

Step 5: Conclusion: We did not find that the mean number of units completed by Statsville College students is more than seven units higher than the mean number of units completed by Wudmoor College students.

So, while we determined that the mean for Statsville is larger than the mean for Wudmoor, we did not find the Statsville mean to be more than seven units higher. Sometimes, simply inferring a difference in two parameters—statistical significance—does not rise to the level of practical significance. Practical significance often needs to be determined by someone knowledgeable in the substantive research field. How much difference matters?

Suppose the problem read, "...determine if the mean number of units completed by Statsville college students is seven units higher... " In that case, the alternative hypothesis would have been $\mu_1 \neq \mu_2 + 7$

Example: Selected residents from Statsville and Tanayatown had their ages (in years) recorded and entered into Minitab. State the conclusion from the hypothesis test.

Two-Sample T-Test and CI: SVAge, TTAge

```
Two-sample T for SVAge vs TTAge

        N  Mean  StDev  SE Mean
SVAge  40  30.6   14.8      2.3
TTAge  40  62.0   45.6      7.2


Difference = mu (SVAge) - mu (TTAge)
Estimate for difference:  -31.50
97% upper bound for difference:  -16.89
T-Test of difference = 0 (vs <): T-Value = -4.16  P-Value = 0.000  DF = 47
```

Answer: p-value (.000) < α (.05), reject H_0. The mean age of all Statsville residents is less than the mean age of all Tanayatown residents.

Since no significance level was given, we use .05. This output is setup the same way as the difference of proportion output from the previous section. "`Difference = mu (SVAge) - mu (TTAge)`" and "`T-Test of difference = 0 (vs <)`" tell us that we are testing H_0: $\mu_1=\mu_2$ versus H_1: $\mu_1<\mu_2$, with Statsville's population mean age being represented by μ_1. The "`Estimate for difference:  -31.50`" is just referring to the difference in the sample means but, of course, that single quantity cannot be used to make an inference about the populations.

We have a significant result here because we rejected H_0. The mean age of Statsville residents in the sample was so much smaller than the mean for the Tanayatown sample that we had enough evidence to infer that the mean age of the entire Statsville population is smaller than the mean for the Tanayatown population. A Type I error for this problem would be concluding that the mean age of all Statsville residents is less than the mean age of all Tanayatown residents, but then it turns out that is not true in reality.

Example: Suppose the first example of this section was changed as shown below. How would this new problem by handled?

Randomly selected Statsville College students were asked how many units they have completed so far. Their responses were:

32 38 44 51 59 38 54 32 47 38 61 54

Of the students surveyed, 42% intend to major in Statistics and the standard deviation of the GPAs for everyone in the sample is .8017 with mean 2.4583. The mean number of units completed by Wudmoor College students is 37.1 with standard deviation 10.7. Test the belief, at the .05 significance level, that Statsville College students have completed a higher mean number of units than Wudmoor College students.

Answer: This time we only have one sample so we would use the method of Section 8.3. The hypotheses would be

$$H_0: \mu = 37.1$$
$$H_1: \mu > 37.1$$

where μ stands for the mean number of units completed by all Statsville College students. The direct translation of H_1 is "the mean number of units completed by all Statsville College students is greater than 37.1" and using words from the problem this is "Statsville College students have completed a higher mean number of units than Wudmoor College students"

Exercises

For Exercises 1) and 2) do the following: a) Write the null and alternative hypotheses using symbols. b) Write the direct translation of H_1. c) Write the translation of H_1 using words from the original problem. d) Write what the conclusion would be if you rejected H_0. e) Write what the conclusion would be if you failed to reject H_0.

1) We consider a sample of men's soccer matches and a sample of women's soccer matches to see if men's and women's soccer matches have the same average amount of added time.

2) Households in Wudmoor are sampled and households in Twin Streams are sampled to see if Wudmoor homes tend to have cheaper utility bills than Twin Streams homes.

3) Refer to the Sparkelz Gems/Javonn Jewelers data from Exercise 1) in Section 7.2. Test the claim that, on average, Javonn Jewelers has diamonds with sizes that are 2mm larger than the diamonds at Sparkelz. Keep the probability of making a Type I error at .02.

4) Refer to the Jerrell Enterprises and Jaden Industries data from Exercise 2) in Section 7.2. Respond to the assertion that the mean absences among Jerrell Enterprises employees is larger than the mean absences for Jaden Industries employees. Make sure that the probability of wrongly concluding that the mean absences among Jerrell Enterprises employees is larger than the mean absences for Jaden Industries employees is .005.

5) There are four furniture stores in a certain city. One is in the north, one is in the south, one is in the east, and one is in the west. We collected random samples of prices (in dollars) from each store and used the appropriate data for the Minitab output below. Is the mean price of items at the north store smaller than the mean price of items at the south store? Use $\alpha = .025$ and answer this question.

Two-Sample T-Test and CI: North, South

```
Two-sample T for North vs South

        N    Mean  StDev  SE Mean
North  91  1508.9   46.2      4.8
South  82  1514.6   48.4      5.3


Difference = mu (North) - mu (South)
Estimate for difference:  -5.66
95% upper bound for difference:  6.26
T-Test of difference = 0 (vs <): T-Value = -0.79  P-Value = 0.217  DF = 167
```

6) Refer to 5).

a) Describe a Type II error for that exercise.

b) What is the probability that we would conclude that the mean price of items sold at the north store is less than the mean price of items sold at the south store if, in fact, that is not true?

8.7 Hypothesis Testing for the Mean of Differences

Conditions: The data is from two dependent simple random samples (dependent in that there is pairing due to a unique relationship between each value in one sample and a value in the other sample). We do not know the standard deviations of the two populations or their paired differences and make no assumptions about them. The number of sample pairs is larger than 30 and/or the population of paired differences is normally distributed. (Assume all conditions are met for each example and exercise.)

Use the methods of this section when...

testing a claim about a mean of differences

Important formula: $t = \frac{\bar{x}_d - \mu_d}{\frac{s_d}{\sqrt{n}}}$ n is the number of pairs

Example: Randomly selected people were studied and had their hand length, foot length, and ear height (all measured in cm) measured. The measurements for each person in the sample are shown below:

Ear	6.4	6.1	5.9	5.7	6.3	5.8	6.3	6.0
Hand	20.9	18.2	15.6	17.3	19.3	17.0	19.7	18.4
Foot	29.5	27.4	21.6	26.0	25.9	22.8	28.8	24.0

At the .005 significance level, see if a person's foot is generally longer than the person's hand.

Answer:

This is a matched pairs problem because the hand and foot measurements came from the same people. The problem does not involve ears so we can ignore that information. Like we saw with matched pairs in the confidence interval discussion, the data we analyze will be the differences in the matched pairs. We organize that information in a table, including a row of squared differences to facilitate calculating the standard deviation of the differences in the sample.

									Totals
Hand	20.9	18.2	15.6	17.3	19.3	17.0	19.7	18.4	
Foot	29.5	27.4	21.6	26.0	25.9	22.8	28.8	24.0	
d	-8.6	-9.2	-6.0	-8.7	-6.6	-5.8	-9.1	-5.6	-59.6
d^2	73.96	84.64	36	75.69	43.56	33.64	82.81	31.36	461.66

We calculate the mean of the sample of differences as −7.45 cm with standard deviation 1.5875 cm.

Traditional method:

Step 1: H_0: $\mu_d = 0$

H_1: $\mu_d < 0$ (we calculated differences as hand−foot, so "foot longer than hand" would indicate a negative mean difference, so "<")

Step 2: Test statistic: $t = \frac{\bar{x}_d - \mu_d}{\frac{s_d}{\sqrt{n}}} = \frac{-7.45-0}{\frac{1.5875}{\sqrt{8}}} = -13.2736$

0 in the formula comes from $\mu_d = 0$ in H_0

Step 3:

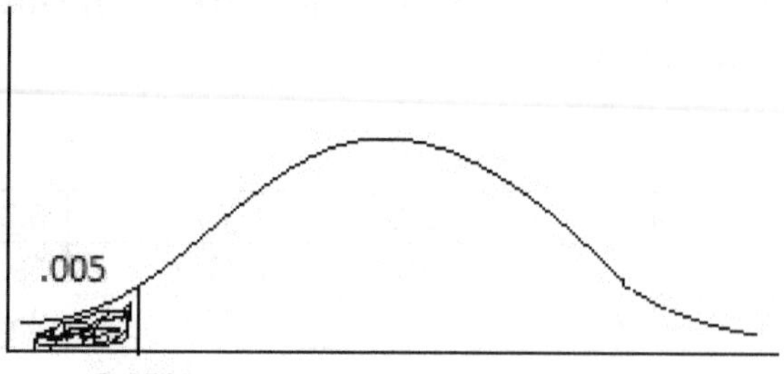

Step 4: The test statistic is in the rejection region; reject H_0

Step 5: Conclusion: A person's foot is generally longer than the person's hand.

P-value method:

Step 1: H_0: $\mu_d = 0$
H_1: $\mu_d < 0$

Step 2: Test statistic: $t = \frac{\bar{x}_d - \mu_d}{\frac{s_d}{\sqrt{n}}} = \frac{-7.45-0}{\frac{1.5875}{\sqrt{8}}} = -13.2736$

Step 3:

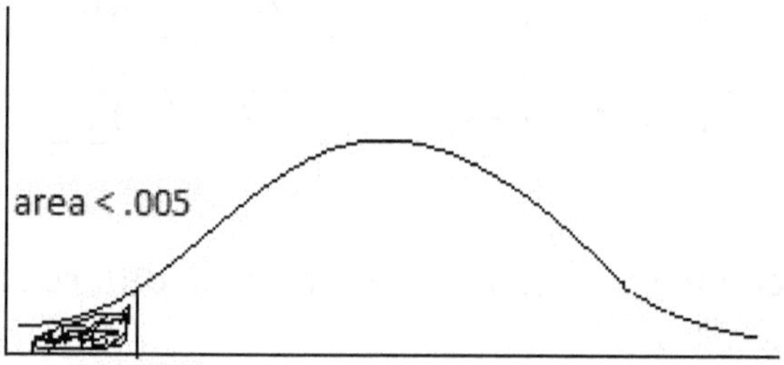

Step 4: total shaded area < .005 so p-value < .005; α given .005; p-value < α; reject H_0

Step 5: Conclusion: A person's foot is generally longer than the person's hand.

To clarify the p-value, remember that Table B only gives areas to the right of positive t-scores. So, we need to find the area to the right of positive 13.2736, knowing that whatever that area is, it's the same as the area to the left of −13.2736. Looking for 13.2736 in df row 7 of the table, we find it's off the table to the left. So, the corresponding area is off the table to the left, making it some number smaller than .005.

Similarly, to get the critical value in the traditional method we observe from df row 7 and column .005 that critical value 3.499 has area .005 to its right. That tells us that critical value −3.499 has area .005 to its left.

μ_d represents the mean of all the differences in people's hand length and foot length. $\bar{x}_d$ is the mean of the differences in people's hand length and foot length in the sample. s_d is the standard deviation of those sample differences. The direct translation of H_1 is "the mean of all the differences in people's hand length and foot length is less than zero." Using words from the problem H_1 is, "a person's foot is generally longer than the person's hand." Had we failed to reject H_1 we would say, "We did not find that a person's foot is generally longer than the person's hand."

Observe the following changes in the research question and the alternative hypothesis that would go with each (of course, the null hypothesis for each would be achieved by replacing the inequality symbol with an equals sign). We are still calculating the differences as hand minus foot.

see if a person's foot is generally shorter than the person's hand

H_1: $\mu_d > 0$

see if a person's foot generally has a different length than the person's hand

H_1: $\mu_d \neq 0$

see if a person's foot, on average, is 2 cm longer than the person's hand

H_1: $\mu_d \neq -2$

see if a person's foot, on average, is 2 cm shorter than the person's hand

H_1: $\mu_d \neq 2$

see if a person's foot, on average, is more than 2 cm longer than the person's hand

H_1: $\mu_d < -2$

As usual, the key is having clearly in mind what you are doing, what everything means, and the order you are using to calculate the differences. Whatever number appears in the hypotheses would be the number used for μ_d when the test statistic is calculated.

If we calculated our differences as foot−hand in the original problem, H_1 would have been $\mu_d > 0$.

Example: 40 homes from Statsville were investigated and the value of the homes (in dollars) before the construction of the new local sports arena were recorded. After the construction of the arena, the value of these same homes was recorded. A local real estate investor believes that Statsville home values increased by more than $5000, on average, after the construction of the arena. Does that seem to be correct? Use the Minitab output with a .10 significance level to answer this question.

Paired T-Test and CI: ValueNow, ValueBefore

```
Paired T for ValueNow - ValueBefore

              N    Mean  StDev  SE Mean
ValueNow     40  203729  41787     6607
ValueBefore  40  187924  37566     5940
Difference   40   15805  29318     4636


96% lower bound for mean difference: 7472
T-Test of mean difference = 5000 (vs > 5000): T-Value = 2.33  P-Value = 0.013
```

> **Answer**: p-value (.013) < α (.10), reject H_0. Statsville home values increased by more than $5000, on average, after the construction of the arena.

In output for a paired differences (μ_d) test, the hypotheses are shown with "`Paired T for ValueNow - ValueBefore`" and "`T-Test of mean difference = 5000 (vs > 5000)`". This is H_0: $\mu_d = 5000$, H_1: $\mu_d > 5000$, where the differences for each pair are calculated as the current value minus the value before the arena. As usual, the output provides the sample results, and even the test statistic (`T-Value = 2.33`), but all we need is the p-value to make our conclusion. As mentioned before, ignore references to upper or lower bounds in this course.

If we wanted to see if home price values increased by $5000 (instead of "more than $5000") we would be testing H_0: $\mu_d = 5000$, H_1: $\mu_d \neq 5000$.

Example: Refer to the previous example. If the sample we got, and the hypothesis test we performed using that sample, led us to an incorrect conclusion about the population, what kind of error would that be? Describe the other type of error for this problem.

> **Answer**: Type I. A Type II error for this problem would occur if we concluded that we did not find that Statsville home values increased by more than $5000, on average, after the construction of the arena if the fact is that they did.

Exercises

For Exercises 1) and 2) do the following: a) Write the null and alternative hypotheses using symbols. b) Write the direct translation of H_1. c) Write the translation of H_1 using words from the original problem. d) Write what the conclusion would be if you rejected H_0. e) Write what the conclusion would be if you failed to reject H_0.

1) Researchers from a consumer group visited several different cities on the same day at the same time. In each city, they observed a fast food restaurant that primarily features chicken and a fast food restaurant that primarily features burgers to see if chicken restaurants generally have longer lines than burger restaurants.

2) A new skin regimen is being tested. Participants in the study had their skin quality measured before using the regimen and then six weeks after using the regimen. The goal of the research is to see if the skin regimen makes a difference in the quality of users' skin.

3) Refer to the Statsville Market and Grocers data from Exercise 1) in Section 7.7. You prefer to shop at Statsville Grocers. Your family thinks something is wrong with you because, in their words, "Statsville Market has much better prices! C'mon!". Perform an appropriate hypothesis test to see if your family's claim (about the prices...) is supported. ($\alpha = .01$).

4) Refer to the Jerrell Enterprises and Jaden Industries data from Exercise 2) in Section 7.2. Suppose you work at Jaden Industries and one of your co-workers swears that the bosses don't like the employees and are constantly trying to, "keep the employees down." At the .05 significance level, see if supervisor performance ratings tend to be smaller than peer performance ratings (for respective employees) at Jaden Industries. How would you respond to your co-worker?

5) There are four furniture stores in a certain city. One is in the north, one is in the south, one is in the east, and one is in the west. We collected random samples of prices (in dollars) from each store and used the appropriate data for the Minitab output below. Use a process with a probability of .07 for making a Type I error to determine if there is a difference, on average, of prices at the north and east stores.

Paired T-Test and CI: North, East

```
Paired T for North - East

              N     Mean  StDev  SE Mean
North        91  1508.94  46.15     4.84
East         91  2005.90  37.18     3.90
Difference   91  -496.95  62.49     6.55


93% CI for mean difference: (-508.97, -484.94)
T-Test of mean difference = 0 (vs not = 0): T-Value = -75.86  P-Value = 0.000
```

6) Refer to 5).

a) Did we detect a significant difference between prices at the north and east stores? Explain.

b) What is the probability of getting sample results like we got if the north and east stores have the same prices (on average)?

8.8 Hypothesis Testing for Two Variances or Standard Deviations

Conditions: The data comes from two independent simple random samples. Each population must be normally distributed. [The normality requirement is very strict for this test.]
(Assume all conditions are met for each example and exercise.)

Use the methods of this section when...

> testing a claim about the relationship between two variances or standard deviations, using one sample from each population

Important formula: $F = \frac{s_1^2}{s_2^2}$

Example: Randomly selected Statsville College students were asked how many units they have completed so far. Their responses were:

32 38 44 51 59 38 54 32 47 38 61 54

Of the students surveyed, 42% intend to major in Statistics and the standard deviation of the GPAs for everyone in the sample is .8017 with mean 2.4583. In a sample of 31 Wudmoor College students, 20 completed less than 50 units. The mean number of units they completed was 37.1 with standard deviation 10.7. At level of significance .025, see if Statsville College students have a smaller standard deviation than Wudmoor College students in their number of units completed.

Answer:
Traditional method:

Step 1: H_0: $\sigma_1 = \sigma_2$
H_1: $\sigma_1 < \sigma_2$ (σ_1 is the standard deviation for Statsville population)

Step 2: Test statistic: $F = \frac{s_1^2}{s_2^2} = \frac{10.1564^2}{10.7^2} = .9010$
(As we have seen before, 10.1564 is calculated from the raw data.)

Step 3:

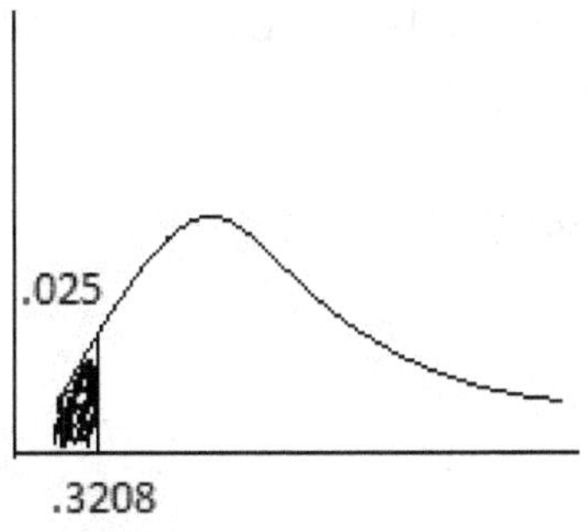

Step 4: The test statistic is not in the rejection region so fail to reject H_0

Step 5: Conclusion: We did not find that Statsville College students have a smaller standard deviation than Wudmoor College students in their number of units completed.

To clarify the critical value, since we are on the left side of the diagram, we are looking for $F_{1-(\alpha/2)}$. We flip the degrees of freedom and use $df_1=30$ and $df_2 = 11$ then use the reciprocal of the critical value we see (on area = .025 page). That gives us $\frac{1}{3.1176} = .3208$

σ_1 represents the standard deviation of the number of units completed by all Statsville College students. σ_2 represents the standard deviation of the number of units completed by all Wudmoor College students. The direct translation of the alternative hypothesis is, "the standard deviation of the number of units completed by all Statsville College students is less than the standard deviation of the number of units completed by all Wudmoor College students." Using words from the problem this is, "Statsville College students have a smaller standard deviation than Wudmoor College students in their number of units completed."

Recall, if this test had required a critical value for the right side we would have retained the original df_1 and df_2 and used the critical value given in the table. If either exact number of degrees of freedom did not appear in the table we would use the critical value for the closest number of degrees provided in the table. If our number of degrees of freedom was right in the middle of two degrees of freedom in the table, we would calculate the critical value that falls right in the middle of the respective critical values for those two degrees of freedom values.

We will only use the traditional method for hypothesis tests requiring the F table due to the limited area information it gives. Of course, if we used Minitab or some other statistical software, we would use the p-value that it provides.

Example: Selected residents from Statsville and Tanayatown had their ages (in years) recorded and entered into Minitab. Is there a difference in how consistent Statsville ages are compared to Tanayatown ages?

Test and CI for Two Variances: SVAge, TTAge

```
Method

Null hypothesis         Sigma(SVAge) / Sigma(TTAge) = 1
Alternative hypothesis  Sigma(SVAge) / Sigma(TTAge) not = 1


Statistics

Variable   N   StDev  Variance
SVAge     40  14.791   218.767
TTAge     40  45.599  2079.279

Ratio of standard deviations = 0.324
Ratio of variances = 0.105


95% Confidence Intervals

                                   CI for
Distribution   CI for StDev       Variance
of Data            Ratio            Ratio
Normal        (0.236, 0.446)  (0.056, 0.199)
Continuous    (0.227, 0.569)  (0.052, 0.324)


Tests

                                                Test
Method                           DF1  DF2  Statistic  P-Value
F Test (normal)                   39   39       0.11    0.000
Levene's Test (any continuous)     1   78      15.10    0.000
```

Answer: p-value (.000) < α (.05), reject H_0. There is a difference in how consistent Statsville ages are compared to Tanayatown ages.

The question is asking for how "consistent" the values are which is another way of asking about the variation or spread in the population. Minitab uses the more technical approach to stating the hypotheses:

```
Null hypothesis         Sigma(SVAge) / Sigma(TTAge) = 1
Alternative hypothesis  Sigma(SVAge) / Sigma(TTAge) not = 1
```

because this ratio is actually what is being tested. However, algebraically these statements are the same: H_0: $\sigma_1 = \sigma_2$, H_1: $\sigma_1 \neq \sigma_2$. After giving other information, all of which is clearly labeled, we find the one piece of information we need to make an inference—the p-value. We will just stick with the "`F Test (normal)`" output, as the other p-value comes from a test that is beyond the level of an introductory Statistics course. No significance level was given so we knew to use .05.

The probability that this hypothesis testing process would lead us to conclude that there is a difference in how consistent Statsville ages are compared to Tanayatown ages, if that is not really the case, is .05 (alpha). The probability of getting sample results like we got if the two towns actually have equally consistent ages is .000 (p-value). Remember that the p-value is never actually exactly zero, but Minitab shows the value to only three decimal places.

Note: Any problem that deals with variation of two populations given a sample from each population could be written in terms of σ_1 and σ_2 OR σ_1^2 and σ_2^2. The test statistic and the entire process remain the same.

Suppose your null hypothesis is something other than both standard deviations (or variances) being equal, like H_0: $\sigma_1 = 3\sigma_2$. The ratio that is actually being tested is H_0: $\frac{\sigma_1}{3\sigma_2} = 1$. The test statistic will be $F = \frac{s_1^2}{(3s_2)^2}$. To calculate the test statistic correctly, you must have the null hypothesis in the form $\frac{variance_1}{variance_2} = 1$ (the right side has to be 1). An example of this is in Exercise 3.

[It is possible to test equality of more than two variances using something called the Bartlett test, which is beyond the level of this course.]

Exercises

For Exercises 1) and 2) do the following: a) Write the null and alternative hypotheses using symbols. b) Write the direct translation of H_1. c) Write the translation of H_1 using words from the original problem. d) Write what the conclusion would be if you rejected H_0. e) Write what the conclusion would be if you failed to reject H_0.

1) Chef County residents were studied, as were Tanayatown residents (one sample from each area). Each person was asked to record the number of miles he or she traveled on vacation so researchers could see if there is more dispersion in the travel distances of Chef County residents compared to Tanayatown residents.

2) A collection of data from randomly selected professional tennis matches and randomly selected amateur tennis matches was studied to determine if there is the same spread in the number of unforced errors in each match for professional tennis matches compared to amateur matches.

3) Refer to the Sparkelz Gems/Javonn Jewelers data from Exercise 1) in Section 7.2. See if the weights of Sparkelz Gems' diamonds have a standard deviation that is more than 1.1 times the standard deviation of the weights of Javonn Jewelers' diamonds.

4) Refer to the Jerrell Enterprises and Jaden Industries data from Exercise 2) in Section 7.2. A Human Resources manager thinks that there is a difference in the variability in supervisor ratings at Jerrell Enterprises and Jaden Industries. Respond to that thought at the .05 significance level.

5) There are four furniture stores in a certain city. One is in the north, one is in the south, one is in the east, and one is in the west. We collected random samples of prices (in dollars) from each store and used the appropriate data for the Minitab output below.

Test and CI for Two Variances: West, South

```
Method

Null hypothesis         Sigma(West) / Sigma(South) = 1
Alternative hypothesis  Sigma(West) / Sigma(South) > 1


Statistics

Variable   N   StDev  Variance
West      73  79.113  6258.907
South     82  48.404  2342.953

Ratio of standard deviations = 1.634
Ratio of variances = 2.671


90% One-Sided Confidence Intervals

              Lower Bound   Lower Bound
Distribution   for StDev  for Variance
of Data          Ratio         Ratio
Normal           1.412         1.993
Continuous       1.316         1.733


Tests

                                              Test
Method                          DF1  DF2  Statistic  P-Value
F Test (normal)                  72   81       2.67    0.000
Levene's Test (any continuous)    1  153      10.26    0.001
```

a) Are west store's prices more varied than the south store's?

b) Describe a Type I error for this problem.

c) Did we get a significant result?

8.9 Hypothesis Testing: A Mixture

We have now considered all of the basic hypothesis tests. As is the case with most math and Statistics books, each section of this text contains similar problems. So, if you're not really paying attention, you could breeze through the exercises for each section because once you figure out how to answer the first one, the others are all pretty similar. Unfortunately, that is not enough to show that you actually know the material. The use and understanding of Statistics is based on knowing what kind of procedure is appropriate for the data you have and the research question at hand. You have to be able to look at the situation, look at the data, and do the right thing with it. To that end, this section includes problems from the previous hypothesis testing sections. There is no order to them. Read the situation, figure out the correct type of hypothesis test to perform, and then do it!

One tip to keep in mind is that the number of samples will match the number of populations involved in your hypothesis test. For example, if the problem involves proportion and there is one sample, the hypotheses will involve p. If there are two separate samples, the hypotheses will involve p_1 and p_2. Before trying the exercises in this section, you may want to go back and take note of how the tests for a single parameter from one population (p, μ, σ, σ^2) used a single sample, whereas the tests for two populations (p_1 and p_2, μ_1 and μ_2, μ_d, σ_1^2 and σ_2^2, σ_1 and σ_2) used two samples.

As always, it is important to read problems carefully to know what information is relevant and what is not relevant to the specific research question at hand. When using statistical methods, it is quite common to do several different things with one set of data to answer several different research questions. As you go from one question to the next, you have to know what information is relevant to each inquiry. The pieces of data that are not relevant to one inquiry do not disappear, and then reappear when needed for another question. All the data stays where it is but you have to decipher what is needed and when.

There are no trick questions. Read carefully and answer what is asked. There's nothing tricky about that.

As you know by now, there are no "key words" in Statistics problems because the English language affords us the ability to express the same research questions using a wide variety of words. It's all about understanding what's happening. If research questions are about the mean, the average, one group being generally larger than another, one group tending to have smaller values than the other, etc. this is addressed using the mean. If a problem is about a proportion, a percentage, a rate, etc. this is addressed using the proportion. If a problem is about variation, dispersion, spread, inconsistency, variability, etc. we can use the standard deviation or the variance to address that. Notice the repeated use of "etc." in this paragraph. Once again, you cannot be fixated on certain words but think about the context of what is being researched—this is not just for "passing this class" but for being able to use Statistics far beyond the class you're currently taking.

For any hypothesis test question, the first step is figuring out which type of hypothesis test will be used and why. Once that is clear to you, then you can proceed to determine the correct formula to use and the relevant information that will be needed for the formula. After that, carefully perform all steps.

The same pattern from our earlier work still holds that certain distributions are matched with certain parameters. If you're working with the mean(s), the t distribution will be used. This is for u, u_1 and μ_2, μ_d. If you are working with the proportion(s), the z distribution will be used. This is for p, p_1 and p_2. If you are testing a single measure of variation (σ or σ^2), χ^2 is your distribution. If your test involves two measures of variation (σ_1^2 and σ_2^2, σ_1 and σ_2), the F distribution is what you use.

In addition to the exercises in this section, you may also find it helpful to randomly pick exercises from the earlier sections in this chapter and try the exercises again without looking at work you already did or similar examples from the section. Approach each problem like you are seeing it for the first time and see if you can figure out what to do. Hopefully, the only help you will need from your notes or the text is the applicable formula.

Speaking of formulas, you should be able to recognize what formula goes with each hypothesis test. You probably don't need to have the formulas memorized (verify this with your instructor) but you should be able to recognize them. For example, when I see

$$t = \frac{\bar{x}-\mu}{\frac{s}{\sqrt{n}}}$$

I immediately know that is the test statistic when hypothesis testing for μ. How? For starters, I see the μ in the formula. There is only one μ, and it does not have a d subscript. Review the other formulas and look for the indications that tell you which formula is used for which type of hypothesis test.

Finally, remember that hypothesis testing is always used to test something about population(s), never sample(s)!!! That's kind of a big deal.

Exercises

Do not use confidence intervals for any exercises.

1) Randomly selected students from Chef County College were surveyed. Nine of them reported liking lacrosse. Six of them were first-generation college students. Their mean household income was $64397.12 with standard deviation $1923.44. For 85% of them both of their parents are still alive. The variance of their GPAs was .6714. These students were also asked their typical commute time to school, typical commute time from school, typical amount of time grooming on a school day, and whether they typically walk to and from school or ride to and from school. Those responses are recorded below:

Student	A	B	C	D	E	F	G	H	I	J	K	L	M
To	30	10	25	15	45	60	5	25	30	20	50	15	12
From	60	10	60	40	60	60	5	20	35	20	60	15	12
Groom	20	40	30	15	90	35	40	20	30	15	10	25	30
Walk/Ride	r	w	r	r	r	r	w	r	r	w	r	w	w

(times are in minutes)

Randomly selected students from Twin Streams College were surveyed. Of the 19 students surveyed, their mean typical commute time to school was 26.27 minutes with variance 265.69 minutes^2, average typical amount of time spent grooming on a school day was 41.32 minutes with standard deviation 20.1 minutes, mean typical commute time from school was 32.89 minutes with variance 338.56 minutes^2, and 57.9% of them typically walk to and from school. Eight of them were first-generation college students. Their mean household income was $73219.34 with standard deviation $1899.37. 31.58% typically spend less than 20 minutes commuting to school. Their mothers averaged 3.84 years of higher education with a median of three years.

a) Test the claim, at the .01 significance level, that Chef County College riders generally take more than 20 minutes longer to get to school than Chef County College students who usually walk to school.

b) See if there is a difference in the percentage of Twin Streams students with a typical commute to school of less than 20 minutes and Chef County students with a typical commute to school of less than 20 minutes. ($\alpha = 0.10$)

c) The County Register reported that Chef County College students have a mean household income of $63,000. Respond to that claim.

d) Are the typical commute times from school for Twin Streams students more consistent (i.e. less varied) than for Chef County students? Perform a test in which the probability of incorrectly concluding that Twin Streams students have more consistent commute times from school than Chef County Students, if they really do not, is only .025.

e) Is the result in d) significant? Explain.

2) Residents from Chef County and residents from Twin Streams were asked if they **r**ent or **o**wn their home (variable: **status**), how long (years) they have lived at their current residence (variable: **residence**), if children live in the household (yes or no, variable: **children**), how long (years) they have worked at their current job (variable: **job**). Renters were also categorized based on how long they have rented their current home as short term (less than a year), moderate term (from one to five years), and long term (over five years) using the variable **rent-type,** with responses coded as s, m, or l. Each respondent's location of residence was recorded in the **location** variable (c: Chef County, t: Twin Streams). **c-residence** and **c-job** are the variables for length of time at current residence and job for only Chef County residents. The following Minitab output was generated using that data.

Test and CI for Two Proportions: status, location

```
Event = r


location   X   N  Sample p
c         14  33  0.424242
t         17  34  0.500000


Difference = p (c) - p (t)
Estimate for difference:  -0.0757576
95% CI for difference:  (-0.313833, 0.162318)
Test for difference = -0.35 (vs not = -0.35):  Z = 2.26  P-Value = 0.024
```

Test and CI for One Proportion: rent-type

```
Test of p = 0.8 vs p < 0.8

Event = s


                                99% Upper    Exact
Variable     X   N  Sample p        Bound  P-Value
rent-type   14  31  0.451613     0.669808    0.000
```

Test and CI for One Variance: job

```
Method

Null hypothesis          Sigma = 4
Alternative hypothesis   Sigma > 4

The chi-square method is only for the normal distribution.
```

```
The Bonett method is for any continuous distribution.


Statistics

Variable   N  StDev  Variance
job       67   4.21      17.7


98% One-Sided Confidence Intervals

                          Lower
                          Bound
                            for   Lower Bound
Variable  Method          StDev  for Variance
job       Chi-Square       3.57          12.7
          Bonett           3.43          11.8


Tests

                               Test
Variable  Method          Statistic  DF  P-Value
job       Chi-Square          73.04  66    0.258
          Bonett                  —   —    0.301
```

Paired T-Test and CI: c-residence, c-job

```
Paired T for c-residence - c-job

              N    Mean  StDev  SE Mean
c-residence  33   5.600  4.816    0.838
c-job        33   6.327  3.150    0.548
Difference   33  -0.727  5.301    0.923


96% CI for mean difference: (-2.703, 1.248)
T-Test of mean difference = 0 (vs not = 0): T-Value = -0.79  P-Value = 0.436
```

a) Test the claim that the percentage of renters in Twin Streams is 35% higher than the percentage of renters in Chef County.

b) Is the standard deviation of the amount of time that Chef County and Twin Streams residents have been at their current job more than 4 years?

3) 30 Statsville residents and 35 Wudmoor residents were asked, "What is the biggest challenge you face when it comes to exercising?" 70% of the Statsville respondents answered, "Time" as did 89% of the Wudmoor respondents. A local gym owner believes that a greater proportion of Wudmoor residents have time as their greatest exercise challenge than Statsville residents who feel this way. Test the gym owner's belief.

4) Statsville residents and Wudmoor residents were asked, "What is the biggest challenge you face when it comes to exercising?" 70% of the 30 Statsville respondents answered, "Time". 89% of Wudmoor residents have time as their biggest challenge to exercise. A local gym owner believes that a greater proportion of Wudmoor residents have time as their greatest exercise challenge than Statsville residents who feel this way. Test the gym owner's belief.

5) Refer to 4)

a) Describe a Type I error for that situation. What is the probability of such an error?

b) Describe a Type II error for that situation.

c) What is the probability of getting sample results like we did if, in fact, Statsville and Wudmoor have the same proportion of residents who see time as their greatest challenge to exercise?

For the following problems, show as much work as needed to get the conclusion.

6) A recent report said that 2% of people have armpits that produce no odor. Some people were tested in order to address this claim and, using the data collected, the p-value was .4563. Does the report appear to be correct? Explain.

7) Randomly selected gym members were asked to perform a deadlift with the maximum weight they could lift, and a bench press with the maximum weight they could lift. The same gym owner from earlier wanted to see if people can generally lift more weight when doing a deadlift compared to bench press. The critical value for the appropriate test was 2.307, and the data yielded test statistic 3.065. State your conclusion.

8) 12 computer monitors had a mean luminosity of 398 candelas per square meter. A technician wanted to see if a certain maintenance process reduces the variation in the luminosity of computer monitors. When performing the appropriate hypothesis test, the test statistic was 2.798. State your conclusion.

9 Advanced Hypothesis Testing

9.1 Foundations

In the remaining sections of the text, we discuss more advanced hypothesis testing. These tests follow the same basic steps as the previous hypothesis tests, but they handle more sophisticated research questions and data. Also, some tests will have certain peculiar features that must be remembered when using them. When deciding on which hypothesis test to use, we need to consider carefully 1) what kind of data we have and 2) what we are trying to test. This has been true with all hypothesis tests heretofore, but we need to think more deeply when deciding which test to use of those considered in this chapter.

For each research context we will need to start formally considering how many variables we have and what kind they are. In probability, the term "random variable" is a technical term with a specific definition relating to functions. When we use the term "variable" in this book, think in terms of a piece of information we are collecting from each member of the sample. For each variable, we also need to know its kind: categorical or quantitative (with the meanings we learned in Chapter 1). For example, we sample 50 people and ask them their favorite animal, their income, and how many siblings they have. That's three variables and they are categorical, quantitative, and quantitative, respectively. The type and number of variables will dictate what hypothesis test we perform with the data. In the examples and exercises of each section, be sure to verify that the variable quantities and types are as they should be for the procedure we will perform. For example, if a research context is supposed to have one categorical variable and one quantitative variable in order to use a certain test, identify those variables before performing the test described in that section.

9.2 ANOVA (One-Way)

Conditions: The data comes from two or more independent simple random samples. Each population is normally distributed and they have equal variances.
(Assume all conditions are met for each example and exercise.)

Use the methods of this section when...

testing a claim about the equality of means

Variable types: one quantitative and one categorical

Important formulas:

$$\text{SSB} = \sum n_i(\bar{x}_i - \bar{x})^2 = n_1(\overline{x_1} - \bar{x})^2 + n_2(\overline{x_2} - \bar{x})^2 + \cdots + n_k(\overline{x_k} - \bar{x})^2$$

$$\text{MSB} = \frac{SSB}{df\ for\ SSB} = \frac{SSB}{k-1}$$

$$\text{SSW} = \sum(n_i - 1)\, s_i^2 = (n_1 - 1)s_1^2 + (n_2 - 1)s_2^2 + \cdots + (n_k - 1)s_k^2$$

$$\text{MSW} = \frac{SSW}{df\ for\ SSW} = \frac{SSW}{n-k}$$

$$\text{F} = \frac{MSB}{MSW}$$

[One-Way ANOVA can be used to test the equality of exactly two means. However, this is not normally done because the simpler t test for two means is usually preferred. In practice, you'll use this test for testing three or more means. The categorical variable will have three or more categories, or levels.]

Suppose a scientist is investigating levels of radioactivity in foods. She wants to test the claim that the mean level of radioactivity in a typical meal in Maryland is 7000 pCi/kg. As you know, she would collect a sample of typical meals in Maryland and then use it to perform hypothesis testing of H_0: $\mu = 7000$ versus H_1: $\mu \neq 7000$. Suppose she wants to test the claim that the mean level of radioactivity is the same in a typical meal in Maryland and a typical meal in Florida. She would get a sample of typical meals in Maryland, a sample of typical meals in Florida, and then use them to perform hypothesis testing of H_0: $\mu_1 = \mu_2$ versus H_1: $\mu_1 \neq \mu_2$. Now suppose she wanted to test the claim that the mean level of radioactivity is the same for typical meals in all 50 states of the United States. Hmmm... Based on what you have learned thus far, she could get a sample of typical meals from each state and then use them to perform hypothesis testing of H_0: $\mu_1 = \mu_2$ versus H_1: $\mu_1 \neq \mu_2$ for each possible pair of states. That would be a lot of tests (1225, in fact)! Besides the sheer number of tests, there is a bigger issue. Remember that, whenever hypothesis testing is performed, there exists the potential for errors—Type I and Type II. This potential for error would compound with each test performed so the results would be highly suspect and untrustworthy. There is a more efficient test for problems like this.

When we want to test the equality of three or more means we use a process called Analysis Of Variance (ANOVA). This process minimizes and consolidates the potential for error so that it is equivalent to the small tolerable error associated with other statistical tests. We collect one sample from each of the k different populations we are testing and the null hypothesis is:

H_0: $\mu_1 = \mu_2 = \cdots = \mu_k$

There are many different ways that this hypothesis could be wrong so we just write the alternative in words as:

H_1: at least one population mean is different from the others

[make that statement specific to the problem at hand]

These would be the scientist's hypotheses:

H_0: $\mu_1 = \mu_2 = \cdots = \mu_{49} = \mu_{50}$

[where each μ stands for the mean level of radioactivity for all typical meals in one of the states]

H_1: at least one mean level of radioactivity is different from the others among all typical meals in the 50 states

[instead of using the word "population" we can use the word "all"; we specify what mean we are considering: "level of radioactivity"; we specify "others": the 50 states]

Among the 50 states there are lots of specific ways that H_1 could be true. States with names having first letters from A–C could have the same mean level of radioactivity for all typical meals, which is different from the mean level of radioactivity for all typical meals in the rest of the states. States with names having first letters from A–M could have the same mean level of radioactivity for all typical meals, which is different from the mean level of radioactivity for all typical meals in the rest of the states. There are many many many ways there could be differences among the individual population means so there is no way we will try to write each of those using symbols. That is why, with ANOVA, we use words for the alternative hypothesis. While we use symbols for the null hypothesis (because that's easy—everything is equal) we certainly know what that says in words: "The mean level of radioactivity is equal among all typical meals in the 50 states." As you would expect, we would need to get a sample from each population in our test.

In the ANOVA procedure, we analyze two types of variation: variation between the sample means and variation within the samples. (In everyday speech, "between" is usually reserved for strictly two things while "among" is used for more than two. However, in ANOVA we refer to the variation among the sample means as "between".) The variation between the sample means is explained by the "factor"—the categorical variable used to differentiate the groups (the scientist's factor would be the categorical variable, State). The variation within the samples is not explained by, or attributable to, the factor. That variation is due to some other factors not measured in our model, or inexplicable chance (called "error"). If we find that more of the variation is explained by the factor (differences between sample means), compared to what is unexplained, that tells us that the differences we observe in all of our individual sample numbers are primarily due to the category (also called "level") of each sample number. (The scientist's factor has 50 different categories or levels.) If there is not just more variation explained by the factor, but there is *a lot* more, we infer that it is also true that the differences in all individual population numbers are primarily due to the category of each population number. This leads to the conclusion that there is variation between the population means (i.e., they are not equal), so we reject H_0. If, on the other hand, there is not a lot more variation explained by the factor relative to the unexplained variation, then we have no reason to conclude that there is variation between the population means, so we fail to reject H_0.

[Remember from earlier in the course that "variation" means "the numbers are different." The more variation a set of data has the more different the numbers are. The less variation a set of data has the less different the numbers are.]

That is why, even though we are ultimately testing population means, the process is called Analysis of *Variance*. The math we are using to make our inference about population means is based on variation.

As you would expect at this point, we will use formulas to quantify these two types of variation (between samples versus within samples). Before we get to those, recall this formula for a sample variance from earlier in the text:

$$s^2 = \frac{\Sigma(x-\bar{x})^2}{n-1}$$

The numerator is called a *sum of squares* because we are squaring individual quantities and then adding them. The denominator is the degrees of freedom. A sum of squares divided by a corresponding degrees of freedom is called a *mean square.* A variance calculation is a mean square. We use this terminology when calculating variation in ANOVA.

ANOVA notation:

k: the number of different populations we are testing
n_i: the sample size for sample i (i goes from 1 to k)
s_i^2: the variance for sample i
$\bar{x}_i$: the mean of sample i
n: the total number of items sampled (the sum of the individual sample sizes)
$\bar{x}$: the mean of all the sample data (also called the grand mean)

The quantification of the variation between the sample means starts with the SSB: Sum of Squares from variation Between sample means.

$$\text{SSB} = \sum n_i(\bar{x}_i - \bar{x})^2 = n_1(\overline{x_1} - \bar{x})^2 + n_2(\overline{x_2} - \bar{x})^2 + \cdots + n_k(\overline{x_k} - \bar{x})^2$$

The corresponding degrees of freedom value for SSB is (k – 1). We are measuring how the k sample means vary around the grand mean.

Next we calculate the MSB: Mean Square from variation Between sample means.

$$\text{MSB} = \frac{SSB}{df\ for\ SSB} = \frac{SSB}{k-1}$$

The MSB is our final measure for the explained variation (the variation between sample means). We store that value and then get to work on quantifying the variation within the samples (the unexplained variation). That begins with the SSW: Sum of Squares due to variation Within samples.

$$\text{SSW} = \sum(n_i - 1)\, s_i^2 = (n_1 - 1)s_1^2 + (n_2 - 1)s_2^2 + \cdots + (n_k - 1)s_k^2$$

[Notice that $(n_i - 1)s_i^2 = (n_i - 1)\frac{\sum(x-\bar{x}_i)^2}{(n_i-1)} = \sum(x - \bar{x}_i)^2$ so each term in the SSW is the sum of squares (numerator) for the variance of each sample.]

The corresponding degrees of freedom value for SSW is:

$$\begin{aligned} \text{df} &= (n_1 - 1) + (n_2 - 1) + \cdots (n_k - 1) \\ &= n_1 + n_2 + \cdots + n_k - 1 - 1 \ldots - 1 \quad \text{(there are } k \text{ of the } -1 \text{ terms)} \\ &= n - k \end{aligned}$$

Next we calculate the MSW: Mean Square due to variation Within samples.

$$\text{MSW} = \frac{SSW}{df\ for\ SSW} = \frac{SSW}{n-k}$$

The MSW is our final measure for the unexplained variation.

Remember, we want to see if a lot more of the variation is explained by the factor compared to what is unexplained. There are many ways this could be done, but the way that would enable us to use a known distribution is by taking the ratio of explained variation to unexplained variation, $\frac{MSB}{MSW}$. From Section 8.8 we learned that the F distribution is applicable when testing the ratio of two variances (or mean squares), which is what we are doing here, so our test statistic is

$$\text{F} = \frac{MSB}{MSW}$$

If there is a lot more variation explained by the factor than what is unexplained, the numerator would be a lot bigger than the denominator (and we know both are positive). So a big value for the test statistic would indicate that there is, in fact, a lot more variation explained by the factor than what is unexplained so we will reject H_0. If there is not a lot more variation explained by the factor than what is unexplained, then the test statistic will be small and we will fail to reject H_0. As usual, a critical value (when using traditional hypothesis testing) will tell you if a test statistic is "big". Since only a big test statistic would make us reject H_0, ANOVA is always a right-tail test. As in Section 8.8, the F table in this text will only allow us to use the traditional method when performing these tests by hand, but if we got a p-value from technology (Minitab, etc.) we would gladly use it!

The key values from ANOVA are organized in a table, typically having this format:

Source	DF	SS	MS	F	P
Between	k – 1	SSB	MSB	$\frac{MSB}{MSW}$	p-value
Within	n – k	SSW	MSW		
Total	n – 1	SST			

The SST, Sum of Squares Total, is simply (SSB + SSW). This would be the $\sum(x - \bar{x})^2$ part of the variance formula if you calculated the variance of all the sample data. ANOVA partitions the total variation into the two types we have been discussing: between and within.

The total degrees of freedom is simply the sum of the degrees of freedom for between and the degrees of freedom for within: (k – 1) + (n – k) = n – 1.

$\frac{SST}{df\ for\ SST} = \frac{SST}{n-1}$ is the variance for all the sample data (the grand variance).

Suppose we perform ANOVA and reject H_0. We infer that at least one population mean is different from the others. The next obvious thing we want to know is where the differences lie. Back to our scientist, she would want to know which states have typical meals with the same mean level of radioactivity, which states have higher levels than other states, etc. ANOVA will not tell us that. Additional steps are required to ascertain that information. While there are formulas we could consider, in this introductory text we will rely on technology to make those distinctions. If you are doing an ANOVA by hand and reject H_0, just state H_1 as the conclusion and stop there. If there is Minitab output given with the problem, use that to state the specific differences, if the ANOVA table indicates that we need to reject H_0.

Computer programs and other texts may have other notation for SSW, SSB, etc. Instead of the label "between" it is common to see the terms "treatment" or "factor". "Factor" is used because the variation between groups is due to the factor. "Treatment" is used because ANOVA is often used in social science, medicine, and other places where different treatments are being tested to compare their effectiveness. There may be a descriptive label for what the factor is, for example "State" in the case of our scientist. Regardless of the label, the first line of the ANOVA table is usually reserved for the between group variation. Instead of the label "within" on the second line of the table, the label "error" is often used since the within group variation is due to those unexplained sources called *error*. Also, just as there are different formulas for s^2, there are algebraically equivalent versions of the formulas given in this section.

The ANOVA discussed in this section is called *one-way* (or *one-factor*) ANOVA due to its having one categorical variable. What if our scientist, in addition to comparing mean radioactivity in meals between the 50 states, at the same time wanted to compare water quality (good, fair, poor)? Now, with two categorical variables (state and water quality) and one quantitative variable (radioactivity) she would have to use a more sophisticated ANOVA model. There is no end to the variables you can add to an ANOVA model. You will discuss more sophisticated models as you advance to other Statistics classes.

Example: Students at three different colleges were sampled and asked how many units they have completed so far. Their responses are provided in the following table. Perform the appropriate statistical procedure to determine if there is a difference in the mean number of units completed by students at these schools.

Statsville College	Wudmoor College	Chef County College
32	38	28
38	41	30
44	33	40
51	29	42
59	44	29
	39	

Answer:

Traditional method

Step 1: H_0: $\mu_1 = \mu_2 = \mu_3$

H_1: at least one population mean number of units completed is different from the others among students at Statsville College, Wudmoor College, and Chef County College

Step 2: Test statistic: $F = \frac{MSB}{MSW} = \frac{158.5023}{59.7641} = 2.6521$

Using Statsville as sample 1, Wudmoor as sample 2, and Chef County as sample 3, and calculating the mean and variance of those samples, and the grand mean, we get the following results:

$$SSB = 5(44.8-38.5625)^2 + 6(37.3333-38.5625)^2 + 5(33.8-38.5625)^2$$
$$= 194.5320313 + 9.06559584 + 113.4070313$$
$$= 317.004658$$

$$MSB = \frac{317.004658}{3-1} = 158.5023$$

$$SSW = (5-1)112.7 + (6-1)29.8667 + (5-1)44.2$$
$$= 776.9335$$

$$MSW = \frac{776.9335}{16-3} = 59.7641$$

We can organize our results in an ANOVA table:

Source	DF	SS	MS	F
Between	2	317.004658	158.5023	2.6521
Within	13	776.9335	59.7641	
Total	15	1093.9382		

Step 3:

(α not given, so using .05)

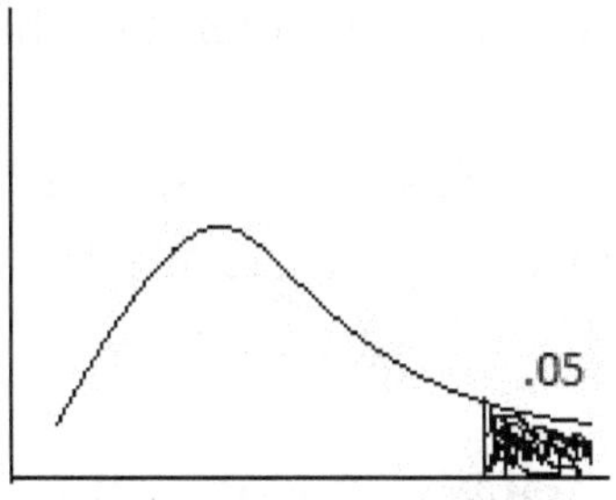

Step 4: The test statistic is not in the rejection region so fail to reject H_0

Step 5: Conclusion: We did not find that at least one population mean number of units completed is different from the others among students at Statsville College, Wudmoor College, and Chef County College.

Even though the directions did not say, "use ANOVA" we knew that to be the correct procedure because we had more than two populations and we wanted to compare means. Also notice that we have one quantitative variable (number of units completed) and one categorical variable (college). From each person in the sample we are getting one quantitative piece of data and one categorical piece. The categorical variable has more than two levels (or categories).

The null hypothesis says that the mean number of units completed by all Statsville College students is equal to the mean number of units completed by all Wudmoor College students which is equal to the mean number of units completed by all Chef County College students.

To clarify the calculations, in SSB we used the sample size and mean for Statsville (5 and 44.8), the sample size and mean for Wudmoor (6 and 37.3333), the sample size and mean for Chef County (5 and 33.8), and the grand mean for all 16 pieces of data (38.5625). For MSB, recall that k, the number of populations we are testing, is 3. In SSW we again needed the sample sizes as well as the variance for each sample: 112.7 for Statsville, 29.8667 for Wudmoor, and 44.2 for Chef County. For MSW, remember that the overall sample size was 16. Be sure to refer to the formulas given earlier and match the values we used to the original values referred to in the general formulas. For our critical value, $df_1 = 2$ and $df_2 = 13$. df_1 is the between df, and df_2 is the within df.

Hold on, wait a minute, in our conclusion we said that we did not find a difference in the means, but we did, didn't we? 44.8, 37.3333, and 33.8 are different, right?! Of course those are different but those are the SAMPLE means!! Our conclusion is indicating that those means were not significantly different. In other words, the sample means were not different enough for us to make the inference that the POPULATION means are different!

Had we rejected H_0, our conclusion would have been: At least one population mean number of units completed is different from the others among students at Statsville College, Wudmoor College, and Chef County College.

Since this test uses the F table, we are unable to get p-value information but, as always, Minitab would be able to calculate a p-value.

Example: For a sample of homes, a real estate broker recorded the current value of each home and the type of home (Cape Cod, Colonial, Cottage, Ranch, or Split Level). Is there a difference in the mean value of these different types of homes? Answer using the Minitab output below.

One-way ANOVA: ValueNow versus HouseType

```
Source     DF           SS          MS     F      P
HouseType   4  10186404289  2546601072  1.54  0.212
Error      35  57914666320  1654704752
Total      39  68101070610


                                Individual 95% CIs For Mean Based on Pooled StDev
Level     N    Mean  StDev     -+---------+---------+---------+--------
cape      8  208277  37203              (---------*---------)
colonial  8  198399  31830          (---------*---------)
cottage   8  218881  57131                 (---------*---------)
ranch     8  175432  37365     (--------*---------)
split     8  217654  34874                 (---------*--------)
                               -+---------+---------+---------+--------
                            150000    180000    210000    240000

Pooled StDev = 40678


Grouping Information Using Tukey Method

HouseType  N    Mean  Grouping
cottage    8  218881  A
split      8  217654  A
cape       8  208277  A
colonial   8  198399  A
ranch      8  175432  A

Means that do not share a letter are significantly different.


Tukey 95% Simultaneous Confidence Intervals
All Pairwise Comparisons among Levels of HouseType
```

```
Individual confidence level = 99.32%


HouseType = cape subtracted from:

HouseType   Lower  Center  Upper  -------+---------+---------+---------+--
colonial   -68413   -9878  48656         (--------*---------)
cottage    -47930   10604  69138            (---------*---------)
ranch      -91380  -32845  25689     (---------*--------)
split      -49158    9376  67910            (---------*--------)
                                  -------+---------+---------+---------+--
                                    -60000          0      60000    120000


HouseType = colonial subtracted from:

HouseType   Lower  Center  Upper  -------+---------+---------+---------+--
cottage    -38052   20482  79017               (--------*---------)
ranch      -81501  -22967  35567      (---------*---------)
split      -39280   19255  77789              (---------*---------)
                                  -------+---------+---------+---------+--
                                    -60000          0      60000    120000


HouseType = cottage subtracted from:

HouseType     Lower  Center  Upper  -------+---------+---------+---------+--
ranch      -101983  -43449  15085  (---------*---------)
split       -59762   -1228  57306         (---------*---------)
                                    -------+---------+---------+---------+--
                                      -60000          0      60000    120000


HouseType = ranch subtracted from:

HouseType   Lower  Center   Upper  -------+---------+---------+---------+--
split      -16313   42222  100756               (---------*---------)
                                   -------+---------+---------+---------+--
                                     -60000          0      60000    120000

stress that the means shown are sample means
```

Answer: p-value (.212) > α (.05), fail to reject H_0. We did not find that at least one population mean current value is different from the others among Cape Cod, Colonial, Cottage, Ranch, and Split Level homes.

When using Minitab output to address a research question involving ANOVA, always, always, always start by considering the p-value (as we do with any other hypothesis test). None of the other output matters until we determine what the p-value is communicating. Since we did not find a difference in the means, we are done. We can ignore the rest of the output.

The null hypothesis would be that all of the five home types have equal mean current values. The alternative hypothesis is that at least one mean current value is different from the others among Cape Cod, Colonial, Cottage, Ranch, and Split Level homes. Obviously, when we say "mean" we're referring to the population mean.

Notice that, for each home in the sample, we got one categorical piece of data (type of home) and one quantitative piece (value of home). The categorical variable had more than two levels (there were five). So, the ANOVA procedure was appropriate. Notice Minitab's label, "**One-way ANOVA**"

Example: For the same scenario described in the previous example, use the following Minitab output to make a conclusion.

One-way ANOVA: ValueNow versus HouseType

```
Source     DF           SS           MS      F      P
HouseType   4  2.52124E+11  63031016800  15.77  0.000
Error      35  1.39848E+11   3995662691
Total      39  3.91972E+11


                                 Individual 95% CIs For Mean Based on Pooled StDev
Level     N    Mean   StDev      +---------+---------+---------+---------
cape      8  208277   37203      (-----*-----)
colonial  8  337104   18389                       (-----*-----)
cottage   8  218881   57131        (----*-----)
ranch     8  404596  117371                                 (-----*----)
split     8  217654   34874        (----*-----)
                                 +---------+---------+---------+---------
                            160000    240000    320000    400000

Pooled StDev = 63211


Grouping Information Using Tukey Method

HouseType  N    Mean  Grouping
ranch      8  404596  A
colonial   8  337104  A
cottage    8  218881    B
split      8  217654    B
cape       8  208277    B

Means that do not share a letter are significantly different.


Tukey 95% Simultaneous Confidence Intervals
All Pairwise Comparisons among Levels of HouseType

Individual confidence level = 99.32%


HouseType = cape subtracted from:

HouseType   Lower  Center   Upper  ---------+---------+---------+---------+
colonial    37868  128827  219786                      (-----*-----)
cottage    -80355   10604  101562              (-----*-----)
ranch      105360  196319  287278                         (-----*-----)
```

```
split      -81582    9376  100335              (-----*-----)
                                    ---------+---------+---------+---------+
                                      -150000         0    150000    300000


HouseType = colonial subtracted from:

HouseType    Lower   Center   Upper  ---------+---------+---------+---------+
cottage   -209182  -118223  -27265        (-----*-----)
ranch      -23467    67492  158451                    (-----*------)
split     -210409  -119451  -28492        (-----*-----)
                                     ---------+---------+---------+---------+
                                       -150000         0    150000    300000


HouseType = cottage subtracted from:

HouseType   Lower  Center   Upper  ---------+---------+---------+---------+
ranch       94757  185715  276674                        (-----*-----)
split      -92186   -1228   89731              (-----*-----)
                                   ---------+---------+---------+---------+
                                     -150000         0    150000    300000


HouseType = ranch subtracted from:

HouseType    Lower   Center   Upper  ---------+---------+---------+---------+
split     -277901  -186943  -95984  (------*-----)
                                     ---------+---------+---------+---------+
                                       -150000         0    150000    300000
```

> **Answer**: p-value (.000) < α (.05), reject H_0. At least one mean current value is different from the others among all Cape Cod, Colonial, Cottage, Ranch, and Split Level homes. Ranch and Colonial homes (group A) have equal mean home values and those values are greater than the mean values for Cottage, Cape Cod, and Split Level homes, which are equal to each other (group B).

Again, we started with the p-value. Since we found at least one difference in means, and since we had the Minitab output to reveal those differences, our conclusion included the specifics of those differences. (When doing ANOVA by hand in this book, if we reject H_0 and state there is a difference, we will stop there. The tedium of doing the multiple comparisons is usually omitted from introductory Statistics courses. However, reading the Minitab output of these comparisons is quite simple.)

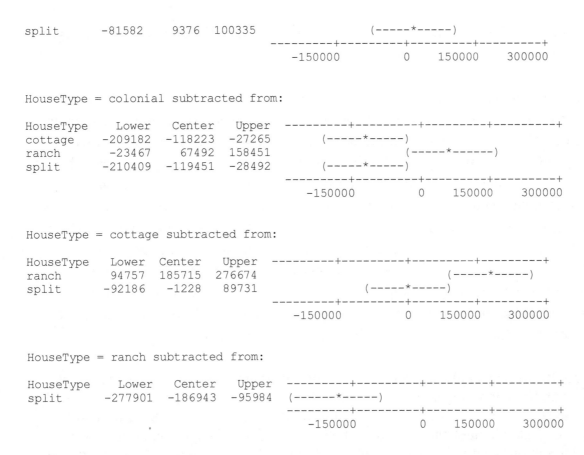

Just below the ANOVA table, Minitab provided confidence intervals for the individual population means. The confidence intervals are shown using dashes and parentheses. Those could be used to look for differences simply by looking for intervals that greatly overlap. In this example, you can easily see that the intervals for Colonial and Ranch greatly overlap. The intervals for Cape Cod, Cottage, and Split Level greatly overlap each other. Under the confidence intervals, there is a number line to get a sense of what values belong in each interval. The Colonial and Ranch intervals are clearly to the right of the intervals for the other three home types, indicating that Colonial and Ranch have a larger mean value than the other three home types.

In practice, it is not always so easy to look at the overlap of confidence intervals and make conclusions from that. It is easier to draw a conclusion by looking at the letter groupings. Tukey is one of several methods that can be used to get these groupings. Categories belonging to Group A have larger population means and, as the groupings progress through the alphabet, the population means are smaller. It is important to note that we do not know the population means—as usual, the means Minitab displays are the sample means. The letter groupings tell us where the significant differences are—in other words, which sample means were different enough for us to conclude that their population means are also different. We would never make an inference by comparing the sample mean numbers. We use the letter groupings. The formulas and theory behind the letter groupings for Tukey and other multiple comparison methods are beyond the scope of an introductory Statistics course. Reading the output, though, is easy.

The rest of the output shows confidence intervals for differences in means. For example, the first confidence interval under the heading "`HouseType = cape subtracted from:`" on the line "`colonial     37868  128827  219786`" is this: $37868 < \mu_1 - \mu_2 < 219786$, where μ_1 is the mean value of all Colonial homes and μ_2 is the mean value of all Cape Cod homes. As we learned back in our study of confidence intervals, since the whole interval is positive, the mean value of all Colonial homes (μ_1) is greater. All of the rest of the confidence intervals for differences in mean home values lead to the same conclusion stated in the answer. (The "center" column on each line gives the difference in sample means.)

In the answer, instead of "equal" you could see "the same" or "there is no difference" in other reports of ANOVA results.

Always be sure to consider the p-value BEFORE looking at letter groupings or confidence intervals. As mentioned previously, when doing multiple tests for one research question, there is a growing opportunity for error. Sometimes the multiple procedures that lead to the letter groupings and confidence intervals can make it appear that there are significant differences when these differences are due to error, as evidenced by the p-value of the ANOVA test being bigger than alpha. The ANOVA test is an overarching test considering everything at once with the minimal acceptable error associated with hypothesis testing. If that test says there is a significant difference (i.e., a difference among the population means), then trust the multiple comparison tests to specify the differences. P-value first, then other stuff later (only if the p-value is less than alpha).

Why do we divide the variation between the means (explained variation) by the variation within the groups (unexplained variation)? Why not just consider the variation between the means and leave it at that?

Consider these two simple scenarios of data. In each scenario there are three samples, each with three measurements, that were taken from their respective populations:

Scenario 1

sample1	sample 2	sample 3
50	100	200
51	101	201
52	102	202

Scenario 2

sample 1	sample 2	sample 3
1	10	14
2	11	53
150	282	536

These are the ANOVA tables for each scenario:

Scenario 1

Source	DF	SS	MS	F	P
Factor	2	35000	17500	17500	0.000
Error	6	6	1		
Total	8	35006			

Scenario 2

Source	DF	SS	MS	F	P
Factor	2	35000	17500	0.45	0.657
Error	6	232942	38824		
Total	8	267942			

Notice that both scenarios have the same variation between the means (MS Factor = 17500). However, in the first scenario it is easily observed that the values within each sample are much closer to each other than are the values within the samples of Scenario 2. So, the variation between the sample means in Scenario 1 is more suggestive of differences between the population means. Nearly all the variation is explained by differences between the samples with very little unexplained variation. In Scenario 2, however, the differences between the sample means are not that impressive because there is still a lot of variation within the samples as well. There is a great deal of variation not explained by the factor used to define the groups. Therefore, in Scenario 1 we reject H_0 (p-value $.000 < \alpha$) and conclude that at least one population mean is different from the others while in Scenario 2 we fail to reject H_0 (p-value $.657 > \alpha$) and we do not find that at least one population mean is different from the others. Dividing the variation between the means (attributable to the factor) by the variation within the samples (attributable to some unknown factor(s) and/or random chance) allows us to contextualize the variation between the means in this way. Otherwise, there would be no way of determining whether or not there is "a lot" of variation between the sample means.

Exercises

1) A sample of transactions from four companies—Jerrell Enterprises, Jaden Industries, Michael Initiatives, and Jordan Developments—were studied and the cost for each transaction was recorded. Test the claim that the companies have different average costs per transaction.

Jerrell	Jaden	Michael	Jordan
20	25	40	25
24	32	38	24
31	27	25	28
38	26	30	26
19	33	33	27
21			30
22			

2) Refer to 1). Suppose the conclusion we got from hypothesis testing does not reflect the true nature of the population mean cost per transaction for the four companies tested. What type of error would have been committed?

3) Is there a difference in the soil quality in Twin Streams, Wudmoor, Statsville, Chef County, and Tanayatown? As part of the investigation of this question, soil samples were randomly collected from each location and the number of earthworms per m^2 was recorded. The nine measurements from Twin Streams had a mean of 31.67 with variance 28.75. The 11 measurements from Wudmoor had a mean of 32.00 with variance 30.20. The 13 measurements from Statsville had a mean of 17.69 with variance 7.73. The 12 measurements from Chef County had a mean of 38.42 with variance 37.36. The 14 measurements from Tanayatown had a mean of 44.93 with variance 34.07. All the readings combined had a mean of 33.17 with variance 118.87. (All means have units earthworms per m^2 and variances have units (earthworms per m^2)^2.) Perform the appropriate test to see if there is a difference in the mean presence of earthworms among these five locations.

4) There are four furniture stores in a certain city. One is in the north, one is in the south, one is in the east, and one is in the west. We collected random samples of prices (in dollars) from each store and used the appropriate data for the Minitab output below. See if there is a difference in mean prices among these four stores.

One-way ANOVA: North, South, East, West

```
Source  DF        SS       MS        F      P
Factor   3  14937570  4979190  1733.43  0.000
```

```
Error    333      956527      2872
Total    336    15894097

Grouping Information Using Tukey Method

         N      Mean   Grouping
East    91   2005.90   A
West    73   1601.81     B
South   82   1514.60       C
North   91   1508.94       C

Means that do not share a letter are significantly different.
```

5) Consider the following partial ANOVA table.

```
Source   DF      SS     MS       F       P
Factor    3   150.6          2.155   0.123
Error    22           23.3
Total    25   663.2
```

a) Using your knowledge of how the numbers in an ANOVA table are related, fill in the missing values.

b) How many populations are being tested?

c) What was the total sample size?

d) What is the variance of all the sample data?

9.3 ANOVA (Two-Way)

Conditions: The data comes from independent simple random samples. The populations have equal variances. For each pair of categories across the two categorical variables, the sample sizes are the same and the populations are normally distributed.
(Assume all conditions are met for each example and exercise.)

Use the methods of this section when...

testing a claim about the equality of means

Variable types: one quantitative and two categorical

Important formulas: none

In the previous section, we looked at examples in which the mean home value was compared for different types of homes. The quantitative variable was home value and the categorical variable (the factor) was type of home. In this section we will add a factor to that scenario: type of windows in the home (single pane or double pane). Now the appropriate hypothesis testing procedure is two-way (or two-factor) ANOVA.

[Note that there are different types of factors that can be used in an ANOVA model: fixed and random. Fixed factors are assumed to include all the possible categories or levels of the factor. Random factors only include some levels that are randomly selected from all the possible levels of the factor. Different situations call for different types of factors, and the treatment of them is different. Anticipate a deeper discussion of that topic in later classes. In this text, all factors are considered to be fixed.]

Unlike the One-Way ANOVA discussed previously, we now have three sets of null and alternative hypotheses. The null hypothesis, H_0, for the first set of hypotheses is labeled H_{01}. The alternative hypothesis, H_1, for the third set of hypotheses is labeled H_{13}, and so on.

Factor 1: H_{01}: The population means in all Factor 1 categories are equal (Factor 1 has no effect on the quantitative variable).
H_{11}: The population mean in at least one Factor 1 category is different from the others (Factor 1 has an effect on the quantitative variable).

Factor 2: H_{02}: The population means in all Factor 2 categories are equal (Factor 2 has no effect on the quantitative variable).
H_{12}: The population mean in at least one Factor 2 category is different from the others (Factor 2 has an effect on the quantitative variable).

Interaction: H_{03}: The population means in all Factor 1 categories are equal for each Factor 2 category (there is no interaction).
H_{13}: The population mean in at least one Factor 1 category is different for at least one of the Factor 2 categories (there is interaction).

The first two listed tests are basically like a one-way ANOVA for each factor. With the interaction test, we are looking to see if Factor 1 and Factor 2 simultaneously have an effect on the quantitative variable, in tandem. When we use the term "effect" in ANOVA we are not necessarily talking about a cause-and-effect, per se, but a difference in the means explained by a factor or factors. [Exactly why there is a difference in the means explained by a factor or factors is a more complex issue that would require more research.] If we say, "Factor 1 has an effect on the quantitative variable" it could also be said that "there is a Factor 1 effect." An effect for either of the factors is called a main effect.

For the two-factor home values example, the hypotheses would look like this:

Factor 1: H_{01}: The population mean home values for Cape Cod, Colonial, Cottage, Ranch, and Split Level homes are equal (the home type has no effect on the home value; there is no home type effect).

H_{11}: The population mean home value for at least one home type is different from the others among Cape Cod, Colonial, Cottage, Ranch, and Split Level homes (the home type has an effect on the home value; there is a home type effect).

Factor 2: H_{02}: The population mean home value for homes with single pane windows is equal to the population mean value for homes with double pane windows (the window type has no effect on the home value; there is no window effect).

H_{12}: The population mean home value for homes with single pane windows is different from the population mean value for homes with double pane windows (the window type has an effect on the home value; there is a window effect).

[The "at least" phrasing here would be a bit awkward because this factor only has two levels.]

Interaction: H_{03}: The population mean home values for Cape Cod, Colonial, Cottage, Ranch, and Split Level homes are equal for homes with single pane windows and homes with double pane windows (there is no interaction).

H_{13}: The population mean home value for at least one home type among Cape Cod, Colonial, Cottage, Ranch, and Split Level homes is different from the others for homes with single pane windows and homes with double pane windows (there is interaction).

For two-way ANOVA we will just rely on technology to provide the results (the formulas are extensions of the one-way ANOVA formulas and they are rather tedious for an introductory text). It is very important that we interpret the test results in the correct order. The first step in interpreting the results is to test the interaction hypotheses. If we find an interaction, we leave the ANOVA table and use the rest of the output to specify the interaction, while ignoring the rest of the ANOVA table. If there is not an interaction, we continue reviewing the ANOVA table to see if there are any main effects. If so, explore and describe any main effects. The term "interaction" only applies to interaction (both factors). For just a single factor, the applicable term is "effect".

The reason we interpret the results this way is because a one-factor test is performed with the assumption that there is no interaction. We are looking for a difference just based on that one factor. If interaction is present, that means that both factors are acting on the quantitative variable together at the same time, so that invalidates the results of the one-factor tests. Even though it will be the third line of the ANOVA table, the interaction test results must be considered first!!

You saw an example of writing out the hypotheses, but I don't require my students to write the hypotheses for two-way ANOVA because that gets a little complicated (it's even more fun if you use symbols for the null hypotheses). When using this text, just be able to interpret the results as demonstrated. We also won't worry about an explicit statement of rejecting or failing to reject H_0. While those rules still apply, when getting the conclusions from two-way ANOVA we will use this shortcut: if the p-value $< \alpha$, something is happening; otherwise it's not. For example, suppose we were looking at the interaction test and the p-value was less than α. We would conclude that the mean home value for at least one type of home is different from the others for at least one of the window types. In other words, there is interaction between the factors home type and window type. So, if we see that the p-value is less than alpha on the interaction line, we know that there is interaction. Otherwise, there is no interaction. The same holds true for the main effect lines. You will see this demonstrated in the following example. Also, varieties in wording will be used so you can become comfortable with that.

Example: What follows are different sets of output for the scenario introduced at the beginning of this section involving home values, types, and windows. How would we interpret each set of output?

a) **General Linear Model: ValueNow versus HouseType, Windows**

```
Factor     Type   Levels  Values
HouseType  fixed       5  cape, colonial, cottage, ranch, split
Windows    fixed       2  double, single


Analysis of Variance for ValueNow, using Adjusted SS for Tests

Source             DF       Seq SS       Adj SS      Adj MS     F      P
HouseType           4    518980376    518980376   129745094  0.28  0.888
Windows             1   1293951750   1293951750  1293951750  2.80  0.105
HouseType*Windows   4   2085635737   2085635737   521408934  1.13  0.362
Error              30  13873507981  13873507981   462450266
Total              39  17772075844


Grouping Information Using Tukey Method and 95.0% Confidence

HouseType  N    Mean  Grouping
colonial   8  336113  A
```

```
ranch      8  335879  A
cottage    8  335692  A
split      8  330693  A
cape       8  327137  A

Means that do not share a letter are significantly different.


Grouping Information Using Tukey Method and 95.0% Confidence

Windows   N    Mean  Grouping
double   20  338790  A
single   20  327415  A

Means that do not share a letter are significantly different.


Grouping Information Using Tukey Method and 95.0% Confidence

HouseType  Windows  N    Mean  Grouping
cottage    double   4  352547  A
colonial   single   4  340614  A
split      double   4  340577  A
ranch      double   4  339743  A
ranch      single   4  332015  A
colonial   double   4  331613  A
cape       double   4  329471  A
cape       single   4  324802  A
split      single   4  320809  A
cottage    single   4  318836  A

Means that do not share a letter are significantly different.
```

Answer: Interaction p-value (.362) > α (.05), no interaction. There is no difference in mean home value for the different home types according to window type.

Home type p-value (.888) > α (.05), no home type effect. There is no difference in mean home value for the different home types.

Windows p-value (.105) > α (.05), no window effect. There is no difference in mean home value for the different window types.

The three different F test statistics, one for each sub-test being performed, are obtained by taking the respective MS divided by the MS for error. This is just like what happened in one-way ANOVA. The main effect tests are conducted in such a way that they assume there is no interaction. Well, if there is interaction then those results are no longer valid. That is why we look for interaction first. If there is interaction then forget about main effects because the two categorical variables are intertwined and connected as it relates to the quantitative variable. The bottom line is: always use the correct order as described earlier when considering two-way ANOVA output.

We are not getting into the intimate details of the test, as that is reserved for higher level courses, but we can certainly read the output. A brief comment, though, regarding the ADJusted versus SEQuential sums of squares. The sequential sums of squares depend on the order in which you enter the factors (like HouseType first and Windows second as opposed to Windows first and HouseType second) when you request the output from Minitab or whatever software you use. The adjusted sums of squares are not dependent on the order. This really becomes a bigger issue as more and more factors are added to the model. We will not go beyond two factors in this course so all we need to focus on is assessing the p-values and performing the steps in the correct order.

For this example, we will totally ignore the rest of the output below the ANOVA table. According to the ANOVA table, there are no differences anywhere, so we're done. As mentioned in the discussion of one-way ANOVA, if the ANOVA table has told us there are no differences, any apparent differences indicated in the rest of the output are attributable to the error that compounds over the course of multiple tests and should be ignored.

The interaction line of this output explicitly shows the two variables that are being tested for interaction. Depending on how you enter the output request in Minitab, the line can be labeled "Interaction" or with a descriptive label showing the factors being tested for interaction. The latter method is necessary when there are more than two factors in the model, since more than one type of interaction is possible.

b) **General Linear Model: ValueNow versus HouseType, Windows**

Factor	Type	Levels	Values
HouseType	fixed	5	cape, colonial, cottage, ranch, split
Windows	fixed	2	double, single

Analysis of Variance for ValueNow, using Adjusted SS for Tests

Source	DF	Seq SS	Adj SS	Adj MS	F	P
HouseType	4	1.00518E+12	1.00518E+12	2.51296E+11	161.27	0.000
Windows	1	35355281	35355281	35355281	0.02	0.881
HouseType*Windows	4	1599182873	1599182873	399795718	0.26	0.903
Error	30	46747322998	46747322998	1558244100		
Total	39	1.05356E+12				

Grouping Information Using Tukey Method and 95.0% Confidence

HouseType	N	Mean	Grouping
ranch	8	623611	A
colonial	8	337104	B
cottage	8	218881	C
split	8	217654	C
cape	8	208277	C

Means that do not share a letter are significantly different.

Grouping Information Using Tukey Method and 95.0% Confidence

Windows	N	Mean	Grouping
double	20	322046	A
single	20	320165	A

Means that do not share a letter are significantly different.

```
Grouping Information Using Tukey Method and 95.0% Confidence

HouseType  Windows  N    Mean  Grouping
ranch      single   4  631422  A
ranch      double   4  615800  A
colonial   double   4  340230    B
colonial   single   4  333979    B
cottage    double   4  227475      C
split      single   4  222706      C
cape       double   4  214123      C
split      double   4  212601      C
cottage    single   4  210288      C
cape       single   4  202432      C

Means that do not share a letter are significantly different.
```

Answer: Interaction p-value (.903) > α (.05), no interaction. We did not find that there is a difference in mean home value in at least one home type for at least one window type.

Home type p-value (.000) < α (.05), home type effect. There is a difference in mean home value in at least one home type.

Windows p-value (.881) > α (.05), no window effect. We did not find that there is a difference in mean home value for at least one window type.

Ranch homes (group A) have the largest mean value followed by Colonial homes (group B). Cottage, Split Level, and Cape Cod homes (group C) have the same mean value and that is less than the mean for Ranch and Colonial.

Again, we follow the same order as described earlier. This time there was a main effect for home type so, after completing the assessment of the p-values, we described the main effect. We only consulted the output that was looking at differences strictly based on home type. Everything else is to be ignored, based on what the ANOVA table told us.

By the way, the word "interaction" only applies to interaction. A statement like, "there is home type interaction" makes no sense because there needs to be at least two categorical variables (factors) involved in an interaction. For a single factor there can be an effect. For multiple factors being considered together, there can be interaction.

If you are still reading this book that means you are still taking the Statistics class that required this book. That being the case, it goes without saying (but, like so many things in our world that go without saying, this will be said anyway) that we cannot draw a conclusion about the differences in mean home values based on the numbers 623611,

337104, 218881, 217654, and 208277. Those are sample means only. Our conclusions, as with all hypothesis tests, are about the entire populations.

c) **General Linear Model: ValueNow versus HouseType, Windows**

```
Factor     Type   Levels  Values
HouseType  fixed       5  cape, colonial, cottage, ranch, split
Windows    fixed       2  double, single


Analysis of Variance for ValueNow, using Adjusted SS for Tests

Source             DF       Seq SS       Adj SS       Adj MS      F      P
HouseType           4  5.94744E+11  5.94744E+11  1.48686E+11  75.54  0.000
Windows             1   9317115236   9317115236   9317115236   4.73  0.038
HouseType*Windows   4  24391221614  24391221614   6097805404   3.10  0.030
Error              30  59053120931  59053120931   1968437364
Total              39  6.87506E+11


Grouping Information Using Tukey Method and 95.0% Confidence

HouseType  N    Mean  Grouping
ranch      8  526286  A
colonial   8  337104    B
cottage    8  218881      C
split      8  217654      C
cape       8  208277      C

Means that do not share a letter are significantly different.


Grouping Information Using Tukey Method and 95.0% Confidence

Windows    N    Mean  Grouping
double    20  316903  A
single    20  286379    B

Means that do not share a letter are significantly different.


Grouping Information Using Tukey Method and 95.0% Confidence

HouseType  Windows  N    Mean  Grouping
ranch      double   4  590085  A
ranch      single   4  462488    B
colonial   double   4  340230      C
colonial   single   4  333979      C D
cottage    double   4  227475        D E
split      single   4  222706          E
cape       double   4  214123          E
split      double   4  212601          E
cottage    single   4  210288          E
cape       single   4  202432          E

Means that do not share a letter are significantly different.
```

Answer: Interaction p-value (.030) < α (.05), there is interaction. The mean home value in at least one home type is different from the others based on window type.

Ranch homes with double pane windows (group A) have the highest mean home value, followed by Ranch homes with single pane windows (group B), followed by Colonial homes with double pane windows (group C). No difference in mean home value was found between Colonial homes with single pane windows (groups C and D) and Colonials with double pane windows (group C) or Cottages with double pane windows (groups D and E). The rest of the home type and window pairings had similar, and the lowest, mean home values (group E).

This time we had interaction so we left the ANOVA table and found the interaction. As you can see, sometimes rows can have multiple letters. If you think back to the confidence intervals for the individual means that will help you see why. For example, confidence intervals would look something like this for the groups listed:

```
colonial double                                    (            )
colonial single                              (           )
cottage double                         (          )
split single                     (          )
cape double                      (          )
```

Colonial single overlaps with colonial double and cottage double, but colonial double does not overlap with cottage double. Cottage double overlaps with colonial single and split single (and many others below it) but colonial single does not overlap with split single. The more categories there are among the factors, the more interesting the letter groupings can become. In the conclusion, after ANOVA has told you there is interaction, just describe what you see. In practice you may often only be interested in a couple of these relationships but, for the purposes of this course, you should describe all the relationships indicated by the letter groupings.

I hope you took the time to identify the two categorical variables (home type and window type) and the one quantitative variable (home value) for this exercise. That's why two-way ANOVA was used. Suppose we wanted to add factors (categorical variables) to our study, like neighborhood quality, quality of schools, and color of home. More advanced methods would be needed. You can look forward to this in later classes!

Bigger picture

The following ANOVA table shows the results from testing one factor, C5.

Source	DF	SS	MS	F	P
C5	3	22035	7345	0.77	0.519
Error	32	305111	9535		
Total	35	327146			

For this same quantitative data, we further categorized it according to a second factor, C6. This is the new ANOVA table.

Source	DF	Seq SS	Adj SS	Adj MS	F	P
C5	3	22035	22035	7345	0.80	0.505
C6	2	9490	9490	4745	0.52	0.602
C5*C6	6	75886	75886	12648	1.38	0.262
Error	24	219735	219735	9156		
Total	35	327146				

Observe that the total SS is the same in both tables (327146). That is the case because we have the same quantitative data so it has a set amount of total variation. Also, both tables show the same amount of variation explained by factor C5 (SS = 22035). However, in the two-factor ANOVA the error, unexplained variation, is smaller than in the one-factor case (SS Error = 219735 for two-way ANOVA, 305111 for one-way ANOVA) . That is because, with the addition of the second categorical variable, there is less variation not explained by our model. Our second model includes more information that explains the variation in the quantitative data. In a study where you use more variables you are more likely to get more useful explanations and inferences. At a point, though, additional variables lose explanatory ability. In later classes you discuss things like variable selection, how to determine which variables are most useful in your model, how many variables you should have (based on research in the substantive field, as well as the statistical methods involved), and so many other related matters. There is so much more for you to learn about Statistics!

Exercises

1) A student earning a degree in Recreation and Leisure was curious about whether or not there is a difference in wait times for different amusement parks during different seasons. She accepts that the summer is the busiest season, so she collects data from fall, spring, and winter visits to two amusement parks: Statsville Amusements and Tanayatown Entertainment. She recorded and entered into Minitab the longest time (minutes) she had to wait to get on a ride during each visit. Write a complete conclusion for each version of Minitab output that follows.

a) Version a

General Linear Model: wait versus season, park

Factor	Type	Levels	Values
season	fixed	3	sp, win, fa
park	fixed	2	amuse, entertain

Analysis of Variance for wait, using Adjusted SS for Tests

Source	DF	Seq SS	Adj SS	Adj MS	F	P
season	2	20.67	20.67	10.33	0.42	0.676
park	1	48.00	48.00	48.00	1.95	0.212
season*park	2	38.00	38.00	19.00	0.77	0.504
Error	6	148.00	148.00	24.67		
Total	11	254.67				

Grouping Information Using Tukey Method and 95.0% Confidence

season	N	Mean	Grouping
fa	4	9.5	A
win	4	9.0	A
sp	4	6.5	A

Means that do not share a letter are significantly different.

Grouping Information Using Tukey Method and 95.0% Confidence

park	N	Mean	Grouping
entertain	6	10.3	A
amuse	6	6.3	A

Means that do not share a letter are significantly different.

Grouping Information Using Tukey Method and 95.0% Confidence

season	park	N	Mean	Grouping
win	entertain	2	12.5	A
fa	entertain	2	12.5	A
sp	amuse	2	7.0	A
fa	amuse	2	6.5	A
sp	entertain	2	6.0	A
win	amuse	2	5.5	A

Means that do not share a letter are significantly different.

b) Version b

General Linear Model: wait versus season, park

Factor	Type	Levels	Values
season	fixed	3	sp, win, fa
park	fixed	2	amuse, entertain

Analysis of Variance for wait, using Adjusted SS for Tests

Source	DF	Seq SS	Adj SS	Adj MS	F	P
season	2	486.50	486.50	243.25	8.27	0.019
park	1	140.08	140.08	140.08	4.76	0.072
season*park	2	1067.17	1067.17	533.58	18.14	0.003
Error	6	176.50	176.50	29.42		
Total	11	1870.25				

```
Grouping Information Using Tukey Method and 95.0% Confidence

season  N  Mean  Grouping
sp      4  22.7  A
fa      4   9.5    B
win     4   9.0    B

Means that do not share a letter are significantly different.


Grouping Information Using Tukey Method and 95.0% Confidence

park          N  Mean  Grouping
amuse         6  17.2  A
entertain     6  10.3  A

Means that do not share a letter are significantly different.


Grouping Information Using Tukey Method and 95.0% Confidence

season  park       N  Mean  Grouping
sp      amuse      2  39.5  A
win     entertain  2  12.5    B
fa      entertain  2  12.5    B
fa      amuse      2   6.5    B
sp      entertain  2   6.0    B
win     amuse      2   5.5    B

Means that do not share a letter are significantly different.
```

c) Version c

General Linear Model: wait versus season, park

```
Factor  Type   Levels  Values
season  fixed       3  sp, win, fa
park    fixed       2  amuse, entertain


Analysis of Variance for wait, using Adjusted SS for Tests

Source       DF   Seq SS   Adj SS   Adj MS       F      P
season        2  38081.2  38081.2  19040.6  288.86  0.000
park          1     80.1     80.1     80.1    1.21  0.313
season*park   2    162.2    162.2     81.1    1.23  0.357
Error         6    395.5    395.5     65.9
Total        11  38718.9


Grouping Information Using Tukey Method and 95.0% Confidence

season  N   Mean  Grouping
win     4  124.7  A
sp      4    5.5    B
fa      4    5.0    B

Means that do not share a letter are significantly different.
```

```
Grouping Information Using Tukey Method and 95.0% Confidence

park        N  Mean  Grouping
amuse       6  47.7  A
entertain   6  42.5  A

Means that do not share a letter are significantly different.


Grouping Information Using Tukey Method and 95.0% Confidence

season  park       N   Mean  Grouping
win     amuse      2  132.5  A
win     entertain  2  117.0  A
sp      entertain  2    6.0    B
fa      amuse      2    5.5    B
sp      amuse      2    5.0    B
fa      entertain  2    4.5    B

Means that do not share a letter are significantly different.
```

d) Consider only the interaction test for Version c. Did that test yield a significant result? Explain.

2) We are interested in the bedding sheet sets produced by four companies: F, G, H, and I. The bedding sheet sets have either solid colors or some sort of design. Randomly selected bedding sheet sets of each color style from each company were weighed (in grams) to see if there are differences in the mean weight of bedding sheet sets based on company and/or color style. With this in mind, interpret the output below.

General Linear Model: premium versus company, color

```
Factor   Type   Levels  Values
company  fixed       4  F, G, H, I
color    fixed       2  design, solid


Analysis of Variance for premium, using Adjusted SS for Tests

Source         DF  Seq SS  Adj SS  Adj MS       F      P
company         3      64      64      21    0.04  0.990
color           1  159800  159800  159800  267.53  0.000
company*color   3     128     128      43    0.07  0.974
Error           8    4779    4779     597
Total          15  164771


Grouping Information Using Tukey Method and 95.0% Confidence

company  N  Mean  Grouping
I        4  1229  A
H        4  1224  A
F        4  1224  A
G        4  1224  A

Means that do not share a letter are significantly different.
```

```
Grouping Information Using Tukey Method and 95.0% Confidence

color     N  Mean  Grouping
design    8  1325  A
solid     8  1125     B

Means that do not share a letter are significantly different.


Grouping Information Using Tukey Method and 95.0% Confidence

company  color   N  Mean  Grouping
F        design  2  1328  A
G        design  2  1325  A
I        design  2  1325  A
H        design  2  1323  A
I        solid   2  1133     B
H        solid   2  1126     B
G        solid   2  1122     B
F        solid   2  1121     B

Means that do not share a letter are significantly different.
```

3) Refer to 2). What is the mean weight of all solid color bedding sheet sets sold by Company H?

9.4 Linear Correlation

Conditions: The data is from two dependent simple random samples (dependent in that there is pairing due to a unique relationship between each value in one sample and a value in the other sample). For each of the x values in the population, the matched y values have a normal distribution, and for each of the y values in the population, the matched x values have a normal distribution.
(Assume all conditions are met for each example and exercise.)

Use the methods of this section when...

testing for a relationship between two variables

Variable types: two quantitative

Important formula:

$$r = \frac{n\sum xy - (\sum x)(\sum y)}{\sqrt{n\sum x^2 - (\sum x)^2}\sqrt{n\sum y^2 - (\sum y)^2}}$$

[It is possible to test for a linear correlation using either a t or r test statistic. We will use the simpler r method.]

We are often interested in knowing whether or not two quantitative variables are related. The simplest and most commonly studied of these relationships is the linear case (increasing values of one variable are associated with increasing values of the other variable, or increasing values of one variable are associated with decreasing values of the other variable). Such a relationship is called a **linear correlation**. We consider questions like, Do older people tend to have less muscle mass? Are more years of education associated with higher incomes? Do students generally get higher scores on exams when they study longer?

A scatterplot is created using the same plotting techniques you learn in algebra. Suppose we collect hand and foot lengths from randomly selected people. The first scatterplot below shows the results when the measurements are in centimeters and the other shows measurements in inches.

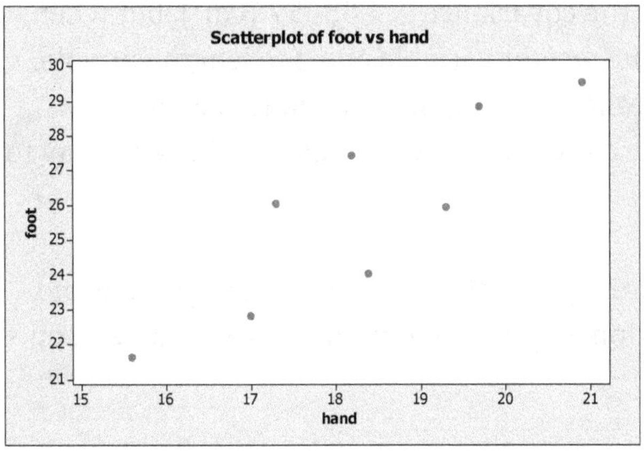

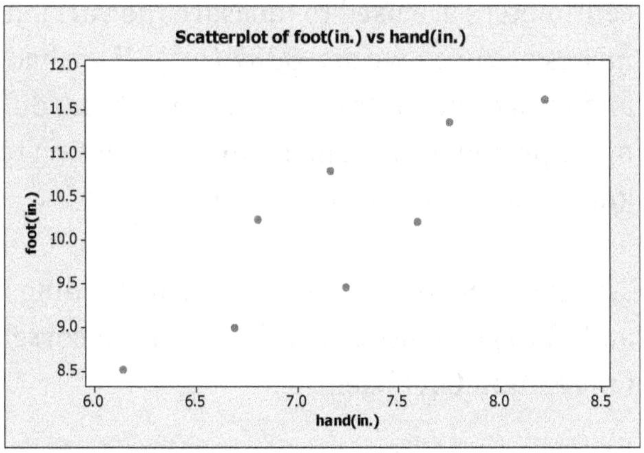

When investigating a possible relationship between two quantitative variables, a scatterplot can be created to see if 1) there are influential points [points that appear to be outside of the general pattern of the points] and/or 2) there is a strong non-linear relationship that seems to exist. In these cases, more sophisticated methods taught in more advanced classes are used. For correlation problems in this course, we will assume that there is no non-linear correlation present.

We cannot rely on a plot of sample data alone to determine whether or not the populations of these variables have a linear correlation. We want to determine if the positive or negative relationship observed in the sample data is strong enough to conclude that these variables have such a relationship in the population from which the (x,y) data pairs were taken. As usual, testing an inference about a population requires calculating a test statistic.

On the way to getting our test statistic, we start with a quantity called the covariance: $cov(x,y) = \frac{\sum(x-\bar{x})(y-\bar{y})}{n-1}$. If there is a negative linear correlation between the variables, most of the terms in the numerator (cross-product deviations) will be negative so the sum will be negative. If there is a positive linear correlation between the variables, most of the terms in the numerator will be positive so the sum will be positive. For the hand and foot data measured in cm, the mean of x (hand) is 18.3 and the mean of y is 25.75. If you plot that point on the graph, and then consider the plotted data points relative to that point, you would see that you would get a positive number if you calculate the covariance.

The covariance is limited as a measure for the strength of a linear correlation. The covariance is greatly influenced by the units used to measure the variables, so one set of data could give different covariance values based on the units. In the hand and foot data, you can see by the two scatterplots that the sample points have the same linear relationship, regardless of which units of measurement are applied. However, when centimeters are used to measure the variables, the covariance is 4.08857 (cm^2), but when inches are used we get .6337 (in^2). If we had one weight variable and one length variable, and varied the units of measurement, the differences among the covariances would be more profound, with no simple conversion to use for comparisons. Also, we have no way to know if a covariance is "big" or "small".

The covariance is standardized (by dividing by the product of the standard deviations of x and y) to get a more useful measure of correlation, not dependent on units, called Pearson's Correlation Coefficient, r.

$$r = \frac{cov(x,y)}{s_x s_y}$$

These are some facts about r:

1) has standard units (like z scores)

"Standard units" essentially means there are no units. For example, in our hand and foot data, r would not have units of cm or inches but it would be a unit-less number. The arithmetic used for the calculation results in the units cancelling. As is the case when we calculate a z-score, the number we get is just applicable to z-scores, no general unit system. Because r is standardized, we can determine if a particular r value is "big" or "small".

2) is always between −1 and 1 (a fact proven in higher math classes)

3) the closer it is to −1 (or 1) the stronger the negative (or positive) correlation

4) only addresses linear (not non-linear) correlation

Other measures are needed to test for non-linear correlation.

5) must be used with variables that both have some variation (otherwise the denominator would be zero)

6) will be the same regardless of which variable is considered "x" and which is considered "y" (although, based on what is coming later, "y" should be the variable about which you may want to make predictions)

This comes from the fact that we are using multiplication, which is commutative.

7) only measures correlation, NOT CAUSATION

Just because we find two variables are related, we cannot jump to a causal claim at that point.

The display that follows shows four scatterplots of sample data. For each of them, the "x" variable is C4. The "y" variables are different for each plot and are labeled C5, C6, C8, and C9, respectively. For the C5 plot r = −.918, for C6 r = −.754, for C8 r = .924, and for C9 r = .759. Observe that displays with a generally negative slope have a negative r value, and displays with a generally positive slope have a positive r value. The C5 and C6 scatterplots both have negative trends but, since C5's is closer to being perfectly linear, its r value is closer to −1. The C8 and C9 scatterplots both have positive trends but, since C8's is closer to being perfectly linear, its r value is closer to 1. In the plots with positive r values, we observe the general trend that, as x increases, y increases (or, as x decreases, y decreases). The variables' quantities are trending in the same direction. This indicates a positive linear correlation in these samples. In the plots with negative r values, we observe the general trend that, as x increases, y decreases (or, as x decreases, y increases). The variables' quantities are trending in opposite directions. This indicates a negative linear correlation in these samples.

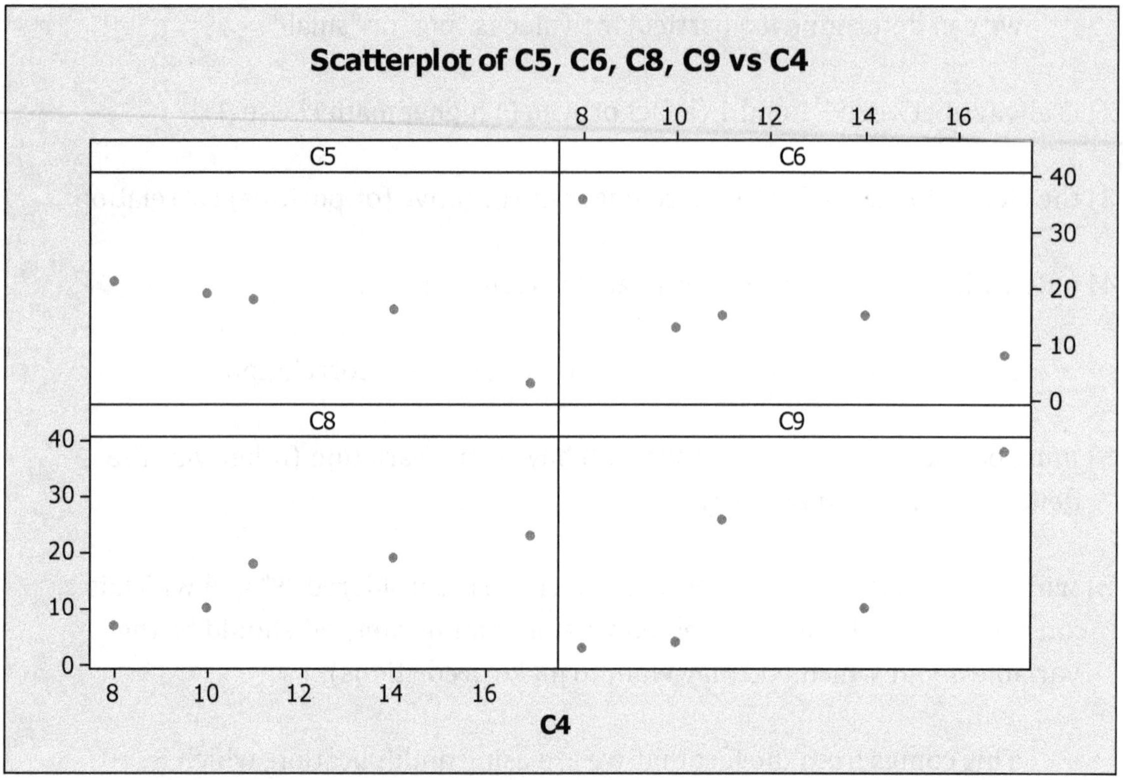

In this scatterplot there is no clear positive or negative trend. Sometimes, small values of x are associated with big values of y. Sometimes, small values of x are paired with small values of y. There is no statement we can make about what happens to values of y as x increases or decreases. There is no clear pattern. There is no linear correlation for this sample and r is approximately zero.

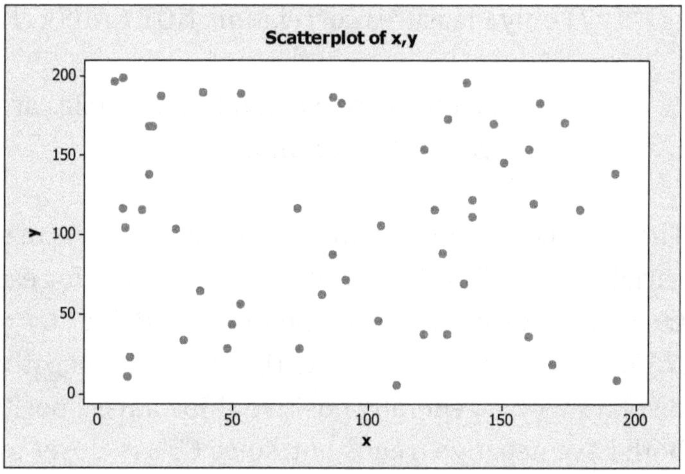

Looking at a scatterplot of sample data will not tell us if the relationship we observe in the sample is convincing enough for us to make the same inference about the population, just like looking at a sample mean does not tell us if it is convincing enough for us to make the same inference about the population mean. This is where hypothesis testing enters.

For tests of linear correlation,

> H_0: There is not a linear correlation between the two variables (in the population)
> H_1: There is a linear correlation between the two variables (in the population)

These are always two-tailed tests because we are actually testing whether or not the linear correlation coefficient for the population, ρ (Greek letter "rho"), is equal to zero. [All of the r values given for samples in previous paragraphs would be ρ values if the data was entire populations. For example, if the C5 scatterplot was for a whole population, $\rho = -.918$.] Rho doesn't appear later in the course, nor in any formulas, so it's easier just to write the hypotheses in words.

r is the test statistic used to test for a linear correlation between two quantitative variables. If r is "big", in absolute value, it will be close to 1 or −1. Since r is standardized, we can determine if a value of r is "big" by using critical values. The bigger r is (i.e., the closer it is to 1 or −1) for the sample, the more compelling the evidence that the linear correlation observed in the sample also exists in the whole population from which that sample was taken. The r table (Table E on page 339) gives the positive critical value, based on the sample size (n, the number of (x,y) pairs). The negative critical value is the opposite of the one shown. The r distribution has a unique unusual symmetric shape that we will not worry about sketching. As usual, we will reject H_0 if the test statistic is beyond the critical values (to the left of the negative critical value or to the right of the positive critical value). Otherwise, we fail to reject H_0. If we infer that there is a linear correlation in the population, the sign of *r* tells what type of linear correlation (positive or negative).

Due to the limited area information in the r table, we will not use the p-value method when doing tests for correlation by hand.

[It is possible to do a directional test (greater than or less than) for correlation but we will use two-tailed tests as is the norm in introductory texts.]

When calculating r by hand, it is advisable to use this algebraically equivalent but less labor-intensive computational formula:

$$r = \frac{n\sum xy - (\sum x)(\sum y)}{\sqrt{n\sum x^2 - (\sum x)^2}\sqrt{n\sum y^2 - (\sum y)^2}}$$

r^2 is called the coefficient of determination. It gives the proportion of the variation in y that is explained by x (stated another way: the proportion of the variation in y that is explained by the linear relationship between x and y). Its value and meaning only apply to the whole population if we infer that there is a linear correaltion in the whole population.

Example: Randomly selected people were studied and had their hand length, foot length, and ear height (all measured in cm) measured. The measurements for each person in the sample are shown below:

Ear	6.4	6.1	5.9	5.7	6.3	5.8	6.3	6.0
Hand	20.9	18.2	15.6	17.3	19.3	17.0	19.7	18.4
Foot	29.5	27.4	21.6	26.0	25.9	22.8	28.8	24.0

See if there is a linear correlation between length of a person's hand and the length of the person's foot.

Answer:

Traditional Method:

Step 1: H_0: there is not a linear correlation between length of a person's hand and the length of the person's foot

H_1: there is a linear correlation between length of a person's hand and the length of the person's foot

Step 2: Test statistic: $r = \frac{8(3798.42)-(146.4)(206)}{\sqrt{8(2698.84)-146.4^2}\sqrt{8(5359.66)-206^2}}$

$$= \frac{228.96}{(12.56025477)(21.00666561)} = .8678$$

Organizing the values in a table provides the needed values:

									totals
Hand (x)	20.9	18.2	15.6	17.3	19.3	17.0	19.7	18.4	146.4
Foot (y)	29.5	27.4	21.6	26.0	25.9	22.8	28.8	24.0	206
x^2	436.81	331.24	243.36	299.29	372.49	289	388.09	338.56	2698.84
y^2	870.25	750.76	466.56	676	670.81	519.84	829.44	576	5359.66
xy	616.55	498.68	336.96	449.8	499.87	387.6	567.36	441.6	3798.42

Step 3:

for $\alpha = .05$, critical values are $\pm .707$

Step 4: The test statistic is in the rejection region so reject H_0

Step 5: Conclusion: There is a linear correlation between length of a person's hand and the length of the person's foot. The correlation is positive.

Be sure to look at the computational formula for r to see how to use the values from the table.

After making our inference that the population of hand and feet measurements has a linear correlation, we further inferred that the correlation is positive based on the sign of r. If we failed to reject H_0, we would simply say, "We did not find that there is a linear correlation between length of a person's hand and the length of the person's foot." There would be no need to say what kind of linear correlation does not exist (think about that...).

Did you identify the two quantitative variables? From each person in the sample we got the length of hand and length of the foot. That's two quantitative pieces of information so the methods of this section are appropriate.

What we are doing in this section is fundamentally different from what we are doing with the mu-d methods from earlier in the text. There we were looking at mean differences, here we are looking for a relationship between the variables. For example, consider the two simple paired data scenarios below:

Scenario 1

x	y
10	11
9	9
8	7

Scenario 2

x	y
10	7
9	9
8	11

In both scenarios, the mean of the differences is zero. However, the first scenario shows a positive linear correlation (r = 1), while the second has a negative linear correlation (r = −1). When working with data, we always have to keep in mind what research question we are trying to address using the data. There will often be multiple procedures you can perform on one set of data, depending on your research question.

Example: Refer to the foot, hand example. What proportion of the variation in foot length is explained by hand length?

Answer: $.8678^2 = .7531$

The answer to the question is the coefficient of determination. In more advanced classes, the r^2 (which becomes R^2 for more advanced procedures) is used quite a bit to compare different statistical models.

Example: Randomly selected persons were given a stress test and were also asked how many minutes they spend doing exercise per week. What does the output below show about the relationship, if any, between stress and exercise?

Correlations: exercise, stress

```
Pearson correlation of exercise and stress = -0.298
P-Value = 0.042
```

> **Answer**: The p-value (.042) < α (.05), reject H_0. There is a negative linear correlation between stress and exercise. Higher levels of stress are associated with smaller amounts of exercise.

The conclusion could also be stated as: Higher levels of exercise are associated with smaller levels of stress, The more a person exercises the smaller amount of stress he or she tends to experience, People who are more stressed tend to exercise less, etc. The way this is stated depends on the situation, the goal of the research, and maybe even the point of view of the researcher. Notice that all these statements are expressing a negative linear correlation—larger values of one variable are associated with smaller values of the other.

The null hypothesis is that there is no linear correlation between stress and exercise. The alternative is that there is a linear correlation between stress (level) and exercise (minutes spent weekly). As usual, rejecting H_0 caused us to conclude H_1, with the added detail of using the sign of r (–.298) to specify which type of correlation there is in the population. If we failed to reject H_0, we would not have found a linear correlation.

It is important to note that, even though we found a correlation, we cannot jump to a cause and effect statement at this point. Maybe exercising more causes people to be less stressed. Maybe being really stressed causes people to not exercise. Maybe some other variable that we did not consider (a genetic factor, diet, family situation, socioeconomic status, etc.) contributes to both higher levels of stress and lower levels of exercise at the same time. There are so many possibilities. To try to make any sort of respectable causal claim, we would next have to engage in much more sophisticated research (which is why students have to take upper division research/Statistics classes) with its associated statistical methods. Establishing a correlation is a very basic step. You will find that many of the more advanced statistical methods have correlations as their foundational component.

What if exercising more makes people less stressed, for a while. However, after reaching a certain amount of exercise, it causes greater stress. Or what if being stressed makes people not want to exercise but then, after reaching a certain stress level, they become motivated to exercise but then, after a brief exercise period, the stress slows down their exercise activity. In those cases we would have some sort of non-linear relationship between stress and exercise. You can probably imagine what graphs would look like for the cases described, but that is something to look forward to in your upper division studies.

Exercises

1) A Public Policy student collects appropriate sample data to test for a relationship between the number of crime incidents and the temperature. He has heard that crimes seem to occur more during warmer periods.

a) State the null and alternative hypotheses for this inquiry.

b) Describe all the possible outcomes for this hypothesis test.

c) What would a Type I error be for this problem?

2) Air quality readings were collected from Statsville, Wudmoor, and Chef County on the same randomly selected days. Perform the appropriate .01 significance level test to see if there is a relationship between air quality readings for Statsville (x) and Wudmoor (y).

Air quality data:

Day	Statsville	Wudmoor	Chef County
1	32	33	35
2	36	40	39
3	36	35	32
4	48	38	42
5	40	36	45
6	37	45	36
7	36	30	36

3) Refer to problem 2). Find and clearly interpret the coefficient of determination. Is this interpretation for the sample only or is it also applicable to the population? Explain.

4) Students at Twin Streams College are given math and reading placement tests. Do students who get higher scores on the math test tend to get higher scores on the reading test? Using the test scores from randomly selected students shown in the following table, perform the appropriate test and address this question.

Major	Math score (x)	Reading score (y)
Criminal Justice	98	57
English	121	81
Math	116	80
Psychology	74	32
English	101	60
Psychology	100	71
Engineering	113	72
Sociology	81	81
Statistics	130	95

5) Refer to problem 4). What proportion of the variation in reading scores is explained by math scores? Is this true for all math and reading placement test scores at Twin Streams College or just this sample? Expound.

6) 40 homes from Statsville were investigated and the value of the homes (in dollars) before the construction of the new local sports arena were recorded. After the construction of the arena, the value of these same homes was recorded. Are higher current values associated with higher before-arena values? Use a .03 significance level when explaining your answer.

Correlations: ValueBefore, ValueNow

```
Pearson correlation of ValueBefore and ValueNow = 0.732
P-Value = 0.000
```

9.5 Regression

Conditions: The data is from two dependent random samples (dependent in that there is pairing due to a unique relationship between each value in one sample and a value in the other sample). For each of the x values in the population, the matched y values have a normal distribution about the regression line and equal standard deviations.
(Assume all conditions are met for each example and exercise.)

Use the methods of this section when...

a) we are instructed to make a regression equation

b) we are instructed to make a prediction and a regression equation is needed for that prediction

Variable types: two quantitative

Important formulas: $\hat{y} = b_o + b_1 x$

$$b_0 = \bar{y} - b_1 \bar{x}$$

$$b_1 = \frac{n\sum xy - (\sum x)(\sum y)}{n\sum x^2 - (\sum x)^2}$$

If there is a linear correlation between two variables, we can make a mathematical model of that relationship and use it to make predictions of y for values of x. This process is called *regression*. (The name comes from an early biological application of the process.) We will learn about the simplest type, ordinary least squares regression.

Using calculus, it can be shown that the regression equation is:

$$\hat{y} = b_o + b_1 x \quad \text{where} \quad b_0 = \bar{y} - b_1 \bar{x} \quad \text{and} \quad b_1 = \frac{n\sum xy - (\sum x)(\sum y)}{n\sum x^2 - (\sum x)^2}$$

The graph of the regression equation is called the regression line. b_0 informs us of the y-intercept of the line, and b_1 is the slope (remember algebra). [In Statistics, we don't care about the x-intercept, only the y-intercept, so the y-intercept is often just called "the intercept".] This is the one line, of all the infinitely many lines, that does the best job of approximating our data. If we can infer there is a linear correlation between the variables in the population (using Section 9.4 techniques), we can also infer that the regression equation is applicable to the population. Based on that, we can use the regression equation to make a prediction about what the value of the y variable would be for some given value of x. We should only use the regression line to make a prediction of a value of y for an x value in the interval of x values in our original data. For x values outside the interval of x values we observed, it might be the case that the relationship between the variables changes (or disappears), thereby making our regression equation worthless.

Even if we cannot infer a relationship between two variables in the population, it is still possible to create a regression equation. However, we would not use that equation to make any predictions because the regression equation would apply strictly to that one sample. In that case, if we need to make a prediction of a value of y for some given value of x, the best prediction would just be $\bar{y}$, the mean of the y values in the sample.

The *x* variable is called the explanatory variable, predictor variable, or independent variable. The *y* variable is called the response or dependent variable. This is what is being predicted. While it does not matter which variable is considered x and what is considered y when working with correlation, it matters greatly when working with regression. When we use the methods of this section, it is said that we are "regressing y on x".

A residual is the difference $(y-\hat{y})$ between the observed value of y and the value (called a fitted value) for y determined by the regression equation, for a value of x in the sample. What makes the line resulting from the regression equation the line of best fit, is that it is the line with minimized residuals (technically, minimized squared residuals). The calculus process used to get the formulas for b_0 and b_1 ensures this.

The scatterplot that follows has x variable "C4" and y variable "C5". The regression line included on the scatterplot has the equation, $\hat{y} = 37.7 - 1.86x$. The leftmost dot on the scatterplot representing an observation from our sample is the point (8, 21). So, the observed value of y when x = 8 is 21. To get the value of y determined by the regression equation when x = 8, we substitute 8 into the regression equation in place of x to get, $\hat{y} = 37.7 - 1.86(8) = 22.82$. So the fitted value (or, "fit") when x = 8 is 22.82. That makes the residual when x = 8: 21 – 22.82 = −1.82. So the regression equation's determination for y is bigger than the actual y value by 1.82 when x = 8.

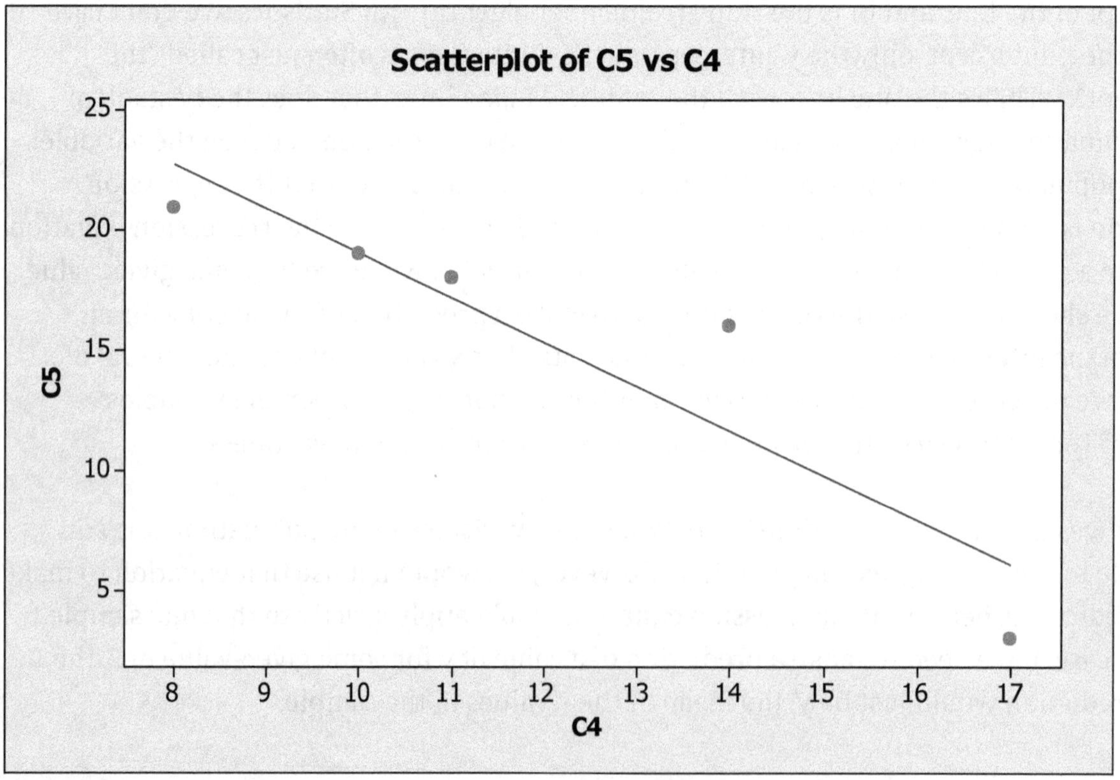

Graphically we can see the residuals for each observed value of x in our sample. Choose a dot representing an observation from our sample. Draw a vertical segment from the dot to the regression line. The length of that segment is the magnitude (absolute value) of the residual. If the vertical segment is below the regression line, the sign on the residual is negative, and positive otherwise. Of course, if the regression line passes right through the center of a dot, the residual would equal zero (the observed y value and the fitted value are the same). Just like in algebra, points on a line are solutions of its equation. So, you could make predictions for non-observed x values graphically by choosing an x value on the x axis, tracing vertically until you hit the regression line, and from there moving horizontally until you hit the y axis. Wherever you land on the y axis, that is your predicted value of y.

In more advanced classes, correlation and regression are handled at the same time instead of in separate steps as is done in this and most other introductory Statistics texts.

Example: Use the hands and feet data from Section 9.4 to write a regression equation. Clearly interpret the slope and y-intercept of the equation. Predict the foot length of a person with a hand length of 18 cm.

Answer:

First part: $b_1 = \frac{8(3798.42)-(146.4)(206)}{8(2698.84)-146.4^2} = \frac{228.96}{157.76} = 1.4513$

$b_0 = 25.75 - 1.4513(18.3) = -.8088$

$\hat{y} = -.8088 + 1.4513x$ (this is the regression equation)

Second part: Slope—For each one cm increase in hand length, foot length is larger by 1.4513 cm, on average.

Intercept—A person that has a hand length of zero cm is predicted to have a foot length of $-.8088$ cm.

Third part: $\hat{y} = -.8088 + 1.4513(18) = 25.3146$ cm

Yes, it is obvious that the interpretation of the intercept makes no sense for this problem. That is because zero is not a possible value for "x" (hand length). The slope always tells you something useful in real world terms, but the intercept's interpretation will only be useful if zero is a possible value for x. In more advanced classes, you talk about things you can do with the data to always get a useful intercept (but the exact meaning of the intercept will change). For this class, though, you should always be able to interpret the slope and the intercept, whether or not the intercept's interpretation is useful. To get the interpretation, we just think back to what the slope and intercept mean in algebra. The slope is the change in y (increase if positive, decrease if negative) per one unit increase in x. The intercept gives you the value of y when x is zero. Think back to your y = mx + b from algebra, where the slope is m and the intercept (y-intercept) is (0, b). The reason for the "on average" in our interpretation of the slope is because the data values are not perfectly linear so the rate of change is not absolute (like it is in a line), but it is the average rate of change.

The interpretation of slope and intercept can be applied to the whole population if we found that there is linear correlation in the population (using methods from the previous section). Otherwise, the interpretation of slope and intercept are restricted to the sample only. Of course, for this data we did infer a linear correlation in the population (in the previous section).

The only reason we are using the regression equation to make the requested prediction is because we found there is a linear correlation between foot length and hand length. If there were no correlation, the best predicted foot length would just be the mean of the foot lengths in the sample (25.75 cm).

To get an even better prediction, you could use something called a prediction interval. As we studied, a confidence interval estimates a parameter. A prediction interval estimates a prediction. So instead of having the single prediction of 25.3146, we would say that the best prediction is between two values created using a prediction interval. In this introductory course, though, we will stick with the best single prediction.

Notice that 18 cm is within the interval of hand length measurements in the collected data. We wouldn't want to try to make a prediction of a foot length based on a hand length smaller than 15.6 cm or larger than 20.9 cm. For hand lengths outside of that interval we have no way of knowing if the same regression model would be applicable.

Side note: What if, instead of only using hand length to predict foot length, we also wanted to use thigh circumference, head width, and torso height to predict foot length? Or what if we wanted to use hand length, thigh circumference, head width, and torso height to simultaneously predict foot length and ear height? Or what if we wanted to use eye color and hair color (categorical variables) to predict foot length and nose width? What if we wanted to use hand length and foot length to predict whether someone is tall or short (after defining those terms)? These are just some of the very cool types of tests you can perform and will learn in later classes. Even if you're of the "I'll never take another math class again" crowd, you may have additional Statistics classes to take but you do not realize that yet. Research methods classes, assessment classes, and other classes with non-math sounding names are actually just cleverly disguised advanced Statistics classes. Any class that uses your current Statistics class as a prerequisite is essentially an advanced Statistics class. There is so much more to learn and there are many fascinating things to discover about Statistics, both in the applications and the rigorous mathematical background!

There are also technical considerations when working with regression models. For one, sometimes there may be influential points—points that are extreme and have a huge effect on the line. If those points were removed the line would drastically change. You may need to get the regression equation with and then without those points. Perhaps a point is a mistake in data collection. This starts to get into the art of Statistics—more about this in your later classes. Also, care needs to be paid to the residuals. For starters, there should not be a discernible pattern among the residuals indicating a systematic over-predicting or under-predicting occurring at certain intervals of x values.

There is so much more to say about regression. In fact, some of the higher level classes you take will consist largely of regression models. Anyway, let's not get ahead of ourselves...

Example: Recall the example from Section 9.4 involving stress and exercise. The same data was used to generate the Minitab output below. What is the best predicted stress level of a person spending 60 minutes per week exercising?

Regression Analysis: stress versus exercise

```
The regression equation is
stress = 134 - 0.0356 exercise


Predictor       Coef  SE Coef      T      P
Constant      133.52    27.62   4.83  0.000
exercise    -0.03562  0.01699  -2.10  0.042


S = 30.5012   R-Sq = 8.9%   R-Sq(adj) = 6.9%


Analysis of Variance

Source          DF       SS      MS     F      P
Regression       1   4091.4  4091.4  4.40  0.042
Residual Error  45  41864.4   930.3
Total           46  45955.8
```

Answer: 134 –.0356(60) = 131.864

Recall that, in Section 9.4, we concluded that there is linear correlation between stress level and weekly exercise time (in the population). That is the reason we used the regression equation to make a prediction. Otherwise, we would have used the mean stress level in the sample as our prediction. (That would have needed to have been given somehow, unless we had the raw data.)

There is a lot of regression output we will not use in this introductory course because there is so much more you can do with regression, particularly as the models include more variables. The only thing of importance to us is the regression equation, clearly labeled with "`The regression equation is`" and the actual variable names instead of $\hat{y}$ and x.

The slope tells us that for each one minute increase in weekly exercise, stress level decreases by .0356, on average. (We know the change is a decrease because the slope is negative.) The best predicted stress level of someone who does no weekly exercise is 134, as indicated by the intercept. Here, the interpretation of the intercept does make sense because there are people who do zero minutes of exercise each week.

Exercises

1) Ages (x, in years) and values (y, in dollars) of randomly selected cars were obtained to create this regression equation: $\hat{y} = 21595 - 1478x$. Interpret the slope and intercept of the equation.

2) Refer to Exercise 2 in Section 9.4. Create a regression equation using the Statsville (x) and Wudmoor (y) data. What is the best predicted air quality reading in Wudmoor on a day that Statsville has an air quality reading of 45?

3) Clearly interpret the slope and intercept of the equation you made in 2). Does the interpretation apply to the sample only or also to all air quality readings for Statsville and Wudmoor? Explain.

4) Refer to Exercise 4 in Section 9.4. Create a regression equation using that data. What is the best predicted reading score for a student with a math score of 107?

5) Clearly interpret the slope and intercept of the equation you made in 4). Does that interpretation apply to the sample only or the whole population? Why?

9.6 Goodness of Fit

Conditions: The frequency data comes from a random sample. The expected frequency for each category is at least five.
(Assume all conditions are met for each example and exercise.)

Use the methods of this section when...

> testing a claim about the proportions of the categories of a categorical variable

Variable type: one categorical

Important formula: $\chi^2 = \sum \frac{(O-E)^2}{E}$ df = k – 1, where k is the number of categories

Some research questions involve theories about the proportions in which the categories of a categorical variable are found in the population. For example, suppose a psychologist believes that, of all people who suffer from social phobia, 20% have moderate social phobia, 40% have marked social phobia, 30% have severe social phobia, and 10% suffer from very severe social phobia. As another example, a demographer theorizes that the age breakdown for people in a certain region is: under 20, proportion = .31; from 20 to 28, proportion = .12; from 29 to 44, proportion = .16; from 45 to 51, proportion = .11; from 52 to 65, proportion = .17; over 65, proportion = .13. In each case, we have one categorical variable because there is one piece of information that the researcher would collect from each person observed (type of social phobia in the first example, age category for the second example).

[Sometimes students confuse the number of variables with the number of categories of a variable. Type of social phobia is one categorical variable—it has four categories. Age category is one categorical variable—it has six categories.]

To address research questions like this, we use a Goodness of Fit test. The idea is that we will collect a sample and, based on the results, see if the theory regarding the value of each proportion seems to be a good fit for the population. If the proportions for each category observed in the sample are reasonably close to the claim about these proportions for the population, we will have no reason to reject that claim. If the sample proportions are quite different from the claim about the population's proportions, then we will reject that claim as it does not appear to be a good fit for the population.

For a Goodness of Fit test, these are the hypotheses in their generic form:

$H_0: p_1 = claim\ for\ category\ 1, p_2 = claim\ for\ category\ 2, \ldots, p_k = claim\ for\ category\ k$
H_1: The categorical variable is not divided into categories according to the proportions in H_0

Of course, these are tailored to the specific research situation.

Notation: O: observed frequency of a category
E: expected frequency of a category
k: number of categories
n: number of trials

When performing this test, we record the quantity of sampled items that belonged to each category (that's O). We compare each of those observed frequencies to the respective quantity of sampled items we expected to belong to each category (that's E), based on the proportions claimed in H_0. If, overall, the respective O's and the E's are reasonably close, then we have no reason to reject the statement that gave us the E's—H_0. The sample proportions are pretty close to the claim about the population proportions, so that claim is plausible. If, however, the O's and the E's are quite different from each other, then we reject the statement that gave us the E's—H_0. The sample proportions are far away from the claim about the population proportions, so that claim is not plausible.

As always, we need some way to quantify "reasonably close" and "quite different". When the conditions of this section are met, the test statistic $\sum \frac{(O-E)^2}{E}$ follows a χ^2 distribution with degrees of freedom (k – 1). By inspection, you can see that, if the O's and E's are quite different from each other, the test statistic will be relatively big due to the squaring of the differences in respective O's and E's. So only a big value of the test statistic would indicate that the O's and E's are far apart, therefore signaling us to reject H_0. For that reason, Goodness of Fit tests are always right-tailed tests.

Be prepared for the fact that the expected frequencies, the E's, will often be decimal numbers. That happens when the sample size you're using is such that the sample could not have the exact proportions stated in H_0 for each category. Keep the decimals when performing the calculations. We are essentially building a theoretical distribution and we are seeing how closely our data conforms to it.

Keep in mind that the null hypothesis is one statement. We are not performing a separate test of each proportion, but we are testing them collectively in one test.

Example: Randomly selected people were asked, "How do you describe your health?" 40 respondents selected "below average", 55 respondents selected "average", and 47 selected "above average". Test the claim that there is an even distribution of people who feel they have below average health, average health, and above average health.

Answer:

Traditional method

Step 1: H_0: $p_1 = p_2 = p_3 = \frac{1}{3}$

H_1: there is not an even distribution of people who feel they have below average health, average health, and above average health

Step 2:

selection	O	E	O−E	$(O-E)^2$	$\frac{(O-E)^2}{E}$
below average	40	(1/3)142=47.3333	−7.3333	53.7773	1.1361
average	55	(1/3)142=47.3333	7.6667	58.7783	1.2418
above average	47	(1/3)142=47.3333	−.3333	.1111	.002347
totals	142			Test statistic χ^2 =	2.3802

Step 3:

(α not given so use .05)

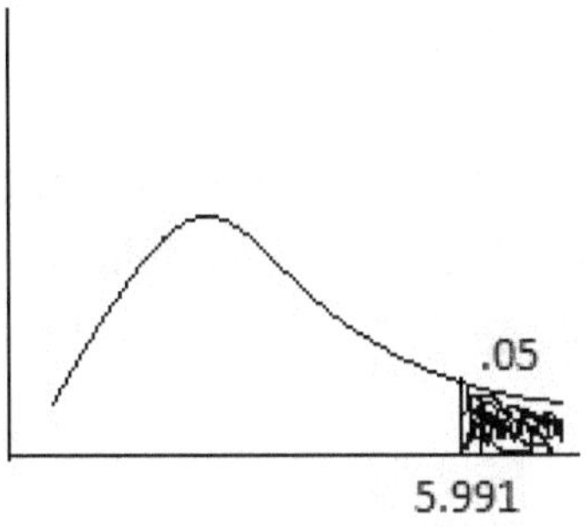

Step 4: Test statistic not in the rejection region; fail to reject H_0

Step 5: Conclusion: We did not find that there is not an even distribution of people who feel they have below average health, average health, and above average health. (The claim that there is an even distribution is plausible.)

P-value method

Step 1: H_0: $p_1 = p_2 = p_3 = \frac{1}{3}$

H_1: there is not an even distribution of people who feel they have below average health, average health, and above average health

Step 2: (same work as shown previously) Test statistic $\chi^2 = 2.3802$

Step 3:

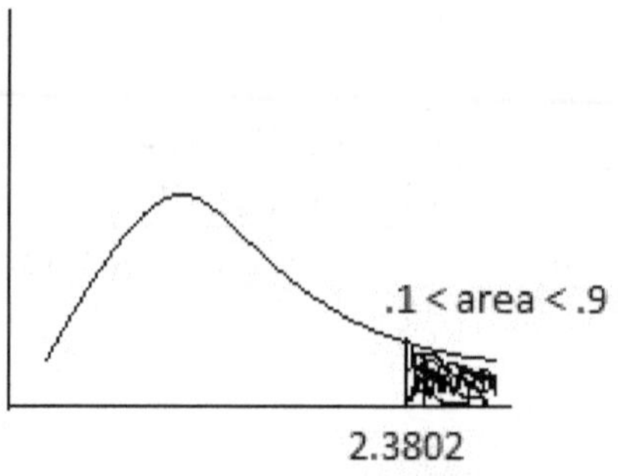

Step 4: p-value is total area so .1<p-value<.9; α not given so use .05; p-value > α; fail to reject H_0

Step 5: Conclusion: We did not find that there is not an even distribution of people who feel they have below average health, average health, and above average health. (The claim that there is an even distribution is plausible.)

We knew a goodness of fit test was needed because we were being asked to test the proportions ("even distribution") for the categories of one categorical variable (view of health). p_1 is the proportion of all people who think they have below average health, p_2 is the proportion of all people who think they have average health, and p_3 is the proportion of all people who think they have above average health. Since we were testing for an even distribution, and we have three categories, the hypothesized proportion for each category would be $\frac{1}{3}$. In general, if we are testing for an even distribution among k categories, the hypothesized proportion for each is $\frac{1}{k}$. Notice that if the problem had been worded, "Test the claim that there is not an even distribution..." we would have performed the test the exact same way. If the test is for an even distribution or not an even distribution, either way the null hypothesis is that all proportions are equal and the alternative is that they are not. The expected frequency for a category is found by taking the hypothesized proportion for that category multiplied by n (the number of trials which is also the sum of the observed frequencies).

Example: Refer to the situation described in the previous problem. Now suppose that a public health administrator believes that 50% of people think they have below average health, 30% think they have average health, and 20% think they have above average health. Test that belief.

Answer:

Traditional method:

Step 1: H_0: $p_1 = .5$, $p_2 = .3$, $p_3 = .2$

H_1: the proportions of the different ways people view their health are not as believed by the public health administrator

Step 2:

Answer	O	E	O−E	$(O−E)^2$	$\frac{(O-E)^2}{E}$
below average	40	(.5)142=71	−31	961	13.5352
average	55	(.3)142=42.6	12.4	153.76	3.6094
above average	47	(.2)142=28.4	18.6	345.96	12.1817
totals	142			Test statistic χ^2 =	29.3263

Step 3:

(α not given so use .05)

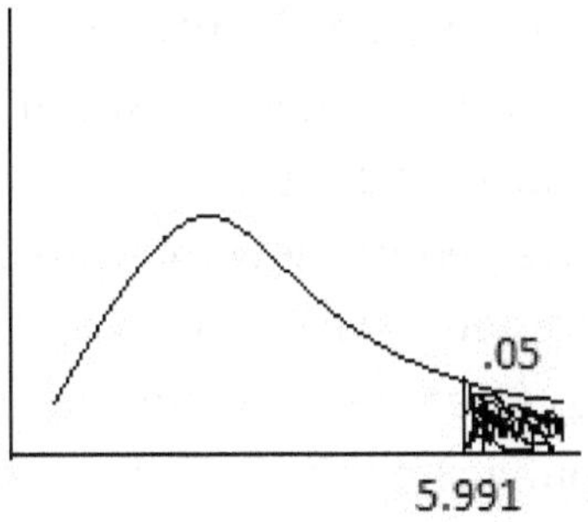

Step 4: Test statistic in the rejection region; reject H_0

Step 5: Conclusion: The proportions of the different ways people view their health are not as believed by the public health administrator.

P-value method

Step 1: H_0: $p_1 = .5$, $p_2 = .3$, $p_3 = .2$

H_1: the proportions of the different ways people view their health are not as believed by the public health administrator

Step 2: (using same work shown above) Test statistic χ^2 = 29.3263

Step 3:

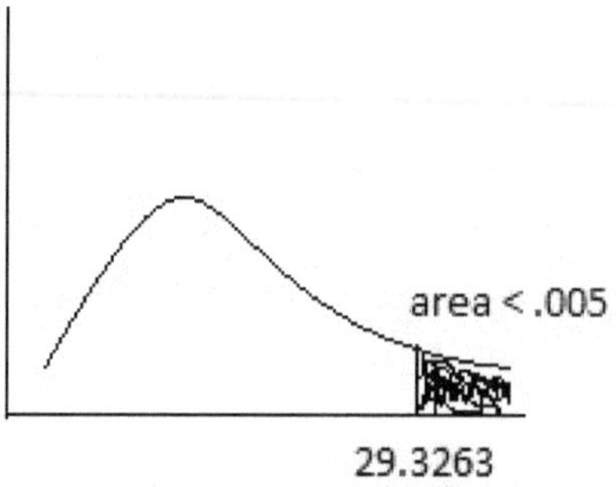

Step 4: Total area < .005 so p-value < .005; α not given so use .05; p-value < α; reject H_0

Step 5: Conclusion: The proportions of the different ways people view their health are not as believed by the public health administrator.

p_1 is the proportion of all people who think they have below average health, p_2 is the proportion of all people who think they have average health, and p_3 is the proportion of all people who think they have above average health. The hypothesized value for each comes from the public health administrator's belief. So, when we reject H_0 and conclude that the true values for the proportions are not as stated in H_0, we are ultimately saying that the proportions of the different ways people view their health are not as believed by the public health administrator. Had we failed to reject H_0, we would not have found that the proportions of the different ways people view their health are not as believed by the public health administrator. So, the public health administrator's belief would have been plausible.

Example: Randomly selected people were asked their favorite news source. We want to see if favorite news sources are equally distributed among ACB, BCN, NCN, and OFX. Use a .10 significance level to address this research question, using the Minitab output below.

Chi-Square Goodness-of-Fit Test for Observed Counts in Variable: Source

Using category names in source

Category	Observed	Test Proportion	Expected	Contribution to Chi-Sq
ACB	65	0.25	75.75	1.52558
BCN	82	0.25	75.75	0.51568
NCN	76	0.25	75.75	0.00083
OFX	80	0.25	75.75	0.23845

N	DF	Chi-Sq	P-Value
303	3	2.28053	0.516

Answer: p-value (.516) > α (.10), fail to reject H_0. We did not find that favorite news sources are not equally distributed among ACB, BCN, NCN, and OFX. It would be plausible to claim that there is an equal distribution of people who prefer these news sources.

The null hypothesis is that the proportion of all people who favor ACB news is equal to the proportion of all people who favor BCN news which is equal to the proportion of all people who favor NCN news which is equal to the proportion of all people who favor OFX news. In symbols, this would be H_0: $p_1 = p_2 = p_3 = p_4 = \frac{1}{4}$, where each respective p stands for the proportions stated in the previous sentence. Having four categories this time, the hypothesized proportion for each is $\frac{1}{4}$. [The "`Test Proportion`" column shows the hypothesized proportion for each category. If we were not testing equal proportions, the numbers in that column would not all be identical.] The alternative hypothesis is that the proportion of all people who favor each news source is not as stated in H_0; in other words, favorite news sources are not equally distributed among ACB, BCN, NCN, and OFX. Had we rejected H_0, our conclusion would have been, "Favorite news sources are not equally distributed among ACB, BCN, NCN, and OFX."

In the Minitab output, notice that we see a table similar to the one we made when doing these Goodness of Fit problems by hand. Once we recognize the output is addressing the research question before us, all we need is the good ol' p-value to get the correct conclusion.

Exercises

1) Air quality index readings were recorded on randomly selected days and categorized as follows: 30-34, 35-39, 40-44, and 45-49. The display below shows how many days had air quality index readings belonging to each category.

reading	30-34	35-39	40-44	45-49
number of days	32	44	32	36

One environmentalist thinks that the air quality index readings occur with equal frequency. Another believes that 10% of the time the air quality is between 30 and 34, 40% of the time the air quality is between 35 and 39, 30% of the time the index is between 40 and 44, and the rest of the time the quality of the air falls between 45 and 49. Perform a separate test for each environmentalist's view and comment on which one seems most likely. (Use α = .01 for each test.)

2) Refer to the Jerrell Enterprises and Jaden Industries data from Exercise 2) in Section 7.2. A Human Resources Generalist theorizes that 52% of Jerrell Enterprises employees are dissatisfied with their current position in the company, 29% are satisfied with their current position in the company, and 19% are neutral on this issue. With a significance level of .025, see if this theory is reasonable.

3) Refer to 2). What is the probability that this hypothesis testing procedure would lead us to conclude that the Generalist's theory is not correct if, in fact, the theory is correct?

4) Randomly selected first-year Statsville College students were asked their favorite time to study (morning, afternoon, or evening) and the area of their major (STEM, Humanities, Social Science, Other). What is this output telling you about all Statsville College students?

Chi-Square Goodness-of-Fit Test for Categorical Variable: major

Category	Observed	Test Proportion	Expected	Contribution to Chi-Sq
Human	51	0.18	34.2	8.25263
Other	22	0.19	36.1	5.50720
Sscience	57	0.28	53.2	0.27143
STEM	60	0.35	66.5	0.63534

N	N*	DF	Chi-Sq	P-Value
190	0	3	14.6666	0.002

9.7 Test of Independence

Conditions: The frequency data comes from a random sample. The expected frequency for each cell is at least five.
(Assume all conditions are met for each example and exercise.)

Use the methods of this section when...

testing for a relationship between two variables

Variable types: two categorical

Important formula: $\chi^2 = \sum \frac{(O-E)^2}{E}$

df = (number of rows – 1)(number of columns – 1)

Suppose we surveyed all residents in Wudmoor, Statsville, Chef County, and Twin Streams and we asked them their favorite weather condition: sunny, cloudy, or rainy. We recorded the number of people who gave the indicated responses in the table.

	Wudmoor	Statsville	Chef County	Twin Streams	Total
Sunny	A				B
Cloudy			E		F
Rainy					
Total	C		G		D

Each rectangle that contains a count for the intersection of two categories is called a *cell.* This table, having categories for one categorical variable on the rows and the categories for another categorical variable on the columns is called a *contingency table* or *two-way table.* The placeholders in the table represent the following:

A: the number of Wudmoor residents who prefer sunny weather
B: the total number of all residents who prefer sunny weather
C: the total number of residents in Wudmoor
D: the total number of residents surveyed (called the *grand total*)
E: the number of Chef County residents who prefer cloudy weather
F: the total number of all residents who prefer cloudy weather
G: the total number of residents in Chef County

In probablility theory, two events, A and B, are independent if and only if P(A and B) = P(A)P(B). For example, P(Wudmoor and Sunny) = P(Wudmoor)P(Sunny)

$$\frac{A}{D} = \left(\frac{C}{D}\right)\left(\frac{B}{D}\right)$$

That equation would be true if and only if the events "resides in Wudmoor" and "prefers sunny weather" are independent. If we solve this equation for A, we have: $A = \frac{BC}{D}$. Notice that B is the row total for the Wudmoor/Sunny cell, C is the column total for the Wudmoor/Sunny cell, and D is the grand total. So the number of residents from Wudmoor who prefer sunny weather would be equal to (the total for that row) times (the total for that column) divided by (the grand total) or $\frac{(row\ total)(column\ total)}{grand\ total}$.

Similarly, if the events "resides in Chef County" and "prefers cloudy weather" are independent, then we would have: P(Chef County and Cloudy) = P(Chef County)P(Cloudy)

$$\frac{E}{D} = \left(\frac{G}{D}\right)\left(\frac{F}{D}\right)$$

Solving this equation for E, we have: $E = \frac{FG}{D}$, which again is $\frac{(row\ total)(column\ total)}{grand\ total}$.

If independence holds for all events across the two variables (Wudmoor/Cloudy, Wudmoor/Rainy, Statsville/Sunny, Statsville/Cloudy, etc.) then the two categorical variables (Location and Weather Preference) would be considered independent.

Now let's think of this in a less technical, more practical way. If location and weather preference are independent,

> Location and weather preference have nothing to do with each other.
> There is no association between location and weather preference.
> A person from one of these locations is not more or less likely to prefer a certain type of weather.
> A person who prefers a certain type of weather is not more or less likely to reside in one of these locations.
> Where a person resides and the person's weather preference are not related.
> There is no difference in a person's weather preference based on location.
> There is no difference in a person's location based on weather preference.

Of course, whenever we refer to "location" we are limited to the specific locations in our study and when we refer to "weather preference" we are limited to the specific weather preferences in our study. If there is no difference in weather preference based on location (i.e., location and weather preference have nothing to do with each other) then the proportion of residents in Wudmoor who prefer sunny weather would equal the proportion of residents in Statsville who prefer sunny weather, which would equal the proportion of residents in Chef County who prefer sunny weather, which would equal the proportion of residents in Twin Streams who prefer sunny weather, which would equal the proportion of all residents who prefer sunny weather. Taking the first and last of those relationships, we have:

proportion of Wudmoor prefer sunny = proportion of all prefer sunny

$$\frac{A}{C} = \frac{B}{D}$$

solving for A we have, $A = \frac{BC}{D} = \frac{(row\ total)(column\ total)}{grand\ total}$

Under this same condition (location and weather preference are not related) the proportion of residents in Chef County who prefer cloudy weather would equal the proportion of all residents who prefer cloudy weather, so

$$\frac{E}{G} = \frac{F}{D}$$

$$\text{and } E = \frac{FG}{D} = \frac{(row\ total)(column\ total)}{grand\ total}$$

If the quantity in each cell is equal to $\frac{(row\ total)(column\ total)}{grand\ total}$ then we know that location and weather preference are not related—those two categorical variables are independent. Of course, if it turns out there is some relationship between location and weather preference (say, people in Twin Streams are more likely to prefer rainy weather but people in Statsville tend to prefer sunny weather) then the counts for each cell would not be $\frac{(row\ total)(column\ total)}{grand\ total}$.

This background leads us to the Test of Independence.

These are the hypotheses for a Test of Independence in their generic form:

H_0: The two categorical variables are independent (they are not related)
H_1: The two categorical variables are not independent (they are related)

Of course, these are tailored to the specific research situation.

When performing this test, we make a two-way table and place in each cell the quantity of sampled items belonging to the indicated categories of the cell. We will again use "O" to symbolize those observed frequencies. We then calculate the total for each row, the total for each column, and then the grand total. The expected frequencies for each cell are the frequencies we would expect under the null hypothesis' claim of independence. Using our earlier discussion, the expected frequency of each cell is given by E = $\frac{(row\ total)(column\ total)}{grand\ total}$.

At this point, the procedure is very similar to the Goodness of Fit test. We again use the test statistic $\chi^2 = \sum \frac{(O-E)^2}{E}$ to determine if the observed and expected frequencies are "reasonably close" or "quite different". Again, only a big test statistic would cause us to conclude that the O's and E's are "quite different" thereby causing us to reject the source of the E's (H_0), so these tests are always right-tailed. The applicable χ^2 distribution has degrees of freedom (number of rows – 1) times (number of columns – 1), based on the number of rows and columns in the two-way table *before* any totals were included. As with the Goodness of Fit test, keep decimal numbers for the expected counts.

Example: A body image counselor asked several people whether or not they ate healthy all day during the prior day. The counselor also asked the people how they feel about their weight, with response options being below average, average, or above average. Of the people who felt they ate healthy the day before, 24 saw themselves as having below average weight, 28 saw themselves as average weight, and 13 believed themselves to be above average weight. Of those who did not see themselves as healthy eaters the day before, 16 felt they were below average weight, 27 considered themselves average weight, and 34 thought their weight was above average. With the probability of making a Type I error being .05, see if there is a relationship between people's perception of their food choices and their view of their weight.

Answer: The first thing we need to do is to create a contingency table so that this data will be more manageable:

	below	average	above
ate healthy	24	28	13
did not eat healthy	16	27	34

Traditional method:

Step 1: H_0: perception of food choices and view of weight are independent (not related)

H_1: perception of food choices and view of weight are not independent (they are related)

Step 2: (To facilitate calculating the test statistic, we'll add the totals to the table, get the expected frequency of each cell, and then perform the calculation.)

	below	average	above	total
ate healthy	24	28	13	65
did not eat healthy	16	27	34	77
total	40	55	47	142

Expected counts:

healthy/below: $\frac{(65)(40)}{142} = 18.3100$

healthy/average: $\frac{(65)(55)}{142} = 25.1761$

healthy/above: $\frac{(65)(47)}{142} = 21.5141$

not healthy/below: $\frac{(77)(40)}{142} = 21.6901$

not healthy/average: $\frac{(77)(55)}{142} = 29.8239$

not healthy/above: $\frac{(77)(47)}{142} = 25.4859$

$$\text{Test statistic } \chi^2 = \frac{(24-18.3100)^2}{18.3100} + \frac{(28-25.1761)^2}{25.1761} + \frac{(13-21.5141)^2}{21.5141} + \frac{(16-21.6901)^2}{21.6901} + \frac{(27-29.8239)^2}{29.8239} + \frac{(34-25.4859)^2}{25.4859}$$

$$= 10.05880$$

Step 3:

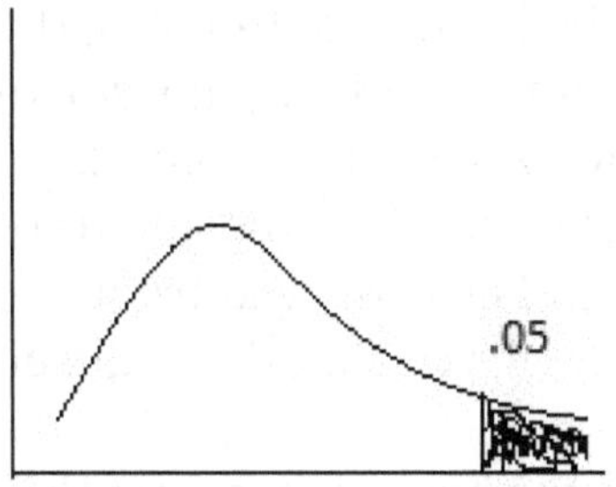

Step 4: Test statistic in rejection region, reject H_0.

Step 5: Conclusion: Perception of food choices and view of weight are not independent (they are related). There is a relationship between people's perception of their food choices and their view of their weight.

P-value method

Step 1: H_0: perception of food choices and view of weight are independent (not related)
H_1: perception of food choices and view of weight are not independent (they are related)

Step 2: (using same work as above) Test statistic $\chi^2 = 10.05880$

Step 3:

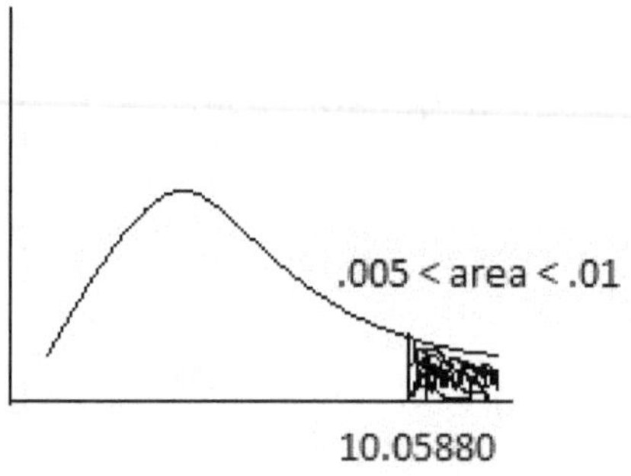

Step 4: total area between .005 and .01 so .005 < p-value < .01; given α .05; p-value < α; reject H_0.

Step 5: Conclusion: Perception of food choices and view of weight are not independent (they are related). There is a relationship between people's perception of their food choices and their view of their weight.

As described earlier, we got the expected frequency for each cell by taking the row total for that cell, times the column total for that cell divided by the grand total. For example, for the "healthy/below" cell we start by looking at that cell, then looking across to the far right of the table to get the total for that row (65). Again, starting in that cell, we go straight to the bottom of the table and see the total for that column (40). We multiply those numbers and divide by the grand total (142) to get the expected frequency of that cell. We followed the same procedure for all six cells. The observed frequency for each cell was given in the problem. The grand total, 142, tells us that the counselor interviewed 142 people.

To get the degrees of freedom, we have to consider the contingency table BEFORE the totals were added. We had two rows and three columns, so the degrees of freedom were $(2-1)(3-1) = 2$.

This question could have been phrased differently, and here are some alternatives:

Is there a difference in perception of food choices based on weight view?

Is there a difference in weight view based on perception of food choices?

Test the claim that weight view and perception of food choices are unrelated.

Are perception of food choices and view of weight independent?

Is view of weight independent of perception of food choices?

See if perception of food choices and view of weight are associated.

As you know by now, research questions can be stated many different ways. You cannot expect situations to only present themselves a certain way. An understanding of the situation, regardless of what English words are used to express it, is necessary. That is not just a lesson for this class—this is how the whole world of Statistics works.

Notice that we got two pieces of categorical data from each respondent for this research question. We were looking for a relationship between two categorical variables: view of weight and perception of food choices. That's what told us to use a test of independence.

Did we execute the test of independence to see if, for the 142 people surveyed, there is some relationship between perception of food choice and view of weight? NO WAY! We don't care about those 142 people (though I am sure they are all lovely). We are making an inference about the entire population from which those 142 people were selected. Among all people (as a group, of course) there is a relationship between people's perception of their food choices and their view of their weight.

We concluded that there is a relationship between perception of food choices and view of weight. However, we cannot make a causal claim at this point. Claims like, "Making healthy food choices makes people feel better about their weight" or "Having a negative view of one's weight will cause a person to eat poorly" are highly inappropriate. One of those statements (or both, to some extent) might be true, but we have not established that at this point. Much more work would need to be performed to make such claims. It could be that some other factor (relationship status, family bonds, occupation, IQ, education, etc.) could affect both perception of food choices and view of weight without those two variables having any direct effect on each other. We have no clue at this point *why* the relationship exists, we just can infer that it exists. We can't even say exactly what the relationship is (statements like, "people of average weight are more likely to perceive that they eat unhealthy than people of below average weight") without additional steps. The other work and research that would be needed is the reason for the higher level research classes you may be (hopefully will be) taking!

Big picture

Suppose, in the previous example, we had the actual numerical weight of respondents instead of their view of their weight? We would then have one categorical variable (perception of food choices) and one quantitative variable (weight). Since there are only two categories or levels for the categorical variable, we could make a confidence interval of $\mu_1 - \mu_2$, or use hypothesis testing for μ_1 and μ_2. Though it's generally only used when there are more than two categories, one-way ANOVA could also be used to address the research question. If the confidence interval did not contain zero, or if the hypothesis tests resulted in rejecting H_0, that would tell us there is some relationship between perception of food choices and weight. If we added a third category to the perception of food choices (like, "mostly healthy") we would have to use one-way ANOVA.

What if we had two quantitative variables? Say we gave each respondent a food choices test which gave them a numerical score and we also measured their weight. Now we would use a test for linear correlation to process the data. Realize that the test of independence and the test for linear correlation do the exact same thing: test for a relationship between two variables. With the test of independence, though, we have two categorical variables. With the test for linear correlation we have two quantitative variables.

Suppose we had the variables weight, perception of food choices, and view of weight. Now we have two categorical variables (perception of food choices and view of weight) with one quantitative (weight). We want to see if there is a difference in mean weight for people in different food choice perception groups and/or a difference in mean weight for people based on their view of their weight. Two categorical variables and one quantitative...two-way ANOVA!

As always in Statistics, the research question and the types of data you have determine the procedure you use. What if you wanted to consider ethnicity, perception of food choices, and view of weight all at the same time? How about weight, age, income, view of weight, and perception of food choices? These situations and many more are considered in upper division classes!

Example: Randomly selected people were asked their favorite news source (ACB, BCN, NCN, or OFX) and how they feel about the state of the nation (unhappy, neutral, or happy). We want to see if there is a difference in feeling regarding state of the nation based on favorite news source. Use a .01 significance level to address this research question, using the Minitab output below.

Chi-Square Test: unhappy, neutral, happy

```
Expected counts are printed below observed counts
Chi-Square contributions are printed below expected counts

        unhappy  neutral  happy  Total
ACB          32       12     21     65
          20.17    24.46  20.38
          6.946    6.344  0.019

BCN          18       44     20     82
          25.44    30.85  25.71
          2.175    5.604  1.268

NCN          19       32     25     76
          23.58    28.59  23.83
          0.889    0.406  0.058

OFX          25       26     29     80
          24.82    30.10  25.08
          0.001    0.558  0.612

Total        94      114     95    303

Chi-Sq = 24.879, DF = 6, P-Value = 0.000
```

Answer: p-value (.000) < α (.01), reject H_0. There is a difference in feeling regarding state of the nation based on favorite news source.

The null hypothesis is that there is no association between feeling regarding state of the nation and favorite news source (or, feeling regarding state of the nation and favorite news source are independent). The alternative hypothesis is that there is an association between feeling regarding state of the nation and favorite news source (or, feeling regarding state of the nation and favorite news source are not independent). To say that there is an association between feeling regarding state of the nation and favorite news source is the same as saying that there is a difference in feeling regarding state of the nation based on favorite news source. A person with a particular favorite news source is more or less likely to have a certain view of the nation.

We do not know why that relationship exists or how it exists. Having a certain view of the nation could make a person choose a certain news source. Getting most of one's news from a certain source could make that person feel a certain way about the nation. Maybe some other factor (political party affiliation, religious affiliation, socioeconomic status, etc.) could make a person choose a certain news source and have a certain view of the nation without those two variables having a direct effect on each other. All we know is there is a difference in feeling regarding state of the nation based on favorite news source but exactly how this relationship works is for additional research to determine. By the way, the conclusion could be worded, there is a difference in favorite news source based on feeling regarding state of the nation, but that was not how the research questions was asked. If a person with a particular favorite news source is more or less likely to have a certain view of the nation, it can also be said that a person having a certain view of the nation is more or less likely to have a particular favorite news source. Either way, there is a difference in one variable based on the other, but we don't know why or how until we do further research.

As usual, everything in the Minitab output is clearly labeled.

Exercises

1) Air quality index readings were recorded for Chef County, Statsville, and Wudmoor on randomly selected days. The readings were categorized as follows: 30-34, 35-39, 40-44, and 45-49. The table below shows how many readings there were for each category by city:

	30-34	35-39	40-44	45-49
Chef County	6	26	6	10
Statsville	12	10	14	12
Wudmoor	14	8	12	14

Is there a difference in air quality based on city? Perform the appropriate test and answer the question.

2) Refer to the Jerrell Enterprises and Jaden Industries data from Exercise 2) in Section 7.2. The Professional Development Coordinator at Jaden Industries feels that participation in professional development helps employees feel better about their position in the company. See if there is a connection between how Jaden Industries employees feel about their position in the company (dissatisfied, satisfied, or neutral) and whether or not they participate in professional development. Use a .10 significance level.

3) Randomly selected first-year Statsville College students were asked their favorite time to study (morning, afternoon, or evening) and the area of their major (STEM, Humanities, Social Science, Other). Do students from particular majors prefer to study during certain times of the day over others? ($\alpha = .025$)

Tabulated statistics: major, study

Rows: major   Columns: study

	Afternoon	Evening	Morning	All
Human	10	28	13	51
Other	7	6	9	22
Sscience	31	14	12	57
STEM	12	7	41	60
All	60	55	75	190

Cell Contents: Count

Pearson Chi-Square = 52.807, DF = 6, P-Value = 0.000

4) What is the probability of getting sample results like we got in 3) if there is no relationship between area of major and when students prefer to study?

9.8 Advanced Hypothesis Testing: A Mixture

By now you are familiar with the mixture sections. This gives you an opportunity to see if you can distinguish among the correct method to use for particular situations, selecting from the body of topics covered in this chapter.

Remember that it is really important to consider the type and quantity of variables in the problem, in order to settle on the correct hypothesis test to perform. The different hypothesis tests we covered in this section all have unique variable type and quantity compositions. You might want to go back and review that information from the introductory page for each section.

Exercises

1) In the fall, selected students who graduated in the recent spring from five different high schools were assessed for their overall academic college readiness. Each student was determined to belong to one of three categories of overall academic college readiness: at college level, one level below college level, or two levels below college level. The research was for the purpose of seeing if there is a difference in overall academic college readiness based on the high school students attended. The test statistic for the appropriate hypothesis test was 16.391. With alpha set at .01, state your conclusion.

2) Moviegoers were asked their favorite soft drink and whether they prefer their popcorn with or without butter. The table below represents how many respondents stated each preference:

	With butter	w/o butter
Popsee	11	3
Diet Popsee	16	1
Lemon Crest	12	10
Orange Kick	7	12

a) At the .10 significance level, see if there is a relationship between the soft drink moviegoers prefer and how they like their popcorn.

b) A manager at a cinema supply company posits that 20% of moviegoers prefer Popsee, 25% prefer Diet Popsee, 30% prefer Lemon Crest, and 25% prefer Orange Kick. Perform the appropriate test to see if the manager appears to be correct.

3) The same busy gym owner from previous studies measured the strength to weight ratio (STWR—amount a person can lift divided by the person's weight) for several people. The STWR was measured for two exercises: squat and power snatch. The results are recorded below:

member	squat	power snatch
Christine	.9	.75
Todd	1.2	.9
Stacy A.	1.1	.81
Angelo	.8	.5
William	1.3	1.11
Armani	.91	.6
Kate	.7	.5
Ted	1	.75
Dave	.6	.49
Stacy M.	1.2	1

What is the best predicted power snatch STWR for someone with a squat STWR of .83?

4) Air quality readings were collected from Statsville, Wudmoor, and Chef County on randomly selected days. The data is shown below. See if there is a difference in air quality for these three areas.

Statsville	Wudmoor	Chef County
32	33	35
36	40	39
36	35	32
48	38	42
40	36	45
37	45	36
36	30	36

5) Residents from Chef County and residents from Twin Streams were asked if they rent or own their home (variable: **status**), how long they have lived at their current residence (variable: **residence**), if children live in the household (yes or no, variable: **children**), how long they have worked at their current job (variable: **job**). Renters were also categorized based on how long they have rented their current home as short term (less than a year), moderate term (from one to five years), and long term (over five years) using the variable **rent-type,** with responses coded as s, m, or l. Each respondent's city of residence was recorded in the **location** variable (c: Chef County, t: Twin Streams). **t-residence** and **t-job** are the variables for length of time at current residence and job for only Twin Streams residents. Participants were also asked their feelings about the job the local government is doing (approve, disapprove, undecided—variable: **feelLG**) and whether or not they plan to move within the next year (variable: **move**). Times were measured in years. Below is Minitab output created using this data.

Tabulated statistics: children, location

```
Rows: children   Columns: location

        c   t  All

n       7   7   14
y       7  10   17
All    14  17   31

Cell Contents:      Count


Pearson Chi-Square = 0.241, DF = 1, P-Value = 0.623
Likelihood Ratio Chi-Square = 0.241, DF = 1, P-Value = 0.623
```

Chi-Square Goodness-of-Fit Test for Categorical Variable: rent-type

```
                             Test              Contribution
Category  Observed  Proportion  Expected      to Chi-Sq
l                5         0.2       6.2       0.232258
m               12         0.3       9.3       0.783871
s               14         0.5      15.5       0.145161


 N  N*  DF   Chi-Sq  P-Value
31   0   2  1.16129    0.560
```

Chi-Square Goodness-of-Fit Test for Categorical Variable: rent-type

```
                             Test              Contribution
Category  Observed  Proportion  Expected      to Chi-Sq
l                5    0.333333   10.3333        2.75269
m               12    0.333333   10.3333        0.26882
s               14    0.333333   10.3333        1.30108


 N  N*  DF   Chi-Sq  P-Value
31   0   2  4.32258    0.115
```

One-way ANOVA: residence versus feelLG

```
Source  DF     SS    MS     F      P
feelLG   2   22.8  11.4  0.42  0.663
Error   33  901.6  27.3
Total   35  924.3

S = 5.227   R-Sq = 2.46%   R-Sq(adj) = 0.00%


                               Individual 95% CIs For Mean Based on
                               Pooled StDev
Level   N   Mean  StDev  ------+---------+---------+---------+---
a      12  8.508  5.740          (-----------*-----------)
d      12  6.567  4.458  (-----------*------------)
u      12  7.667  5.398      (------------*-----------)
                         ------+---------+---------+---------+---
                             5.0       7.5      10.0      12.5

Pooled StDev = 5.227


Grouping Information Using Tukey Method

feelLG   N   Mean  Grouping
a       12  8.508  A
u       12  7.667  A
d       12  6.567  A

Means that do not share a letter are significantly different.


Tukey 95% Simultaneous Confidence Intervals
All Pairwise Comparisons among Levels of feelLG

Individual confidence level = 98.04%
```

```
feelLG = a subtracted from:

feelLG   Lower  Center  Upper    -+---------+---------+---------+--------
d       -7.177  -1.942  3.294    (--------------*--------------)
u       -6.077  -0.842  4.394        (--------------*--------------)
                                 -+---------+---------+---------+--------
                               -7.0      -3.5       0.0       3.5


feelLG = d subtracted from:

feelLG   Lower  Center  Upper    -+---------+---------+---------+--------
u       -4.136   1.100  6.336              (--------------*--------------)
                                 -+---------+---------+---------+--------
                               -7.0      -3.5       0.0       3.5
```

General Linear Model: residence versus feelLG, move

```
Factor  Type   Levels  Values
feelLG  fixed       3  a, d, u
move    fixed       2  n, y


Analysis of Variance for residence1, using Adjusted SS for Tests

Source       DF  Seq SS  Adj SS  Adj MS     F      P
feelLG        2   22.75   22.75   11.38  0.41  0.667
move          1   14.57   14.57   14.57  0.53  0.474
feelLG*move   2   56.10   56.10   28.05  1.01  0.375
Error        30  830.90  830.90   27.70
Total        35  924.32


S = 5.26276   R-Sq = 10.11%   R-Sq(adj) = 0.00%


Unusual Observations for residence1

Obs  residence1      Fit  SE Fit  Residual  St Resid
  5     20.0000   8.2000  2.1485   11.8000      2.46 R
 25     21.3000  10.0667  2.1485   11.2333      2.34 R

R denotes an observation with a large standardized residual.


Grouping Information Using Tukey Method and 95.0% Confidence

feelLG   N   Mean  Grouping
a       12  8.508  A
u       12  7.667  A
d       12  6.567  A

Means that do not share a letter are significantly different.


Grouping Information Using Tukey Method and 95.0% Confidence

move   N   Mean  Grouping
y     18  8.217  A
n     18  6.944  A
```

```
Means that do not share a letter are significantly different.


Grouping Information Using Tukey Method and 95.0% Confidence

feelLG  move  N    Mean  Grouping
u       y     6  10.067  A
a       n     6   8.817  A
a       y     6   8.200  A
d       n     6   6.750  A
d       y     6   6.383  A
u       n     6   5.267  A

Means that do not share a letter are significantly different.
```

Descriptive Statistics: t-residence, t-job

```
Variable      N  N*   Mean  SE Mean  StDev  Minimum     Q1  Median      Q3
t-residence  34   0  4.935    0.829  4.833    0.200  0.775   4.000   7.175
t-job        34   0  7.871    0.850  4.958    1.900  4.300   6.600  12.300

Variable     Maximum
t-residence   21.300
t-job         20.000
```

Correlations: t-residence, t-job

```
Pearson correlation of t-residence and t-job = -0.011
P-Value = 0.953
```

Regression Analysis: t-job versus t-residence

```
The regression equation is
t-job = 7.92 - 0.011 t-residence


Predictor       Coef  SE Coef      T      P
Constant       7.924    1.244   6.37  0.000
t-residence  -0.0108   0.1813  -0.06  0.953


S = 5.03457   R-Sq = 0.0%   R-Sq(adj) = 0.0%


Analysis of Variance

Source          DF      SS     MS     F      P
Regression       1    0.09   0.09  0.00  0.953
Residual Error  32  811.10  25.35
Total           33  811.19


Unusual Observations

Obs  t-residence   t-job    Fit  SE Fit  Residual  St Resid
 19          8.9  20.000  7.828   1.124    12.172     2.48R
 22         16.3  18.200  7.748   2.234    10.452     2.32RX
 23         21.3   4.500  7.693   3.091    -3.193    -0.80 X
```

```
R denotes an observation with a large standardized residual.
X denotes an observation whose X value gives it large leverage.
```

a) Among renters in Chef County and Twin Streams, is there an equal proportion who are short, long, and moderate term?

b) For all Chef County and Twin Streams residents who approve the job the local government is doing and plan to move within the next year, what is the mean time they have lived in their current residence?

c) What is the best predicted years at the current job for a Twin Streams resident who has been at the current residence for five years?

d) If there is any output that you did not use to answer a) b) and c), consider what inferences you can draw from that output.

6) Volume and price data was collected from 13 new refrigerators. The mean volume of the refrigerators was 23.94 cubic feet with standard deviation 3.2665 cubic feet. The mean price of the refrigerators was $1285.92 with variance 586889.2916 dollars^2. Pearson's Correlation Coefficient for the sample was .3907 and the regression equation created using that data was: $\hat{y} = -907.9683 + 91.6412x$, where the predictor variable was volume. The volume of the smallest refrigerator in the sample was 18 cubic feet, with the largest volume being 28.15 cubic feet.

What is the best predicted price of a refrigerator that has a volume of 23.1 cubic feet?

7) In a sample of 15 consumers, 67.24% of the variation in consumer spending is explained by total wealth. Can we infer that there is a linear correlation between consumer spending and total wealth among all consumers? Explain.

9.9 Looking Ahead

Congratulations on making it this far! This text, and whatever course you are taking in conjunction with it, have a lot of material, most of which is brand new. As much as you think you have learned in this course, you have just barely become introduced to the world of Statistics. There is so much more to learn. You can earn a degree in Statistics, all the way to the doctoral level because there is so much great information to learn, so many applications of the procedures, and so many unique technical problems and challenges yet to be tackled. A degree in Statistics qualifies you for many jobs in teaching, the private sector, and government, as a person who has good quantitative skills and knows how to work with data is a very attractive job candidate.

What else is there to learn about Statistics? Throughout the text, you were given some indications of how increasing the complexity of the research situation would require statistical methods beyond the level of this text and the course you are taking. On Minitab output you will often see things that are unfamiliar now, but you can learn about the advanced methods to which they refer. Per the instructions, you assumed all the problems in the book met required conditions for the statistical methods being discussed. However, in practice, you should always test the conditions because if there are gross departures (and for tests dealing with variance or standard deviation the departures don't need to be that big) from them, your statistical results might be useless. How do you rigorously test all those conditions and what other methods should you employ if your data does not meet conditions? That's more stuff to learn! In a perfect world, when you get data from a sample, every member of the sample provides every piece of data you need. Often, this does not happen (people don't answer all the questions on a survey, data is taken in stages and someone is unavailable at a stage, people drop out in the middle of a study, etc.). What are the options for dealing with this missing data, and which specific one should you use in a specific research context? Suppose you want to measure something like "happiness" or "success". What are legitimate measures for those unobservable (called, *latent*) variables? There are lots of procedures (factor analysis, structural equation modeling, etc.) for questions like that. What if you have nested data, like children in classrooms, in schools, in districts, and you want to make inferences? Hierarchical linear modeling can be used for that. Some of these procedures can also make it easier to make more trustworthy causal claims, which we are unable to do based on the simple methods we considered. Suppose you want to assess changes over time for a particular phenomenon? Time series methods work for that. Like most introductory Statistics texts, this one presents frequentist (or classical) Statistics in which we get a sample and make an inference about the population based on it. There is another type of Statistics, Bayesian Statistics, in which prior information is worked into the inference process. And what about the theory behind all the distributions (normal, chi-square, etc.), how they are formed, the math used to show their relationships, and how to prove which is appropriate for the various statistical procedures? There are classes and texts for that. There are many other topics to learn that can't even be described here because a higher level of math and Statistics (probability as well) background is needed to even describe the topic. The point is that there is a lot to learn about both the theory and practice of Statistics! As you move on to upper division courses, you may be required to take classes about some of this material, but look for electives as well.

Besides learning about statistical methods, it is also important to learn more about how to obtain data and what kind to obtain. For example, in section 9.7 there was an example dealing with respondents' self reports of their eating habits and their weight. Depending on the goal of the research and the resources of the researchers, it might be more helpful to have researchers perform some objective observations of diet and weight. Self-report data is often flawed based on factors such as who is asking the questions (race, age, sex, etc. of the researcher), whether the answers are given orally or in writing, whether answers are anonymous are not, how questions are worded, order of questions, and so on. Issues like these are tackled in the later research classes that you may take. Quality of data is a huge concern in research. You can have a solid understanding of every statistical process in the world but, if you have poor data, you have nothing.

Be aware that different books use different notation. For example, $\bar{x}_d$ in this book is $\bar{d}$ in other books. Where this and many other books use p for the population proportion some use π (not to be confused with the constant 3.14...) because they strictly use Greek letters to symbolize population parameters. Once you discover the notation a book uses, just go with it.

You may not ever use Minitab again, but all statistical output is basically the same. The same in that it's all labeled. If you need a confidence interval or hypothesis test results, they will be labeled. As warned earlier in the text, some software gives the p-value as the significance (or sig.). When you use a certain statistical program regularly, you very quickly pick up its style and notation. While all output is easy to read, all software is not equally easy to use. Some software is of the menu driven point and click variety, but others require you to code your requests, like when using a computer programming language. Some software, like Minitab, allows both methods of entering instructions.

It is necessary to oversimplify word problems in a text because we don't have several weeks or more to complete each problem. Along with verifying conditions before using a statistical procedure, when doing actual research you should always start by looking at your data, plotting it, and seeing if anything looks "weird". On the most basic level, if you know you interviewed 77 people and you have 76 pieces of data, ummm...what's the deal?! Was there an error in data collection or do you already have an explanation for that? When you collect the data yourself and are familiar with the research area, you'll know something "weird" when you see it—values in the data that are not possible, values you know you did not observe, values representing an element in the sample that may require additional investigation, etc.. Address that before doing any analysis as that might help you find errors in the data or may help you consider research questions that weren't apparent to you before. It might even help you realize another piece of data you should collect to complete the research. None of this is done in an introductory class or text due to time constraints, but are necessary first steps in real research.

One more thing...

9.10 Inferential Statistics: A Mixture

This section contains a variety of problems from the confidence interval, early hypothesis testing, and later hypothesis testing material. This will really help you see how well you mastered the inferential statistics topics.

Recall that some questions can be answered using hypothesis tests or confidence intervals. Sometimes the results can be different depending on which approach you use (this is particularly true when dealing with proportions). In practice, hypothesis testing is preferred over confidence intervals (when there is a choice). Of course, for problems in the text or on an exam, follow whatever directions you are given.

Exercises

1) Ornithologists studied a sample of peregrine falcons, golden eagles, and gyrfalcons. They recorded maximum horizontal speed, diving speed, and airspeed for each bird. The mean and standard deviation for each sample are shown below. (all measurements in mph)

	Number of birds	Horizontal $(\bar{x}, s)$	Diving $(\bar{x}, s)$	Airspeed $(\bar{x}, s)$
peregrine	11	67.1, 1.26	200.8, 2.77	239.6, 3.95
golden	13	79.4, 3.47	150, 3.98	189.2, 3.05
gyrfalcons	8	87.2, 2.98	125.7, 2.06	130, 4.8

a) In the sample, it is easily observed that the peregrine falcons had a larger mean maximum airspeed than the golden eagles, but create a 90% confidence interval to see if we can make that inference about all such birds.

b) At the .05 significance level, test the claim that golden eagles and gyrfalcons have the same standard deviation in their maximum horizontal speeds.

c) Is the mean maximum diving speed of gyrfalcons less than 80 mph smaller than that of peregrine falcons? ($\alpha = .025$)

2) A mail order company claims that 95% of orders are received by its customers within two business days. In a sample of orders from this company, only 93.4% of the customers received their orders within two business days. The following confidence interval was made for the proportion of orders received by this company's customers within two business days:

$$.8913 < p < .9769$$

Respond to the company's claim.

3) A baseball manager wanted to test the claim that the average in-game life of a major league baseball is 6.3 pitches. The p-value for the appropriate test was .2273 and the critical value was 1.664. State your conclusion.

4) Two placement exams are being studied to see which one does a better job of placing students in appropriate classes. Hypothesis testing was used to determine if there is a difference in the proportion of students appropriately placed when using Exam 1 versus the proportion of students appropriately placed when using Exam 2. The test statistic was −2.635 and alpha was .08. State your conclusion.

5) Randomly selected teenagers were asked how they prefer to spend their free time: partying, reading, listening to music, or exercising. The number of responses for each activity are recorded below. Are these activities preferred equally among teenagers? Explain using the appropriate statistical procedure.

Party	Read	Music	Exercise
20	6	13	7

6) Air quality readings were collected from Statsville, Wudmoor, and Chef County on the same randomly selected days. The readings are in the table below. Perform the appropriate .01 significance level test to see if Wudmoor generally has better air quality than Statsville. (Note that lower readings indicate better air quality.)

Statsville	Wudmoor	Chef County
32	33	35
36	40	39
36	35	32
48	38	42
40	36	45
37	45	36
36	30	36

7) Residents from Chef County and residents from Twin Streams were asked if they rent or own their home (variable: **status**), how long they have lived at their current residence (variable: **residence**), if children live in the household (yes or no, variable: **children**), how long they have worked at their current job (variable: **job**). Renters were also categorized based on how long they have rented their current home as short term (less than a year), moderate term (from one to five years), and long term (over five years) using the variable **rent-type,** with responses coded as s, m, or l. Each respondent's city of residence was recorded in the **location** variable (c: Chef County, t: Twin Streams). **c-residence** and **c-job** are the variables for length of time at current residence and job for only Chef County residents. Participants were also asked their feelings about the job the local government is doing (approve, disapprove, undecided—variable: **feelLG**) and whether or not they plan to move within the next year (variable: **move**). All times were measured in years. The Minitab output below was generated using that data.

Tabulated statistics: children, location

```
Rows: children   Columns: location

         c   t  All

n        7   7   14
y        7  10   17
All     14  17   31

Cell Contents:      Count


Pearson Chi-Square = 0.241, DF = 1, P-Value = 0.623
Likelihood Ratio Chi-Square = 0.241, DF = 1, P-Value = 0.623
```

Test and CI for Two Variances: job vs location

```
Method

Null hypothesis         Sigma(c) / Sigma(t) = 0.5
Alternative hypothesis  Sigma(c) / Sigma(t) not = 0.5
Significance level      Alpha = 0.1


Statistics

location   N  StDev  Variance
c         33  3.150     9.925
t         34  4.958    24.582

Ratio of standard deviations = 0.635
Ratio of variances = 0.404


90% Confidence Intervals

                                  CI for
```

```
Distribution    CI for StDev      Variance
of Data            Ratio            Ratio
Normal         (0.474, 0.852)  (0.225, 0.726)
Continuous     (0.464, 0.897)  (0.215, 0.805)


Tests

                                                 Test
Method                          DF1  DF2  Statistic  P-Value
F Test (normal)                  32   33       1.62    0.176
Levene's Test (any continuous)    1   65       1.67    0.201
```

One-way ANOVA: residence versus feelLG

```
Source  DF     SS    MS     F      P
feelLG   2   22.8  11.4  0.42  0.663
Error   33  901.6  27.3
Total   35  924.3

S = 5.227   R-Sq = 2.46%   R-Sq(adj) = 0.00%


                               Individual 95% CIs For Mean Based on
                               Pooled StDev
Level   N   Mean  StDev  ------+---------+---------+---------+---
a      12  8.508  5.740         (-----------*-----------)
d      12  6.567  4.458  (-----------*------------)
u      12  7.667  5.398     (------------*-----------)
                         ------+---------+---------+---------+---
                             5.0       7.5      10.0      12.5

Pooled StDev = 5.227


Grouping Information Using Tukey Method

feelLG   N   Mean  Grouping
a       12  8.508  A
u       12  7.667  A
d       12  6.567  A

Means that do not share a letter are significantly different.


Tukey 95% Simultaneous Confidence Intervals
All Pairwise Comparisons among Levels of feelLG

Individual confidence level = 98.04%


feelLG = a subtracted from:

feelLG   Lower  Center  Upper   -+---------+---------+---------+--------
d       -7.177  -1.942  3.294   (--------------*--------------)
u       -6.077  -0.842  4.394       (--------------*--------------)
                                -+---------+---------+---------+--------
                              -7.0      -3.5       0.0       3.5


feelLG = d subtracted from:
```

```
feelLG   Lower  Center  Upper   -+---------+---------+---------+--------
u       -4.136   1.100  6.336            (--------------*--------------)
                                -+---------+---------+---------+--------
                              -7.0      -3.5       0.0       3.5
```

Test and CI for One Proportion: rent-type

```
Test of p = 0.8 vs p < 0.8

Event = s


                                   99% Upper    Exact
Variable     X   N  Sample p          Bound  P-Value
rent-type   14  31  0.451613       0.669808    0.000
```

One-Sample T: residence

```
Variable    N   Mean  StDev  SE Mean        95% CI
residence  67  5.263  4.800    0.586  (4.092, 6.433)
```

Paired T-Test and CI: c-residence, c-job

```
Paired T for c-residence - c-job

              N    Mean  StDev  SE Mean
c-residence  33   5.600  4.816    0.838
c-job        33   6.327  3.150    0.548
Difference   33  -0.727  5.301    0.923


96% CI for mean difference: (-2.703, 1.248)
T-Test of mean difference = 0 (vs not = 0): T-Value = -0.79  P-Value = 0.436
```

a) *Twin Streams Monthly* magazine stated that less than 80% of renters in Chef County and Twin Streams are short term renters. Test this claim using $\alpha = .005$.

b) Is a person more or less likely to have children in the household based on where the person lives (Chef County or Twin Streams)? (use alpha = .12)

c) Is it reasonable to conclude that Twin Streams and Chef County residents have been at their current residence for an average of 7.8 years?

8) For the following problems you will need to refer to the Minitab output in 7). The output will not have complete output to answer the research questions, but it contains numbers you will need to use as you complete the exercises by hand.

a) Use hypothesis testing to test the claim that the variance of the time Twin Streams residents have been at their current job is less than 40 years^2.

b) Make a 93.4% confidence interval for the difference in the proportion of Chef County residents who have children living in the household and the proportion of Twin Streams residents who have children living in the household. Make a logical inference based on that interval.

c) Suppose areas where the residents have been at their current residence for an average of 10 years are considered "stable areas". Use a 95% confidence interval to see if Chef County should be considered a stable area.

9) A consumer group is comparing insurance premiums for insurance Company I and insurance Company II for customers in Chef County, Statsville, Twin Streams, and Wudmoor. Is there a difference (on average) in insurance premium based on company and/or area? Fully explain and use a .01 significance level.

General Linear Model: premium versus area, company

```
Factor   Type   Levels  Values
area     fixed       4  ChefCounty, Statsville, TwinStreams, Wudmoor
company  fixed       2  I, II


Analysis of Variance for premium, using Adjusted SS for Tests

Source        DF  Seq SS  Adj SS  Adj MS         F      P
area           3    8378    8378    2793    893.68  0.000
company        1  160000  160000  160000  51200.00  0.000
area*company   3      13      13       4      1.33  0.330
Error          8      25      25       3
Total         15  168416


Grouping Information Using Tukey Method and 95.0% Confidence

area          N  Mean  Grouping
TwinStreams   4  1299  A
ChefCounty    4  1247    B
Wudmoor       4  1246    B
Statsville    4  1246    B

Means that do not share a letter are significantly different.


Grouping Information Using Tukey Method and 95.0% Confidence

company  N  Mean  Grouping
I        8  1359  A
II       8  1159    B

Means that do not share a letter are significantly different.


Grouping Information Using Tukey Method and 95.0% Confidence

area          company  N  Mean  Grouping
TwinStreams   I        2  1399  A
```

```
Wudmoor        I      2  1347    B
ChefCounty     I      2  1347    B
Statsville     I      2  1345    B
TwinStreams    II     2  1199      C
Statsville     II     2  1147        D
ChefCounty     II     2  1147        D
Wudmoor        II     2  1145        D

Means that do not share a letter are significantly different.
```

10) Randomly selected people were asked questions about their personal and professional lives. One question was about the distance (in miles) they travel to work, and another question asked the type of work they do. Suppose people who travel more than 50 miles to work (one way) are said to have a "stretch commute". The table below shows how many people of each job type were in the sample and how many of them had stretch commutes:

type of work	manufacturing/construction	professional/technical	sales/service	other
stretch commuters	118	219	157	67
total in sample	11675	21667	15532	6689

a) Suppose we want to test the claim that 21% of stretch commuters do manufacturing or construction work, 39% of stretch commuters do professional or technical work, 28% of stretch commuters do sales or service work, and the rest do some other kind of work. What numbers would we use from the table? How would those numbers be used?

b) Suppose we wanted to estimate the proportion of sales or service workers who have stretch commutes. What numbers would we use from the table? How would those numbers be used?

11) For each of the following situations, state the type of statistical procedure that would be performed to analyze the data in order to address the research question.

a) We get the age of a sample of people and the number of calories they consume each day. Is there a relationship between age and calorie consumption?

b) We categorize a sample of people as teen, young adult, middle adult, and older adult, and categorize their daily calorie consumption as low, moderate, and high. Is there a relationship between age and calorie consumption?

c) We categorize a sample of people as teen, young adult, middle adult, and older adult, and get the number of calories they consume each day. Is there a relationship between age and calorie consumption?

d) We categorize a sample of people as teen, young adult, middle adult, and older adult. We also categorize their daily activity level as light or heavy. We get the number of calories they consume each day. Is there a relationship among age, activity level, and calorie consumption?

e) We categorize a sample of people as teen, young adult, middle adult, and older adult. In our population, are 20% of people teens, 30% young adults, 30% middle adults, and 20% older adults?

f) We categorize a sample of people as teen, young adult, middle adult, and older adult, and get the number of calories they consume each day. Is the mean number of calories consumed by middle adults greater than 2000?

g) We categorize a sample of people as teen, young adult, middle adult, and older adult, and get the number of calories they consume each day. Is there greater variation in the calories consumed by teens than there is for older adults?

h) We categorize a sample of people as teen, young adult, middle adult, and older adult, and categorize their daily calorie consumption as low, moderate, and high. Estimate the proportion of young adults who have high calorie consumption.

i) We categorize each person in the whole population as teen, young adult, middle adult, and older adult, and get the number of calories each person in the whole population consumes each day. Is the mean calorie consumption for young adults different from the mean calorie consumption for older adults?

Answers to Exercises

Chapter 1

1) population: all perishable items at the market, parameter: 12%, sample: none, statistic: none 2) population: all perishable items at the market, parameter: unknown value (percentage of all perishable items having expiration dates that fall within the next two days), sample: the cart-load of perishable items collected, statistic: 12% 3) population: all minor league pitchers, parameter: unknown value (the average walks per game given up by all minor league pitchers), sample: the few minor league pitchers studied, statistic: 3.72 walks per game 4) population: all minor league pitchers, parameter: 3.72 walks per game, sample: none, statistic: none 5) quantitative, continuous 6) quantitative, discrete 7) quantitative, continuous (the average could have any number of decimal places) 8) categorical 9) quantitative, discrete, 10) quantitative, continuous 11) categorical

Chapter 2

1) 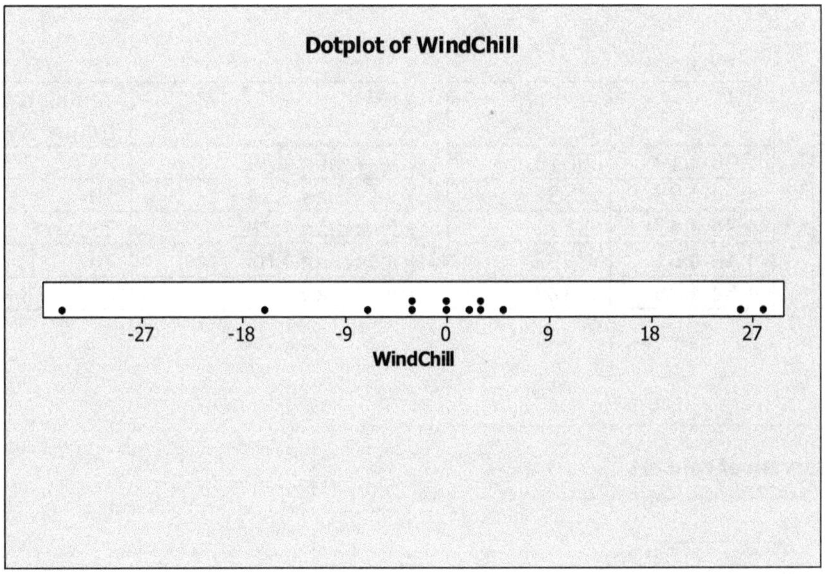

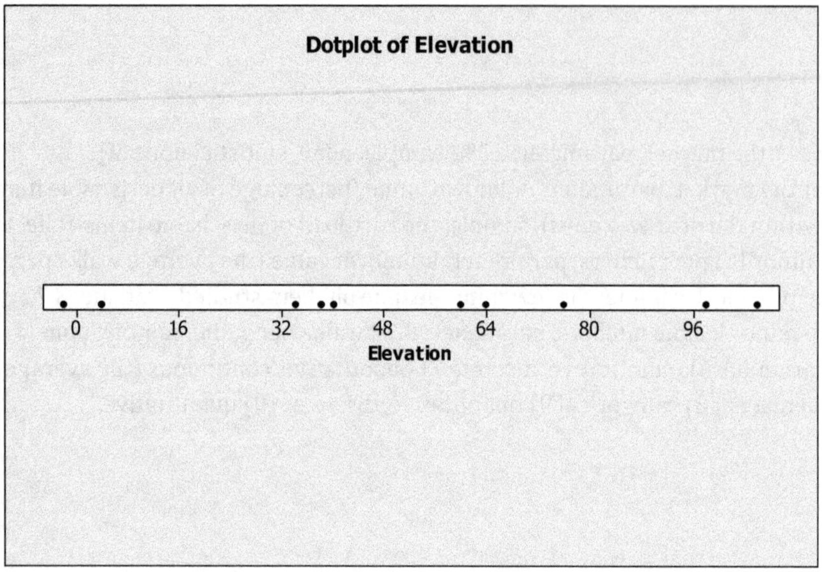

2)

3) a)

pH	frequency
4.06-4.19	2
4.20-4.33	8
4.34-4.47	11
4.48-4.61	9
4.62-4.75	1

b)

pH	relative frequency
4.06-4.19	.0645
4.20-4.33	.2581
4.34-4.47	.3548
4.48-4.61	.2903
4.62-4.75	.0323

c)

pH	cumulative frequency
less than 4.20	2
less than 4.34	10
less than 4.48	21
less than 4.62	30
less than 4.76	31

d)

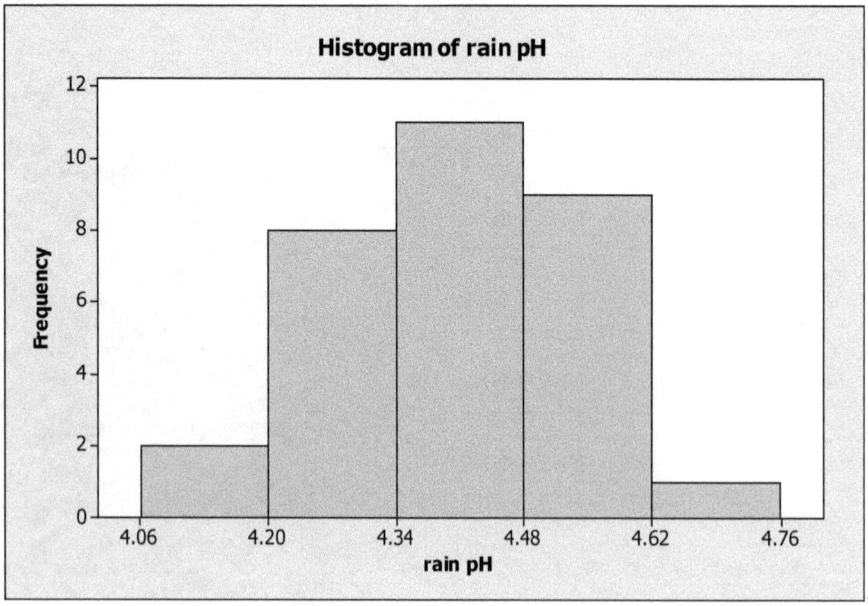

e)

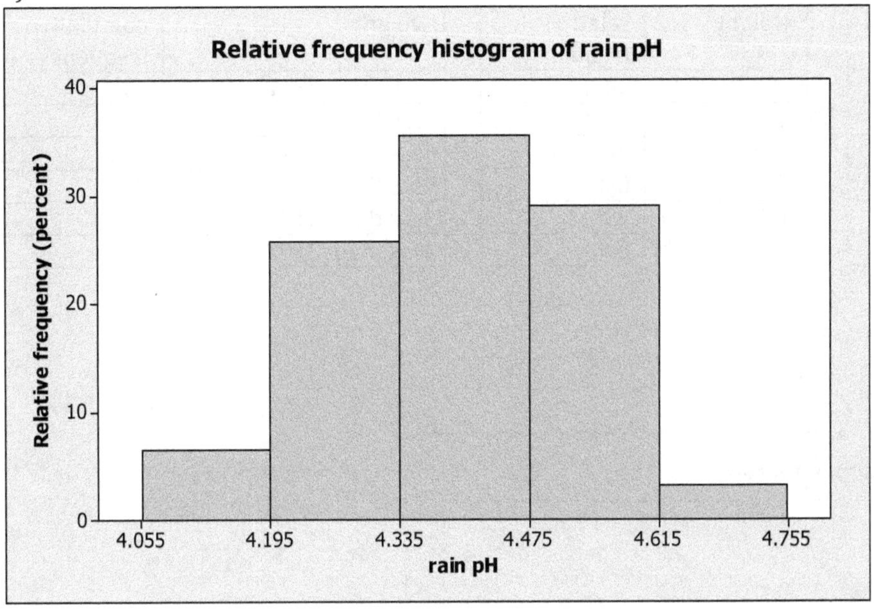

f)

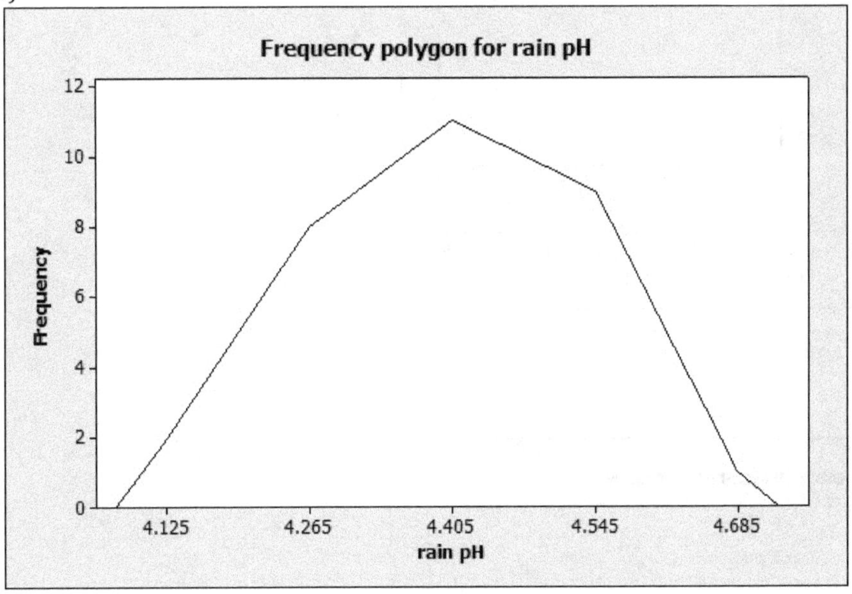

g) stem leaf

stem	leaf
4.0	7
4.1	7
4.2	0 0 6 7 8 9
4.3	0 1 4 5 6 8 9
4.4	0 2 2 4 5 6 9
4.5	0 1 2 3 4 5
4.6	0 0 2

h) Yes, the data is approximately normal. The frequencies start low, increase to a maximum frequency, then decrease to a low frequency. Also, the frequencies are approximately symmetric.

4) a)

weight	number of cans
.7760-.7831	11
.7832-.7903	3
.7904-.7975	7
.7976-.8047	0
.8048-.8119	2
.8120-.8191	2

b)

weight	relative frequency
.7760-.7831	44%
.7832-.7903	12%
.7904-.7975	28%
.7976-.8047	0%
.8048-.8119	8%
.8120-.8191	8%

c)

weight	cumulative frequency
less than .7832	11
less than .7904	14
less than .7976	21
less than .8048	21
less than .8120	23
less than .8192	25

d)

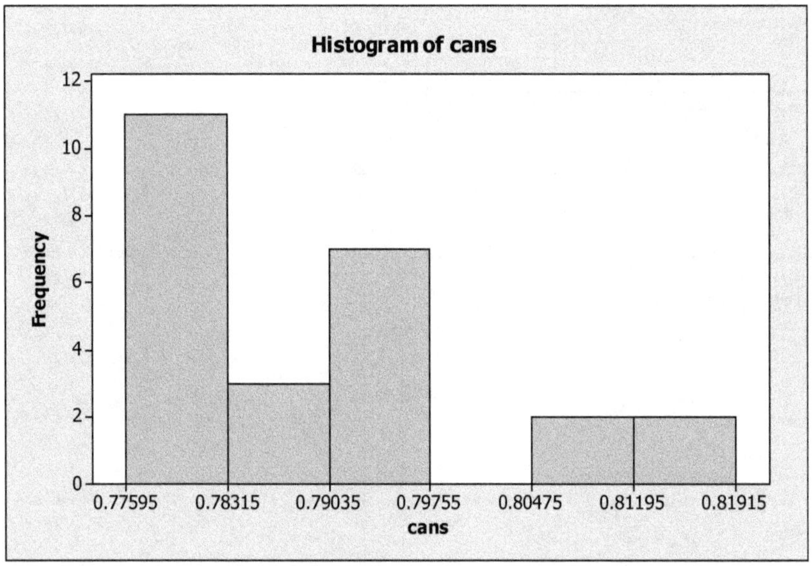

e)

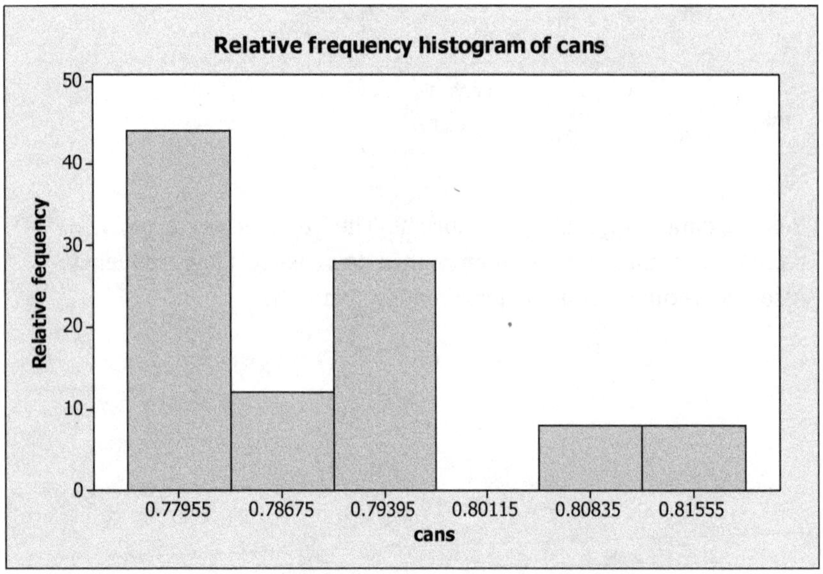

f)

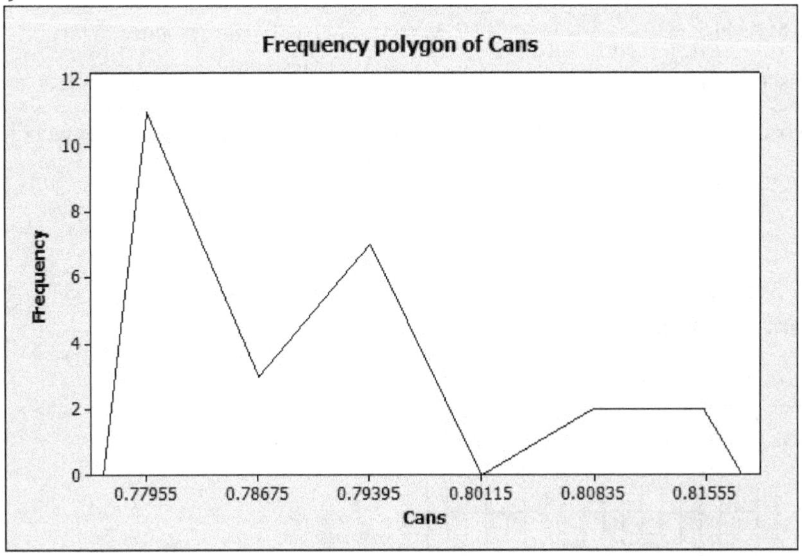

g) stem leaf

stem	leaf
.77	62 73 84 97
.78	03 04 14 17 19 23 27 41 55 87
.79	14 27 38 41 50 60 62
.80	77
.81	02 23 70

h) No, the data is not approximately normal. The frequencies do not start low, increase to a maximum frequency, then decrease to a low frequency. That's enough to make the data not approximately normal. (Also, the frequencies are not symmetric.)

5) 2 (134.5 would be the midpoint of class 130-139, and the dot has a height of 2)

Chapter 3

Section 3.1

1) mean = .3077 °F, median = 0 °F, mode = -3 °F, 0 °F, 3 °F (this is multimodal data). These would also be the measures of center if this were a population. 2) mean = 51.4 m, median = 50 m, no mode 3) mean = 102.8 m, median = 100 m, no mode 4) mean = 53.4 m, median = 52 m, no mode

5) The mean tensile strength for this selection of bolts is $\bar{x} = \frac{6654284.4}{312} = 21327.8346$ foot pounds.

tensile strength	number of bolts (f)	class midpoint (x)	fx
21300-21309.9	22	21304.95	468708.9
21310-21319.9	49	21314.95	1044432.55
21320-21329.9	116	21324.95	2473694.2
21330-21339.9	78	21334.95	1664126.1
21340-21349.9	36	21344.95	768418.2
21350-21359.9	11	21354.95	234904.45
total	312		6654284.4

Section 3.2

1) standard deviation = 15.7712 °F, variance = 248.7308 °F^2, range = 62 °F. If this were a population the measures of dispersion would be: standard deviation = 15.1525 °F, variance = 229.5976 °F^2, range = 62 °F. 2) standard deviation = 36.04380 m, variance = 1299.1556 m^2, range = 107 m 3) standard deviation = 72.08760 m, variance = 5196.6222 m^2, range = 214 m 4) standard deviation = 36.04380 m, variance = 1299.1556 m^2, range = 107m 5) standard deviation = 5.2165 pounds, variance = 27.2119 pounds^2, range = 13.41 pounds

Section 3.3

1) September: $z = \frac{124.2-101.6}{28.3} = .7986$, March: $z = \frac{129.4-83.6}{31.7} = 1.4448$, March score is better; z-score is larger. Neither exam score is unusual, both z-scores between -2 and 2. 2) Martin: $z = \frac{163.0-180}{51.2} = -.3320$, Gina: $z = \frac{134.2-125}{62.9} = .1463$, Gina's distance is more noteworthy because her z-score is higher (also her distance is above average while Martin's is below average)

Section 3.4

1) a) -34, -5, 0, 4, 28 b) cutoffs are -18.5 and 17.5 for mild outliers, -32 and 31 for extreme outliers; so -34 $^{\circ}$F is an extreme outlier, 26 $^{\circ}$F and 28 $^{\circ}$F are mild outliers

c)

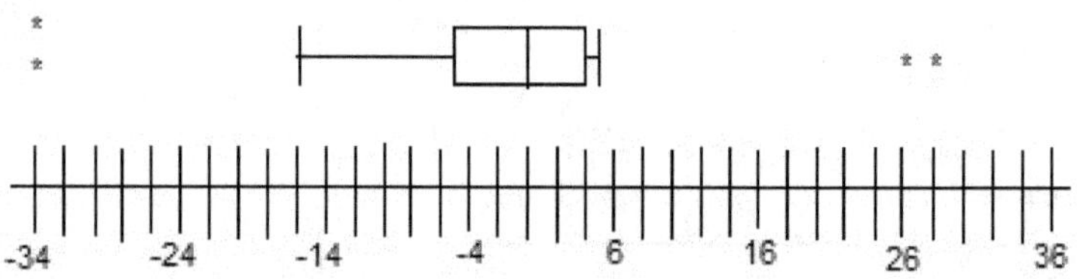

2) a) -2, 34, 50, 76, 105 b) cutoffs are -29 and 139 for mild outliers, -92 and 202 for extreme outliers; there are no outliers

c)

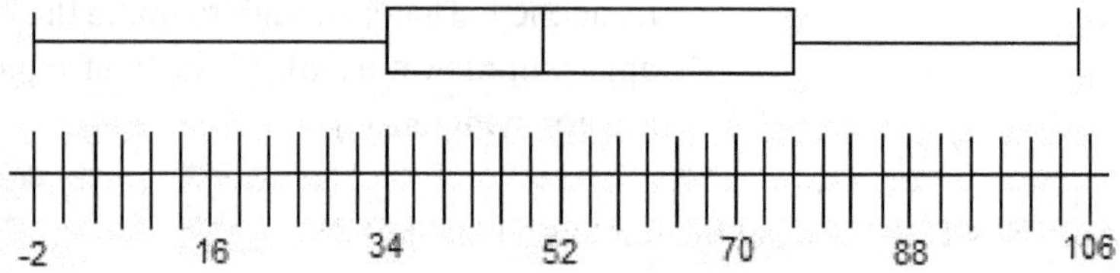

Section 3.5

1) a) mean = \$1508.90, median = \$1512.40, (mode not provided in output) b) standard deviation = \$48.40, variance = (\$48.40)^2 = 2342.56 squared dollars, range = \$210.10 c) biggest value is 2095.1. using z-score: $z = \frac{2095.1-2005.9}{37.2} = 2.3978$ which is outside the interval from -2 to 2 so \$2095.10 is an outlier (based on z-score method). using IQR: cutoffs for extreme outliers are 1807.4 and 2203.6, cutoffs for mild outliers are 1892.3 and 2118.7 so \$2095.10 is not an outlier (based on IQR method) d) 73 2) 113 minutes 3) "1.3 standard deviations below the mean" gives us z-score = -1.3. using z-score formula: $-1.3 = \frac{x-11.4}{4.92}$, solve for x to get answer: 5.004 years. (4.92 is $\sqrt{24.2064}$)

4) a) $\bar{x} = \frac{180.6}{28} = 6.45$ hours b)

hours of sleep	number of residents
3.4 – 4.2	1
4.3 – 5.1	7
5.2 – 6.0	5
6.1 – 6.9	4
7.0 – 7.8	2
7.9 – 8.7	4

c) (1.58 hours)^2 = 2.4964 hours^2

5) a) $\bar{x} = 4.6$ tons, s = .5850 ton, $z = \frac{5.1-4.6}{.5850} = .8547$, so 5.1 is .8547 standard deviation from (specifically above, since it's positive) the mean. b) 3.5, 4.3, 4.65, 5.1, 5.2

Chapter 5

Section 5.1

1) .9904 2) 78.23% 3) .2454 4) .0540 5) .67 6) −1.23 7) 1.555 8) 2.575 9) .0001 10) 2.22 11) .9968 12) .299 13) (The problem is asking for the percentage of z-scores less than −2.6 or more than 2.6.) answer: .94% 14) .8665 15) 2.33 16) −.81 17) −2.05 18) .8617 19) (Your sketch should have three regions with area .003 on the left, .994 in the middle, and .003 on the right.) answer: ± 2.75 20) .74 21) 1 (Every z-score is greater than −3.42 or less than 1.61 so the whole diagram is shaded.) 22) a) 68.26% b) 95.44% c) 99.74% 23) a) $\frac{2-1}{8}[f(1)+f(1.125)+f(1.25)+f(1.375)+f(1.5)+f(1.625)+f(1.75)+f(1.875)] = .1487$ b) $\frac{1}{2}\cdot\frac{2-1}{8}[f(1)+2f(1.125)+2f(1.25)+2f(1.375)+2f(1.5)+2f(1.625)+2f(1.75)+2f(1.875)+f(2)] = .1361$ c) $\frac{1}{3}\cdot\frac{2-1}{8}[f(1)+4f(1.125)+2f(1.25)+4f(1.375)+2f(1.5)+4f(1.625)+2f(1.75)+4f(1.875)+f(2)] = .1359$ d) .9772 − .8413 = .1359

Section 5.2

1) a) .7058 b) 79.94mg c) .0047 d) 79.661mg 2) a) 156.748 minutes b) 63.436 minutes c) .1977 d) 34.83% 3) a) $\frac{104-96}{8}[f(96)+f(97)+f(98)+f(99)+f(100)+f(101)+f(102)+f(103)] = .05379$ b) $\frac{1}{2}\cdot\frac{104-96}{8}[f(96)+2f(97)+2f(98)+2f(99)+2f(100)+2f(101)+2f(102)+2f(103)+f(104)] = .06079$ c) $\frac{1}{3}\cdot\frac{104-96}{8}[f(96)+4f(97)+2f(98)+4f(99)+2f(100)+4f(101)+2f(102)+4f(103)+f(104)] = .06060$ d) $z = \frac{104-116}{8} = -1.5, z = \frac{96-116}{8} = -2.5$, .0668 − .0062 = .0606

Section 5.3

1) a) .4207 b) .0154 c) 1.54% d) 79.9277 mg 2) a) .6038 b) .1016 3) a).4078 b) The sample size is small, and we are not told that the population is normally distributed. The Central Limit Theorem is not applicable. The tools we need for this problem are beyond the level of this text.

Chapter 7

Section 7.2

1) $E = 1.81\sqrt{\frac{\left(\frac{8}{17}\right)\left(\frac{9}{17}\right)}{17}} = .2191$; .2515 < p < .6897; With 93% confidence we can say that the proportion of all Javonn Jewelers diamonds that are color H is between .2515 and .6897. Interval contains .43 so the claim is plausible. 2) $E = 2.575\sqrt{\frac{\left(\frac{50}{157}\right)\left(\frac{107}{157}\right)}{157}} = .09574$; .2227 < p < .4142; With 99% confidence we can say that the proportion of all Jaden Industries employees who are satisfied with their current position in the company is between .2227 and .4142. Whole interval is less than .5 so we can infer that less than half of Jaden Industries employees are satisfied with their current position in the company. 3) $n = \frac{2.17^2(.25)}{.04^2} = 735.7656$, answer: 736 air fresheners 4) $n = \frac{1.555^2(.17)(.83)}{.05^2} = 136.4733$, answer: 137 students 5) With 95% confidence we can say that the proportion of all Drug A users who experience nausea is between .146301 and .403448. Since the whole interval is less than .5, the researchers' belief appears to be correct. 6a) That proportion is just for the SAMPLE and cannot by itself be used to make a conclusion about the population. b) Same error as in a). c) The sample proportion will always be in the interval for the population proportion. That does not give any information for making a conclusion. d) The whole purpose of confidence intervals is to give information (an estimate) for the whole population based on sample results. e) The confidence interval is for a proportion, not an amount or quantity. f) The question and output are about nausea (Event = n) not fatigue.

Section 7.3

1) $E = 1.761\frac{1.4219}{\sqrt{15}} = .6465$; 6.2335 < μ < 7.5265; With 90% confidence we can say that the mean size of all Sparkelz Gems diamonds is between 6.2335 and 7.5265 mm. The whole interval is not below 7 so it does not appear that the current Sparkelz' diamonds have a smaller mean size than last year. 2) $E = 1.984\frac{31}{\sqrt{142}} = 5.1613$; 152.1387 < μ < 162.4613; With 95% confidence we can say that the mean of all supervisor ratings at Jerrell Enterprises is between 152.1387 and 162.4613. Whole interval is below 170 so the assertion appears to be incorrect. 3) With 99% confidence we can say that the average price of all items at the north store is between $1496.21 and $1521.68. Since $1500 is in the interval, the north store appears to be a high end store. 4) a) 99% is the confidence level used to create the interval and has nothing to do with the percentage of prices in the store in a certain interval. b) $1508.94 is just the mean price of items in the sample and is useless for making a statement about the whole population (all prices at the store). c) The sample mean is always found in the confidence interval for the population mean, so that information is not helpful for making an inference. d) The confidence interval is for the average (or mean) price of items, not for each individual price. e) Use either mean OR average, not both. f) The confidence interval is estimating the population value, NEVER the sample value. 5) $\frac{1521.68-1496.21}{2} = 12.735$ dollars 6) Any $\hat{y}$ in the middle (1 – α) proportion of $\hat{y}$ values is such that $y_p - t_{\frac{\alpha}{2}} s_{\hat{y}} < \hat{y} < y_p + t_{\frac{\alpha}{2}} s_{\hat{y}}$ Using the algebra described in this section and in 7.2, you have several algebra steps concluding with: $\hat{y} - t_{\frac{\alpha}{2}} s_{\hat{y}} < y_p < \hat{y} + t_{\frac{\alpha}{2}} s_{\hat{y}}$

Section 7.4

1) $\frac{(17-1).6708^2}{32} < \sigma^2 < \frac{(17-1).6708^2}{5.812}$; $.2250 < \sigma^2 < 1.2387$; With 98% confidence we can say that the variance of the weights of all Javonn Jewelers' diamonds is between .2250 and 1.2387 carats². OR $.4743 < \sigma < 1.1130$; With 98% confidence we can say that the standard deviation of the weights of all Javonn Jewelers' diamonds is between .4743 and 1.1130 carats. (The answer could be either interval since the problem did not specify, and both estimate how inconsistent, or varied, the population is.) 2) $\frac{(157-1)28^2}{179} < \sigma^2 < \frac{(157-1)28^2}{133.8}$; 26.1393 < σ < 30.2338; With 80% confidence we can say that the standard deviation of all peer performance ratings at Jaden Industries is between 26.1393 and 30.2338. Since the whole interval is greater than 23.5, the peer performance ratings at Jaden Industries are more dispersed than at the other company. (Again, we could have used the interval for the variance and compared it to 23.5².) 3) $\frac{(20-1)1513.74}{30.144} < \sigma^2 < \frac{(20-1)1513.74}{10.117}$; $954.1222 < \sigma^2 < 2842.8447$. With 90% confidence we can say that the variance of the amount of time all Tanayatown residents spend acquiring news is between 954.1222 and 2842.8447 minutes². 4) With 97% confidence we can say that the standard deviation of the prices of items at the south store is between 41.3 and 58.3 dollars. Since the whole interval is greater than $32, it appears that the south store has prices with a larger standard deviation than reported by the furniture magazine.

Section 7.5

1) $E = 1.88\sqrt{\frac{\left(\frac{5}{15}\right)\left(\frac{10}{15}\right)}{15} + \frac{\left(\frac{8}{17}\right)\left(\frac{9}{17}\right)}{17}} = .3227$; $-.4600 < p_1 - p_2 < .1854$; With 94% confidence we can say that the difference in the percentage of all Sparkelz Gems' diamonds that are VS1 and the percentage of all Javonn Jewelers' diamonds that are VS1 is between –46.00% and 18.54%. Since the interval contains zero, neither source has a higher percentage of VS1 diamonds (there is no difference in the percentages). 2) $E = 1.6\sqrt{\frac{\left(\frac{15}{32}\right)\left(\frac{17}{32}\right)}{32} + \frac{\left(\frac{25}{49}\right)\left(\frac{24}{49}\right)}{49}} = .1816$; $-.2231 < p_1 - p_2 < .1401$; With 89% confidence we can say that the difference in the proportion of all neutral employees who participate in professional development at Jerrell Enterprises and the proportion of all neutral employees who participate in professional development at Jaden Industries is between –.2231 and .1401. Since the interval contains zero, there is no difference in the proportion of

neutral employees who participate in professional development at Jerrell Enterprises compared to Jaden Industries. 3) With 92% confidence we can say that the difference in the proportion of all Drug A users who experience nausea and the proportion of all Drug B users who experience nausea is between −.408674 and −.0713263. Since the whole interval is negative, the proportion of all Drug B users who experience nausea is larger than the proportion of all Drug A users who experience nausea. 4) $\frac{-.0713263-(-.408674)}{2} = .1687$ 5) Increase the sample of Drug A users and the sample of Drug B users. In other words, have more people try Drug A and more people try Drug B.

Section 7.6

1) $E = 2.977\sqrt{\frac{.8009^2}{15}+\frac{.6708^2}{17}} = .7833$; $-1.0831 < \mu_1 - \mu_2 < .4835$; With 99% confidence we can say that the difference in the mean weight of all Sparkelz diamonds and the mean weight of all Javonn diamonds is between −1.0831 and .4835 carats. Since the interval contains zero, it is not true that Sparkelz diamonds typically weigh less than Javonn diamonds. (There is no difference in mean weight.) 2) $E = 2.364\sqrt{\frac{31^2}{142}+\frac{42^2}{157}} = 10.03051$; $-28.8305 < \mu_1 - \mu_2 < -8.7695$; With 98% confidence we can say that the difference in the mean of all supervisor ratings at Jerrell Enterprises and the mean of all supervisor ratings at Jaden Industries is between −28.8305 and −8.7695. Since the whole interval is negative, Jaden Industries has the larger mean supervisor rating. 3) a. We have no idea! The interval estimates the difference in the population means, NOT the individual population means. b. Since the whole interval is negative, Wudmoor (the second population) had the larger mean tax return amount. The interval tells us that Wudmoor's mean was larger by somewhere between $203.97 and $984.35. 4) With 95% confidence we can say that the difference in the mean price of north store products and south store products is between −19.89 and 8.57. Since the interval contains zero, there is no difference in the prices of products from the north and south stores, on average.

Section 7.7

1) $E = 1.895\frac{.1115}{\sqrt{8}} = .07470$; $-.1460 < \mu_d < .00345$; With 90% confidence we can say that the mean of the differences of prices for all identical items at Statsville Market and Statsville Grocers is between −$0.1460 and $0.00345. (you could round to two decimal places) Since the interval contains zero, there is no difference (on average) between the price of items at one store versus the price of the same items at the other store. 2) $E = 1.290\frac{51.8}{\sqrt{142}} = 5.6076$; $5.4924 < \mu_d < 16.7076$; With 80% confidence we can say that the mean of the differences in supervisor performance ratings and peer performance ratings for all Jerrell Enterprises employees is between 5.4924 and 16.7076. Since the whole interval is positive, supervisor performance ratings are generally larger than peer performance ratings. The employee's conjecture is apparently incorrect. 3)a) We can say that the mean of the differences in incomes of all Tanayatown siblings who do not have a college degree and those who have a college degree is between −11925 and −5961 dollars. (No confidence level was given so our interpretation does not include that part.) Since the whole interval is negative, siblings with a college degree tend to have larger incomes. b) The relevant sample mean would be $\bar{x}_d$, the mean of the "sibling with no degree minus sibling with degree" income differences for the sample of matched pairs. $\bar{x}_d = \frac{-11925+(-5961)}{2} = -8943$ dollars. 4) a) The interval is for a mean of differences (μ_d), not a difference of means ($\mu_1 - \mu_2$). b) The groups are backwards: no degree is first and with degree is second. c) The interval is for a mean of differences. It does not give any information about the population mean for one population. d) Again the interval is for a mean of *differences*. e) Confidence intervals never estimate a sample value. Never. f) ???? Any statement you write needs to make sense. 5) With 93% confidence we can say that the mean of the differences for all prices at the north store and prices at the east store is between

−508.97 and −484.94 dollars. Since the whole interval is negative, the east store tends to sell more expensive items.

Section 7.8

1) $\frac{1.4219^2/1.1221^2}{2.7875} < \frac{\sigma_1^2}{\sigma_2^2} < \frac{1.4219^2/1.1221^2}{.3391}$; $.5760 < \frac{\sigma_1^2}{\sigma_2^2} < 4.7353$; With 95% confidence we can say that the ratio of the variance of the sizes of all Sparkelz diamonds and the variance of the sizes of all Javonn diamonds is between .5760 and 4.7353. Interval contains 1 so there is not less spread in the sizes for Sparkelz diamonds compared to Javonn diamonds. (no difference in the variance) 2) $\frac{46^2/28^2}{1.380} < \frac{\sigma_1^2}{\sigma_2^2} < \frac{46^2/28^2}{.7225}$; $1.9558 < \frac{\sigma_1^2}{\sigma_2^2} < 3.7356$; With 95% confidence we can say that the ratio of the variance of all peer ratings at Jerrell Enterprises and the variance of all peer ratings at Jaden Industries is between 1.9558 and 3.7356. Whole interval is bigger than 1 so peer ratings at Jerrell Enterprises have more variability than the peer ratings at Jaden Industries. 3) With 90% confidence we can say that the ratio of the standard deviation of the prices of all items at the west store and the standard deviation of the prices of all items at the south store is between 1.354 and 1.978. Since the whole interval is greater than one, the west store has more inconsistent pricing among its items. [Note: CI for Variance could also be used to come to this same conclusion.] 4) About 270. [from .90 times 300]

Section 7.9

1) a) (using the "to" and "from" data from Chef County for μ_d interval) $E = 2.179\frac{13.2530}{\sqrt{13}} = 8.009396$; $-16.8556 < \mu_d < -.8368$; With 95% confidence we can say that the mean of the differences in typical commute times to school and from school for all Chef County College students is between −16.8556 and −.8368 minutes. Since the whole interval is negative, students generally have a longer commute from school than to school. b) $E = 1.81\sqrt{\frac{(.579)(.421)}{19} + \frac{\left(\frac{5}{13}\right)\left(\frac{8}{13}\right)}{13}} = .3189$; $-.1245 < p_1 - p_2 < .5133$; With 93% confidence we can say that the difference in the percentage of all Twin Streams College students who generally walk to and from school and the percentage of all Chef County College students who generally walk to and from school is between −12.45% and 51.33%. Since the interval contains zero, there is no difference in the percentage of Twin Streams College students who generally walk to and from school and the percentage of Chef County College students who generally walk to and from school. c) $E = 2.05\sqrt{\frac{\left(\frac{6}{13}\right)\left(\frac{7}{13}\right)}{13}} = .2834$; $.1781 < p < .7449$; With 96% confidence we can say that the percentage of all Chef County College students who are first generation college students is between 17.81% and 74.49%. Since 47% is in the interval, that claim is reasonable. d) $\frac{(19-1)20.1^2}{28.869} < \sigma^2 < \frac{(19-1)20.1^2}{9.390}$; $251.9027 < \sigma^2 < 774.4601$; With 90% confidence we can say that the variance in the typical grooming times of all Twin Streams College students is between 251.9027 and 774.4601 minutes^2. Since the whole interval is less than 1043.29 minutes^2, the variance in the typical grooming times of Turner College students, there is less dispersion in grooming times among Twin Streams College students compared to Turner. This contradicts the claim. [We squared 32.3 to get 1043.29. Of course, we could have used the confidence interval for standard deviation, (15.8714, 27.8291) and compared that to 32.3.] e) $\frac{32.3^2/20.1^2}{2.3421} < \frac{\sigma_1^2}{\sigma_2^2} < \frac{32.3^2/20.1^2}{.3931}$ [.3931 came from $\frac{1}{2.5436}$], $1.1026 < \frac{\sigma_1^2}{\sigma_2^2} < 6.5692$, With 90% confidence we can say that the ratio of the variance of typical grooming times of all Turner College students and the variance of typical grooming times of all Twin Streams College students is between 1.1026 and 6.5692. Since the whole interval is greater than one, there is more dispersion in grooming times among Turner College students compared to Twin Streams College students. This contradicts the claim. f) Using the result from part a), with 95% confidence we can say that number would be between −16.8556 and −.8368. 2) We can say that the ratio of the standard deviation of the hours worked by all Statsville residents last week

and the standard deviation of the hours worked by all Tanayatown residents last week is between .4987 and .8402. [We have no confidence level so we omit that part of the interpretation.] Since the whole interval is less than one, this supports the claim that Tanayatown residents have greater variability in the amount of time they spend working (assuming last week was a typical week). 3) We can say that the mean amount of time all Americans watch TV is between 18 and 22 hours per week. [We have no confidence level so we omit that part of the interpretation.] Since the interval contains 19.7, the journal's claim is supported. 4) The interval in 2) must be positive because it's estimating a ratio of two positive numbers (standard deviations). That fact tells us nothing about which standard deviation is bigger. The interval in 3) must be positive because it's estimating an average amount of time. That fact tells us nothing about the journal's claim. 5) a) We can say that the difference in the proportion of all Chef County residents who would continue working if they suddenly got enough money to live as comfortably as they would like for the rest of their lives and the proportion of all Wudmoor residents who would continue working if they suddenly got enough money to live as comfortably as they would like for the rest of their lives is between .1568 and .2128. [We have no confidence level so we omit that from the interpretation.] Since the whole interval is positive, Chef County has a larger proportion than Wudmoor of residents who would continue working if they suddenly got enough money to live as comfortably as they would like for the rest of their lives. b) margin of error is $\frac{.2128-.1568}{2} = .028$ c) $\hat{p}_1 - \hat{p}_2$ is in the middle of the interval so that equals $\frac{.2128+.1568}{2} = .1848$ d) We have no idea! The interval only estimates the difference of the population proportions, not the individual population proportions. 6) $E = 2.575\sqrt{\frac{\left(\frac{14}{72}\right)\left(\frac{58}{72}\right)}{72}} = .1201$; $.07434 < p < .3145$; With 99% confidence we can say that the proportion of all moviegoers who prefer Popsee is between .07434 and .3145. 7) $E = 1.771\frac{\sqrt{600.3}}{\sqrt{14}} = 11.5968$; $134.0032 < \mu < 157.1968$; With 90% confidence we can say that the mean amount of time moviegoers who prefer Popsee stay at the movie theater on a single visit is between 134.0032 and 157.1968 minutes. 8) $n = \frac{1.51^2(.25)}{.07^2} = 116.3316$, so the answer is 117 moviegoers 9) a) With 99% confidence we can say that the difference in the mean amount of time all Chef County residents have been at their current residence and the mean amount of time all Twin Streams residents have been at their current residence is between −2.47 and 3.79 years. Since the interval contains zero, there is no difference in the mean amount of time Chef County residents have been at their current residence compared to Twin Streams residents. b) With 98% confidence we can say that the proportion of all residents in Chef County and Twin Streams who rent their homes is between .3195 and .6104. Since the whole interval is less than .84, it appears that the proportion of renters is less than that claimed by the local paper. c) 95%

Chapter 8

Section 8.2

1) a) H_0: p = .3, H_1: p ≠ .3 b) The proportion of all Americans who lived in a large city when they were 16 years old is not equal to .3. c) The percentage of Americans who lived in a large city when they were 16 years old is different from 30%. d) The percentage of Americans who lived in a large city when they were 16 years old is different from 30%. e) We did not find that the percentage of Americans lived in a large city when they were 16 years old is different from 30%. 2) a) H_0: p = .4, H_1: p < .4 b) The proportion of all Americans who lived with their own father and a stepmother on their 16th birthday is less than .4. c) The proportion of Americans who lived with their own father and a stepmother on their 16th birthday is less than .4. d) The proportion of Americans who lived with their own father and a stepmother on their 16th birthday is less than .4. e) We did not find that the proportion of Americans who lived with their own father and a stepmother on their 16th birthday is less than .4. 3) H_0: p = .71, H_1: p ≠ .71 (direct translation of H_1: the proportion of all Sparkelz Gems diamonds that are VS1 or VVS1 is not equal to .71; using words from the problem: the percentage of Sparkelz

Gems diamonds that are VS1 or VVS1 is not 71%) , $z = \frac{\frac{9}{15} - .71}{\sqrt{\frac{(.71)(.29)}{15}}} = -.9389$, critical values ±2.05, p-value = .3472, fail to reject H_0, We did not find that the proportion of Sparkelz Gems' diamonds that are VS1 or VVS1 is different from .71. The claim that 71% of Sparkelz Gems' diamonds are VS1 or VVS1 is plausible. 4) ("8 out of 10" is .8) H_0: p = .8, H_1: p < .8 (direct translation of H_1: the proportion of all satisfied employees who participate in professional development at Jerrell Enterprises is less than .8; using words from the problem: the proportion of satisfied employees who participate in professional development is smaller at Jerrell Enterprises than what is reported by *ProfDev.*), $z = \frac{\frac{34}{46} - .8}{\sqrt{\frac{(.8)(.2)}{46}}} = -1.03209$, critical value −1.88, p-value = .1515, fail to reject H_0, We did not find that the proportion of satisfied employees who participate in professional development is smaller at Jerrell Enterprises than what is reported by *ProfDev.* 5) p-value (.000) < α (.025) so reject H_0. Less than half of all Drug A users experience nausea. 6) .000 (the p-value). .025 (the significance level, α).

Section 8.3

1) a) H_0: μ = 140, H_1: μ > 140 b) The mean number of people who work at the locations where all employed Tanayatown residents work is greater than 140. c) Over 140 people, on average, work at the locations where employed Tanayatown residents work. d) Over 140 people, on average, work at the locations where employed Tanayatown residents work. e) We did not find that over 140 people, on average, work at the locations where employed Tanayatown residents work. 2) a) H_0: μ = 10, H_1: μ ≠ 10 b) The mean weight of all filled shopping bags is not equal to 10 pounds. c) The average weight of filled shopping bags is different from 10 pounds. d) The average weight of filled shopping bags is different from 10 pounds. e) We did not find that the average weight of filled shopping bags is different from 10 pounds. 3) H_0: μ = 8, H_1: μ < 8 (direct translation of H_1: the mean size of all diamonds at Javonn Jewelers is less than 8 mm; using words from the problem: Javonn Jewelers is a mid-size supplier), $t = \frac{7.4824-8}{\frac{1.1221}{\sqrt{17}}} = -1.9019$, critical value −1.337, .025 < p-value < .05, reject H_0. Javonn Jewelers is a mid-size supplier. 4) H_0: μ = 160, H_1: μ > 160 (direct translation of H_1: the mean of all supervisor ratings at Jaden Industries is greater than 160; using words from the problem: Jaden Industries is employee friendly), $t = \frac{176.1-160}{\frac{42}{\sqrt{157}}} = 4.8032$, critical value 2.345, p-value < .005, reject H_0. Jaden Industries is employee friendly. 5) p-value (.068) > α (.01), fail to reject H_0. We did not find that the mean price of items at the north furniture store is different from $1500. 6) a) Type II b) The probability of getting a sample of prices like we got from the north store, if the mean price of all items at the north store is $1500, is .068.

Section 8.4

1) a) H_0: σ = 22.7, H_1: σ < 22.7 b) The standard deviation of the number of hours per week worked by all married people's spouses is less than 22.7 hours. c) There has been a reduction in the standard deviation of the hours per week typically worked by spouses. d) There has been a reduction in the standard deviation of the hours per week typically worked by spouses. e) We did not find that there has been a reduction in the standard deviation of the hours per week typically worked by spouses. 2) a) H_0: σ^2 = 27.9, H_1: σ^2 ≠ 27.9 b) The variance of the number of days all people usually work overtime is not equal to 27.9 days^2. c) There has been a change in the dispersion of the number of days per month people usually work overtime. d) There has been a change in the dispersion of the number of days per month people usually work overtime. e) We did not find that there has been a change in the dispersion of the number of days per month people usually work overtime. 3) H_0: σ = .58, H_1: σ > .58 (direct translation of H_1: the standard deviation of the weight of all Sparkelz diamonds is greater than .58; using words from the problem: the diamonds have more variation in their weights), $\chi^2 = \frac{(15-1).8009^2}{.58^2} = 26.6949$, critical value 23.685, .01 < p-value < .025, reject H_0. [α is .05 since α was not given.] The diamonds have more variation in their weights. 4) H_0: σ^2 = 2218.41, H_1: σ^2 ≠ 2218.41 (direct translation of H_1: the variance of all peer ratings at Jerrell Enterprises is not equal to 2218.41; using

words from the problem: there has been a change in the variance of peer ratings), $\chi^2 = \frac{(142-1)46^2}{2218.41} =$ 134.4909, critical values 67.328, 140.169, fail to reject H_0. We did not find that there has been a change in the variance of peer ratings. 5) p-value (.000) < α (.03), reject H_0. The standard deviation of all prices from the south store is different from $32. 6) Yes because we rejected H_0. 7) Type I

Section 8.5

1) a) H_0: $p_1 = p_2$, H_1: $p_1 \neq p_2$ b) The proportion of all Wudmoor residents who believe the attorney's client is guilty is not equal to the proportion of all Chef County residents who believe the attorney's client is guilty. c) Among all Wudmoor and Chef County residents there is a difference in the proportion who believe the attorney's client is guilty. d) Among all Wudmoor and Chef County residents there is a difference in the proportion who believe the attorney's client is guilty. e) We did not find that, among all Wudmoor and Chef County residents, there is a difference in the proportion who believe the attorney's client is guilty. 2) a) H_0: $p_1 = p_2$, H_1: $p_1 < p_2$ b) The proportion of all Deli 1 customers who prefer turkey and Swiss is less than the proportion of all Deli 2 customers who prefer turkey and Swiss. c) A smaller proportion of Deli 1 customers prefer turkey and Swiss sandwiches compared to Deli 2 customers. d) A smaller proportion of Deli 1 customers prefer turkey and Swiss sandwiches compared to Deli 2 customers. e) We did not find that a smaller proportion of Deli 1 customers prefer turkey and Swiss sandwiches compared to Deli 2 customers. 3) H_0: $p_1 = p_2$, H_1: $p_1 < p_2$ (direct translation of H_1: the proportion of all Sparkelz Gems diamonds that are color H is less than the proportion of all Javonn Jewelers diamonds that are color H; using words from the problem: a smaller percentage of Sparkelz Gems diamonds are color H compared to Javonn Jewelers diamonds),

$\bar{p} = \frac{5+8}{15+17} = .4063$, $z = \frac{\left(\frac{5}{15}-\frac{8}{17}\right)-0}{\sqrt{\frac{(.4063)(.5937)}{15}+\frac{(.4063)(.5937)}{17}}} = -.7889$, critical value −2.575, p-value = .2148, fail to reject H_0.

We did not find that a smaller percentage of Sparkelz Gems diamonds are color H compared to Javonn Jewelers diamonds. 4) H_0: $p_1 = p_2$, H_1: $p_1 > p_2$ (direct translation of H_1: the proportion of all Jerrell Enterprises employees who are dissatisfied with their current position at the company is greater than the proportion of all Jaden Industries employees who are dissatisfied with their current position at the company; using words from the problem: there is a higher proportion of Jerrell Enterprises employees dissatisfied with their current position at the company than there is at Jaden Industries), $\bar{p} = \frac{64+58}{142+157} = .4080$, $z = \frac{\left(\frac{64}{142}-\frac{58}{157}\right)-0}{\sqrt{\frac{(.4080)(.5920)}{142}+\frac{(.4080)(.5920)}{157}}} =$ 1.4280, critical value 1.28, p-value = .0764, reject H_0. There is a higher proportion of Jerrell Enterprises employees dissatisfied with their current position at the company than there is at Jaden Industries. The belief appears valid. 5) H_0: $p_1 = p_2 + .15$, H_1: $p_1 \neq p_2 + .15$ (direct translation of H_1: the proportion of all employed Americans who are employed by private employers is not equal to the proportion of all employed non-Americans who are employed by private employers plus .15; using words from the problem: the proportion of employed Americans who are employed by private employers is different from .15 higher than the proportion of employed non-Americans who are employed by private employers),

$z = \frac{\left(.622047-\frac{8}{21}\right)-.15}{\sqrt{\frac{(.622047)(.377953)}{127}+\frac{\left(\frac{8}{21}\right)\left(\frac{13}{21}\right)}{21}}} = .7965$, critical values ±1.96, p-value = .4238, fail to reject H_0. We did not find that the proportion of employed Americans who are employed by private employers is different from .15 higher than the proportion of employed non-Americans who are employed by private employers. It's a plausible claim that the proportion of employed Americans who are employed by private employers is .15 higher than the proportion of employed non-Americans who are employed by private employers. 6) p-value (.339) > α (.04), fail to reject H_0. We did not find that the percentage of users who experience nausea with Drug B is more than 20% higher than for Drug A. 7) The probability that we would conclude that the percentage of users who experience nausea with Drug B is more than 20% higher than for Drug A, if that is not actually true, is .04.

Section 8.6

1) a) H_0: $\mu_1 = \mu_2$, H_1: $\mu_1 \neq \mu_2$ b) The mean amount of added time for all men's soccer matches is not equal to the mean amount of added time for all women's soccer matches. c) Men's and women's soccer matches have a different average amount of added time. d) Men's and women's soccer matches have a different average amount of added time. e) We did not find that men's and women's soccer matches have a different average amount of added time. 2) a) H_0: $\mu_1 = \mu_2$, H_1: $\mu_1 < \mu_2$ b) The mean utility bill amount for all Wudmoor households is less than the mean utility bill amount for all Twin Streams homes. c) Wudmoor homes tend to have cheaper utility bills than Twin Streams homes. d) Wudmoor homes tend to have cheaper utility bills than Twin Streams homes. e) We did not find that Wudmoor homes tend to have cheaper utility bills than Twin Streams homes. 3) H_0: $\mu_1 = \mu_2+2$, H_1: $\mu_1 \neq \mu_2+2$ (direct translation of H_1: the mean size of all Javonn Jewelers diamonds is not equal to the mean size of all Sparkelz Gem diamonds plus 2; using words from the problem: on average, Javonn Jewelers has diamonds with sizes that are not 2mm larger than the diamonds at Sparkelz), $t = \frac{(7.4824-6.8800)-2}{\sqrt{\frac{1.1221^2}{17}+\frac{1.4219^2}{15}}} = -3.05818$, critical values ±2.624, p-value < .01, reject H_0. On average, Javonn Jewelers has diamonds with sizes that are not 2mm larger than the diamonds at Sparkelz. 4) H_0: $\mu_1 = \mu_2$, H_1: $\mu_1 > \mu_2$, (direct translation of H_1: the mean absences among all Jerrell Enterprises employees is greater than the mean absences for all Jaden Industries employees; using words from the problem: the mean absences among Jerrell Enterprises employees is larger than the mean absences for Jaden Industries employees), $t = \frac{(3.71-3.26)-0}{\sqrt{\frac{3.42^2}{142}+\frac{2.58^2}{157}}} = 1.2740$, critical value 2.626, p-value > .10, fail to reject H_0. We did not find that the mean absences among Jerrell Enterprises employees is larger than the mean absences for Jaden Industries employees. 5) p-value (.217) > α (.025), fail to reject H_0. We did not find that the mean price of items at the north store is smaller than the mean price of items at the south store. 6) a) Concluding that we did not find that the mean price of items at the north store is smaller than the mean price of items at the south store when, in fact, the mean at the north store is smaller. b) .025.

Section 8.7

1) a) H_0: $\mu_d = 0$, H_1: $\mu_d > 0$ b) The mean of the differences of all chicken restaurant line lengths and burger restaurant line lengths in the visited cities is greater than zero. c) Chicken restaurants generally have longer lines than burger restaurants. d) Chicken restaurants generally have longer lines than burger restaurants. e) We did not find that chicken restaurants generally have longer lines than burger restaurants. 2) a) H_0: $\mu_d = 0$, H_1: $\mu_d \neq 0$ b) The mean of the differences of all skin quality measurements for people before using the skin regimen and six weeks after using the regimen is not equal to zero. c) The skin regimen makes a difference in the quality of users' skin. d) The skin regimen makes a difference in the quality of users' skin. e) We did not find that the skin regimen makes a difference in the quality of users' skin. 3) H_0: $\mu_d = 0$, H_1: $\mu_d < 0$ (direct translation of H_1: the mean of the differences of all prices for identical items at Statsville Market and Statsville Grocers is less than 0; using words from the problem: Statsville Market has much better prices), $t = \frac{-.07125-0}{\frac{.1115}{\sqrt{8}}} = -1.8074$, critical value −2.998, .05 < p-value < .10, fail to reject H_0. We did not find that Statsville Market has much better prices. 4) H_0: $\mu_d = 0$, H_1: $\mu_d < 0$ (direct translation of H_1: the mean of differences for all supervisor performance ratings and peer performance ratings (for respective employees) at Jaden Industries is less than 0; using words from the problem: supervisor performance ratings tend to be smaller than peer performance ratings (for respective employees) at Jaden Industries), $t = \frac{-4.3-0}{\frac{49.1}{\sqrt{157}}} = -1.09733$, critical value −1.653, p-value > .10, fail to reject H_0. We did not find that supervisor performance ratings tend to be smaller than peer performance ratings (for respective employees) at Jaden Industries. Co-worker appears to be wrong. 5) p-value (.000) < α (.07), reject H_0. There is a difference, on average, of prices at the north and east stores. 6) a) Yes, there was a significant difference because we rejected H_0. b) .000 (the p-value)

Section 8.8

1) a) H_0: $\sigma_1 = \sigma_2$, H_1: $\sigma_1 > \sigma_2$ b) The standard deviation of the number of miles traveled on vacation by all Chef County residents is greater than the standard deviation of the number of miles traveled on vacation by all Tanayatown residents. c) There is more dispersion in the travel distances of Chef County residents compared to Tanayatown residents. d) There is more dispersion in the travel distances of Chef County residents compared to Tanayatown residents. e) We did not find that there is more dispersion in the travel distances of Chef County residents compared to Tanayatown residents. 2) a) H_0: $\sigma_1 = \sigma_2$, H_1: $\sigma_1 \neq \sigma_2$ b) The standard deviation in the number of unforced errors for all professional tennis matches is not equal to the standard deviation in the number of unforced errors for all amateur tennis matches. c) There is different spread in the number of unforced errors in each match for professional tennis matches compared to amateur matches. d) There is different spread in the number of unforced errors in each match for professional tennis matches compared to amateur matches. e) We did not find that there is different spread in the number of unforced errors in each match for professional tennis matches compared to amateur matches.
3) H_0: $\sigma_1 = 1.1\sigma_2$, H_1: $\sigma_1 > 1.1\sigma_2$ (direct translation of H_1: the standard deviation of the weights of all Sparkelz Gems' diamonds is greater than 1.1 times the standard deviation of the weights of all Javonn Jewelers' diamonds; using words from the problem: the weights of Sparkelz Gems' diamonds have a standard deviation that is more than 1.1 times the standard deviation of the weights of Javonn Jewelers' diamonds), $F = \frac{.8009^2}{(1.1 \cdot .6708)^2} = 1.1781$, critical value 2.3522, [alpha not given so use .05] fail to reject H_0. We did not find that the weights of Sparkelz Gems' diamonds have a standard deviation that is more than 1.1 times the standard deviation of the weights of Javonn Jewelers' diamonds. 4) H_0: $\sigma_1 = \sigma_2$, H_1: $\sigma_1 \neq \sigma_2$ (direct translation of H_1: the standard deviation of all supervisor performance ratings at Jerrell Enterprises is not equal to the standard deviation of all supervisor performance ratings at Jaden Industries; using words from the problem: there is a difference in the variability in supervisor ratings at Jerrell Enterprises and Jaden Industries), $F = \frac{31^2}{42^2} = .5448$, critical values .6980, 1.4327 [using 120 for df_1 and 120 for df_2], reject H_0. There is a difference in the variability in supervisor ratings at Jerrell Enterprises and Jaden Industries. 5) a) p-value (.000) < α (.05), reject H_0. [use $\alpha = .05$ since it was not given] West store's prices more varied than the south store's. b) Concluding that west store's prices are more varied than the south store's if that is not actually true. c) Yes because we rejected H_0.

Section 8.9

1) a) (We first need to separate the "r" and "w" samples and calculate the mean and standard deviation of each) H_0: $\mu_1=\mu_2+20$, H_1: $\mu_1>\mu_2+20$, $t = \frac{(35-12.4)-20}{\sqrt{\frac{15.1186^2}{8}+\frac{5.5946^2}{5}}} = .4405$, critical value 3.747, p-value > .10, fail to reject H_0. We did not find that Chef County College riders generally take more than 20 minutes longer to get to school than Chef County College students who usually walk to school. b) H_0: $p_1 = p_2$, H_1: $p_1 \neq p_2$,
$z = \frac{\left(.3158-\frac{5}{13}\right)-0}{\sqrt{\frac{(.3438)(.6562)}{19}+\frac{(.3438)(.6562)}{13}}} = -.4025$, where $\bar{p} = \frac{6+5}{19+13} = .3438$, critical values ±1.645, p-value = .6892, fail to reject H_0. We did not find that there is a difference in the percentage of Twin Streams students with a typical commute to school of less than 20 minutes and Chef County students with a typical commute to school of less than 20 minutes. c) H_0: $\mu = 63000$, H_1: $\mu \neq 63000$, $t = \frac{64397.12-63000}{\frac{1923.44}{\sqrt{13}}} = 2.6189$, critical values ±2.179, .02 < p-value < .05, reject H_0 [α not given so use .05]. Chef County College students do not have a mean household income of $63,000. d) H_0: $\sigma_1 = \sigma_2$, H_1: $\sigma_1 < \sigma_2$, $F = \frac{338.56}{504.4744} = .6711$, critical value $\frac{1}{2.7689} = .3612$, fail to reject H_0. We did not find that the typical commute times from school for Twin Streams students are more consistent (i.e. less varied) than for Chef County students. e) No. We did not reject H_0. 2)

a) p-value (.024) < α (.05), reject H_0. the percentage of renters in Twin Streams is not 35% higher than the percentage of renters in Chef County. b) p-value (.258) > α (.05), fail to reject H_0. We did not find that the standard deviation of the amount of time that Chef County and Twin Streams residents have been at their current job is more than 4 years. 3) H_0: $p_1 = p_2$, H_1: $p_1 < p_2$, $z = \frac{(.70-.89)-0}{\sqrt{\frac{(.8)(.2)}{30}+\frac{(.8)(.2)}{35}}} = -1.9091$ with $\bar{p} = \frac{21+31}{30+35} = .8$, critical value −1.645, p-value = .0281, reject H_0. A greater proportion of Wudmoor residents have time as their greatest exercise challenge than Statsville residents who feel this way. The gym owner's claim is plausible. 4) H_0: p = .89, H_1: p < .89, $z = \frac{.7-.89}{\sqrt{\frac{(.89)(.11)}{30}}} - 3.3260$, critical value −1.645, p-value = .0004, reject H_0. A greater proportion of Wudmoor residents have time as their greatest exercise challenge than Statsville residents who feel this way. The gym owner's claim is plausible. 5) a) Concluding that a greater proportion of Wudmoor residents have time as their greatest exercise challenge than Statsville residents who feel this way if, in truth, that is not the case. Probability .05 (the α we are using since none was provided). b) Concluding that we did not find that a greater proportion of Wudmoor residents have time as their greatest exercise challenge than Statsville residents who feel this way if, actually, the proportion is greater for Wudmoor. c) .0004 (p-value) 6) In words, H_1 is "the percentage of people having armpits that produce no odor is different from 2%", p-value (.4563) > α (.05), fail to reject H_0. We did not find that the percentage of people having armpits that produce no odor is different from 2%. The report's claim is likely correct. 7) In words, H_1 is "people can generally lift more weight when doing a deadlift compared to bench press", the given test statistic is in the rejection region determined by the given critical value, reject H_0. People can generally lift more weight when doing a deadlift compared to bench press. 8) In words, H_1 is "a certain maintenance process reduces the variation in the luminosity of computer monitors", the χ^2 critical value is 4.575 for this left-tail test with α (not given) = .05, .005 < p-value < .01, reject H_0 The maintenance process reduces the variation in the luminosity of computer monitors. (The given information about the mean is not relevant to addressing the research question.)

Chapter 9

Section 9.2

1) H_0: $\mu_1 = \mu_2 = \mu_3 = \mu_4$, H_1: at least one population mean cost per transaction is different from the others among Jerrell Enterprises, Jaden Industries, Michael Initiatives, and Jordan Developments, SSB = $7(25-28)^2+5(28.6-28)^2+5(33.2-28)^2+6(26.6667-28)^2 = 210.6661$, MSB = $\frac{210.6661}{4-1} = 70.2220$, SSW = $(7-1)48.6667+(5-1)13.3+(5-1)36.7+(6-1)4.6667 = 515.3337$, MSW = $\frac{515.3337}{23-4} = 27.1228$, F = $\frac{70.2220}{27.1228} = 2.5890$, critical value 3.1274,fail to reject H_0. We did not find that at least one population mean cost per transaction is different from the others among Jerrell Enterprises, Jaden Industries, Michael Initiatives, and Jordan Developments. The companies do not have different average costs per transaction.

Source	DF	SS	MS	F
Between	3	210.6661	70.2220	2.5890
Within	19	515.3337	27.1228	
Total	22	725.9998		

2) Type II. 3) H_0: $\mu_1=\mu_2=\mu_3=\mu_4=\mu_5$, H_1: at least one population mean number of earthworms per m^2 is different from the others among soil in Twin Streams, Wudmoor, Statsville, Chef County, and Tanayatown. SSB=$9(31.67-33.17)^2+11(32-33.17)^2+13(17.69-33.17)^2+12(38.42-33.17)^2+14(44.93-33.17)^2=5417.4195$, MSB=$\frac{5417.4195}{5-1} = 1354.3549$, SSW=$(9-1)28.75+(11-1)30.2+(13-1)7.73+(12-1)37.36+(14-1)34.07=1478.63$, MSW=$\frac{1478.63}{59-5} = 27.3820$, F = $\frac{1354.3549}{27.3820} = 49.4615$, critical value 2.5252, reject H_0. At least one population mean number of earthworms per m^2 is different from the others among soil in Twin Streams, Wudmoor, Statsville,

Chef County, and Tanayatown. There is a difference in the mean presence of earthworms among these five locations.

Source	DF	SS	MS	F
Between	4	5417.4195	1354.3549	49.4615
Within	54	1478.63	27.3820	
Total	58	6896.0495		

4) p-value (.000) < α (.05), reject H_0. At least one population mean price is different from the others among items at the north, south, east, and west stores. The east store (group A) has the highest mean price, followed by the west store (group B). The south and north stores (group C) have equal mean prices and their means are lower than both of the other stores. 5) a) MS Factor = $\frac{150.6}{3}$ = 50.2, SS Error = (22)(23.3)=512.6 (OR 663.2−150.6) b) 4 (add 1 to DF Factor) c) 26 (add 1 to DF Total) d) $\frac{663.2}{25} = 26.528$

Section 9.3

1) a) Interaction p-value (.504) > α (.05); no interaction. There is no difference in mean wait time for the different parks according to the different seasons. Season main effect p-value (.676) > α (.05); no season effect. There is no difference in mean wait time for the different seasons. Park main effect p-value (.212) > α (.05); no park effect. There is no difference in mean wait time for the different parks. b) Interaction p-value (.003) < α (.05); there is interaction. There is a difference in mean wait time for the different parks according to the different seasons. Statsville Amusements during the spring (group A) has a greater mean wait time than all other season and park pairs (group B). c) Interaction p-value (.357) > α (.05); no interaction. There is no difference in mean wait time for the different parks according to the different seasons. Season main effect p-value (.000) < α (.05); there is a season effect. There is a difference in mean wait time for the different seasons. Park main effect p-value (.313) > α (.05); no park effect. There is no difference in mean wait time for the different parks. The mean wait time during the winter (group A) is greater than for the other seasons (group B). d) No. We failed to reject H_0. 2) Interaction p-value (.974) > α (.05); no interaction. There is no difference in mean weight for the different companies according to color style. Company main effect p-value (.990) > α (.05); no company effect. There is no difference in mean weight for the different companies. Color main effect p-value (.000) < α (.05); there is a color effect. There is a difference in mean weight for the different color styles. Bedding sheet sets with some sort of design (group A) have a greater mean weight than bedding sheet sets with just a solid color (group B). 3) We have no idea. The means shown are just for the samples, not for the populations ("all").

Section 9.4

1) a) H_0: There is no linear correlation between number of crime incidents and high temperature. H_1: There is a linear correlation between number of crime incidents and high temperature. b) (i) Reject H_0 with positive r value: There is a positive linear correlation between number of crime incidents and high temperature. Crimes tend to occur more during warmer periods. (ii) Reject H_0 with negative r value: There is a negative linear correlation between number of crime incidents and high temperature. Crimes tend to occur more during cooler periods. (iii) Fail to reject H_0 with positive r value: We did not find that there is a linear correlation between number of crime incidents and high temperature. Crime incidents and temperature are unrelated. (iv) Fail to reject H_0 with negative r value: We did not find that there is a linear correlation between number of crime incidents and high temperature. Crime incidents and temperature are unrelated. c) A Type I error would be concluding that there is a linear correlation between number of crime incidents and high temperature if, in reality, there is no such correlation. 2) H_0: there is not a linear correlation between Statsville air quality readings and Wudmoor air quality readings, H_1: there is a linear correlation between Statsville air quality readings and Wudmoor air quality readings (H_1 using words from the problem: there is a relationship between air quality readings for Statsville and Wudmoor), $r = \frac{7(9765)-(265)(257)}{\sqrt{7(10185)-265^2}\sqrt{7(9579)-257^2}} =$ .2412, critical values ±.875, fail to reject H_0. We did not find that there is a relationship between air quality readings for Statsville and Wudmoor. 3) $r^2 = (.2412)^2 = .05818$, 5.818% of the variation in Wudmoor air

quality readings is explained by Statsville air quality readings. This only applies to this sample of Statsville and Wudmoor air quality readings because we found no relationship between the population (all) of Statsville and Wudmoor air quality readings in Exercise 2). 4) H_0: there is not a linear correlation between math placement test scores and reading placement test scores for Twin Streams College students, H_1: there is a linear correlation between math placement test scores and reading placement test scores for Twin Streams College students, $r = \frac{9(67242)-(934)(629)}{\sqrt{9(99608)-934^2}\sqrt{9(46645)-629^2}} = .7329$, critical values ±.666 [α not given so use .05], reject H_0. There is a linear correlation between math placement test scores and reading placement test scores for Twin Streams College students. Since r is positive, the linear correlation is positive telling us that students who get higher scores on the math test tend to get higher scores on the reading test. 5) $r^2 = (.7329)^2 = .5371$ is the proportion of the variation in reading scores that is explained by math scores. Since we concluded that there is a positive linear correlation between math scores and reading scores in the population, this proportion also applies to the population of all Twin Streams math and reading placement test scores. 6) p-value (.000) < α (.03), reject H_0. There is a positive linear correlation between current home values and before-arena home values. Higher current values are associated with higher before-arena values.

Section 9.5

1) Slope: For each year that elapses, a car's value decreases by $1478, on average. Intercept: The best predicted value of a new car is $21595. ["new" because it has zero years of age] 2) $b_1 = \frac{7(9765)-(265)(257)}{7(10185)-265^2} = .2336$, $b_0 = 36.7143-(.2336)(37.8571) = 27.8709$, $\hat{y} = 27.8709 + .2336x$. Since we found no linear correlation for this data's population in Section 9.4, we use $\bar{y}$ as our prediction, which is 36.7143. 3) slope: For each one-unit increase in air quality reading in Statsville, the respective Wudmoor air quality reading increases by .2336 unit, on average. intercept: When the air quality reading in Statsville is zero units, the predicted air quality reading in Wudmoor is 27.8709 units. The interpretations apply only to the sample data because we found no linear correlation for this data's population in Section 9.4. 4) $b_1 = \frac{9(67242)-(934)(629)}{9(99608)-934^2} = .7336$, $b_0 = 69.8889-.7336(103.7778) = -6.2425$, $\hat{y} = -6.2425 + .7336x$. Since we did find a linear correlation for this data's population in Section 9.4 our prediction is $\hat{y} = -6.2425 + .7336(107) = 72.2527$. 5) slope: For each one-point increase in math score, reading score increases by .7336 point, on average. intercept: A student with a math score of zero points is predicted to have a reading score of −6.2425 points. The interpretations apply to the population because we found a linear correlation for this data's population in Section 9.4.

Section 9.6

1) (i) H_0: $p_1=p_2=p_3=p_4=\frac{1}{4}$, H_1: air quality index readings do not occur with equal frequency,

air quality	O	E	O−E	$(O-E)^2$	$\frac{(O-E)^2}{E}$
30-34	32	.25(144) = 36	32−36 = −4	16	16/36=.4444
35-39	44	.25(144) = 36	44−36 = 8	64	64/36=1.7778
40-44	32	.25(144) = 36	32−36 = −4	16	16/36=.4444
45-49	36	.25(144) = 36	36−36 = 0	0	0/36=0
total	144		test stat	χ^2=	2.6666

critical value 11.345, .1 < p-value < .9, fail to reject H_0. We did not find that air quality index readings do not occur with equal frequency. (ii) H_0: $p_1=.1$, $p_2=.4$, $p_3=.3$, $p_4=.2$ [note: we knew p_4 had to be .2 because the proportions have to sum to 1], H_1: air quality readings do not occur according to the proportions stated in H_0,

air quality	O	E	O−E	$(O-E)^2$	$\frac{(O-E)^2}{E}$
30-34	32	.1(144)=14.4	32−14.4=17.6	309.76	309.76/14.4=21.5111
35-39	44	.4(144)=57.6	44−57.6=−13.6	184.96	184.96/57.6=3.2111
40-44	32	.3(144)=43.2	32−43.2=−11.2	125.44	125.44/43.2=2.9037
45-49	36	.2(144)=28.8	36−28.8=7.2	51.84	51.84/28.8=1.8
total	144		test stat	χ^2=	29.4259

critical value 11.345, p-value < .005, reject H_0. Air quality readings do not occur according to the proportions stated in H_0. The idea that that 10% of the time the air quality is between 30 and 34, 40% of the time the air quality is between 35 and 39, 30% of the time the index is between 40 and 44, and the rest of the time the quality of the air falls between 45 and 49 appears to be incorrect. (iii) The first environmentalist's view seems most likely. 2) H_0: p_1=.52, p_2=.29, p_3=.19, H_1: the proportions of Jerrell Enterprises employees' feelings about their position in the company are not as stated in H_0,

feeling	O	E	O−E	(O−E)²	$\frac{(O-E)^2}{E}$
dissatisfied	64	.52(142)=73.84	64−73.84=−9.84	96.8256	96.8256/73.84=1.3113
satisfied	46	.29(142)=41.18	46−41.18=4.82	23.2324	23.2324/41.18=.5642
neutral	32	.19(142)=26.98	32−26.98=5.02	25.2004	25.2004/26.98=.9340
total	142		test stat	χ^2=	2.8095

critical value 7.378, .1 < p-value <.9, fail to reject H_0. We did not find that the proportions of Jerrell Enterprises employees' feelings about their position in the company are not as stated in H_0. The theory is reasonable. 3) .025 (alpha). 4) H_0: p_1=.18, p_2=.19, p_3=.28, p_4=.35, H_1: the proportions for all Statsville College students who are Humanities, Other, Social Science, and STEM majors are not according to the respective proportions indicated in H_0, p-value (.002) < α (not given, so .05), reject H_0. For all Statsville College students, the following is NOT the distribution of majors: 18% Humanities, 19% Other, 28% Social Science, 35% STEM.

Section 9.7

1) H_0: air quality and city are independent (not associated), H_1: air quality and city are not independent (they are associated). These are the $\frac{(O-E)^2}{E}$ terms for the indicated cells: $\frac{(6-10.6667)^2}{10.6667} = 2.04169$ (Chef,30-34), $\frac{(26-14.6667)^2}{14.6667} = 8.7575$ (Chef,35-39), $\frac{(6-10.6667)^2}{10.6667} = 2.04169$ (Chef,40-44), $\frac{(10-12)^2}{12} = .3333$ (Chef,45-49), $\frac{(12-10.6667)^2}{10.6667} = .1667$ (Stats,30-34), $\frac{(10-14.6667)^2}{14.6667} = 1.4849$ (Stats,35-39), $\frac{(14-10.6667)^2}{10.6667} = 1.04164$ (Stats,40-44), $\frac{(12-12)^2}{12} = 0$ (Stats,45-49), $\frac{(14-10.6667)^2}{10.6667} = 1.04164$ (Wud,30-34), $\frac{(8-14.6667)^2}{14.6667}$ 3.03033 (Wud,35-39), $\frac{(12-10.6667)^2}{10.6667} = .1667$ (Wud,40-44), $\frac{(14-12)^2}{12} = .3333$ (Wud,45-49). Add those to get test statistic χ^2 = 20.4394. critical value 12.592, p-value < .005, reject H_0. Air quality and city are associated. There is a difference in air quality based on city. 2) H_0: how Jaden Industries employees feel about their position in the company and whether or not they participate in professional development are independent (those variables are unrelated), H_1: how Jaden Industries employees feel about their position in the company and whether or not they participate in professional development are not independent (those variables are related). Two-way table with totals:

	dissatisfied	neutral	satisfied	total
no	28	24	26	78
yes	30	25	24	79
total	58	49	50	157

taking all the no's first and going from left to right, we have test statistic $\chi^2 = \frac{(28-28.8153)^2}{28.8153} + \frac{(24-24.3439)^2}{24.3439} + \frac{(26-24.8408)^2}{24.8408} + \frac{(30-29.1847)^2}{29.1847} + \frac{(25-24.6561)^2}{24.6561} + \frac{(24-25.1592)^2}{25.1592} = .1630$, critical value 4.605, .90 < p-value < .95, fail to reject H_0. We did not find that how Jaden Industries employees feel about their position in the company and whether or not they participate in professional development are related. There is no connection between how Jaden Industries employees feel about their position in the company and whether or not they participate in professional development. 3) p-value (.000) < α (.025), reject H_0. Favorite time to study and area of major are related. Students from particular majors prefer to study during certain times of the day over others. 4) .000 (p-value, as it appears in Minitab)

Section 9.8

1) Two categorical variables, looking for association, use test of independence. df is (5−1)(3−1) = 8. H_1 would say there is a difference in overall academic college readiness based on the high school students attended (those variables are related). critical value 20.090, .025 < p-value < .05, fail to reject H_0. We did not

find that there is a difference in overall academic college readiness based on the high school students attended. (obviously, the conclusion just relates to the five high schools in the study) 2) a) H_0: soft drink preference and popcorn preference among moviegoers are independent (there's no relationship), H_1: soft drink preference and popcorn preference among moviegoers are not independent (there is a relationship), moving across the cells from left to right and from the top down test stat $\chi^2 = \frac{(11-8.9444)^2}{8.9444} + \frac{(3-5.05556)^2}{5.05556} + \frac{(16-10.8611)^2}{10.8611} + \frac{(1-6.1389)^2}{6.1389} + \frac{(12-14.05556)^2}{14.05556} + \frac{(10-7.9444)^2}{7.9444} + \frac{(7-12.1389)^2}{12.1389} + \frac{(12-6.8611)^2}{6.8611} = 14.8984$, critical value 6.251, p-value < .005, reject H_0. There is a relationship between the soft drink moviegoers prefer and how they like their popcorn. b) H_0: p_1=.2, p_2=.25, p_3=.3, p_4=.25, H_1: the proportions for moviegoers' soft drink preferences are not found according to those stated in H_0, critical value 7.815, .975 < p-value < .99, fail to reject H_0. We did not find that the proportions for moviegoers' soft drink preferences are not found according to those stated in H_0. The manager is likely correct. (The claim is plausible.)

	O	E	O−E	$(O-E)^2$	$\frac{(O-E)^2}{E}$
Popsee	14	.2(72)=14.4	14−14.4=−.4	.16	.16/14.4=.01111
Diet Popsee	17	.25(72)=18	17−18=−1	1	1/18=.05556
Lemon Crest	22	.3(72)=21.6	22−21.6=.4	.16	.16/21.6=.007407
Orange Kick	19	.25(72)=18	19−18=1	1	1/18=.05556
total	72		test stat	χ^2=	.1296

3) [power snatch is "y" because that is what is being predicted] sum of x = 9.71, sum of x^2 = 9.9081, sum of y = 7.41, sum of y^2 = 5.9233, sum of xy = 7.629, $r = \frac{10(7.629)-(9.71)(7.41)}{\sqrt{10(9.9081)-9.71^2}\sqrt{10(5.9233)-7.41^2}} = .9526$, critical values ± .632. There is a linear correlation so need regression for the prediction. $b_1 = \frac{10(7.629)-(9.71)(7.41)}{10(9.9081)-9.71^2} = .9045$, b_0 = .741−.9045(.971) = −.1373, $\hat{y}$ = −.1373 + .9045x. Best prediction is $\hat{y}$ = −.1373+.9045(.83)= .6134. 4) H_0: $\mu_1=\mu_2=\mu_3$, H_1: at least one population mean air quality reading is different from the rest among Statsville, Wudmoor, and Chef County, SSB = $7(37.8571-37.4762)^2 + 7(36.7143-37.4762)^2 + 7(37.8571-37.4762)^2$ = 6.09463, MSB = $\frac{6.09463}{3-1}$ = 3.04732, SSW = (7−1)25.4762 + (7−1)23.9048 + (7−1)19.8095 = 415.143, MSW = $\frac{415.143}{21-3}$ = 23.0635, F = $\frac{3.04732}{23.0635}$ = .1321, critical value 3.5546, fail to reject H_0.

Source	DF	SS	MS	F
Between	2	6.09463	3.04732	.1321
Within	18	415.143	23.0635	
Total	20	421.2376		

We did not find that at least one population mean air quality reading is different from the rest among Statsville, Wudmoor, and Chef County. There is no difference in air quality for these three areas. 5) a) p-value (.115) > α (.05), fail to reject H_0. We did not find that there is not an equal proportion of renters who are short, long, and moderate term. It is quite likely that, among renters in Chef County and Twin Streams, there is an equal proportion who are short, long, and moderate term. b) We have no idea. Any means shown on the output are sample means, not population ("all residents") means. c) correlation p-value (.953) > α (.05), fail to reject H_0. We did not find that there is a linear correlation between years at current job for Twin Streams residents and years at the current residence. Best predicted years at current job will be $\bar{y}$, sample mean years at current job for Twin Streams residents. Using the sample mean of `t-job` the answer is 7.871 years. 6) H_0: there is not a linear correlation between volume and price of refrigerators, H_1: there is a linear correlation between volume and price of refrigerators, given r = .3907, critical values ± .553, fail to reject H_0. We did not find that there is a linear correlation between volume and price of refrigerators. Best predicted price is the sample mean of prices, $1285.92. 7) We are given the coefficient of determination, r^2, .6724. Taking the square root of that we know r is either .82 or −.82. Either way, yes there is a linear correlation in the population of all consumers because critical values are ± .514.

Section 9.10

1) a) $E = 1.812\sqrt{\frac{3.95^2}{11}+\frac{3.05^2}{13}} = 2.6470$; $47.753 < \mu_1-\mu_2 < 53.047$; With 90% confidence we can say that the difference in the mean maximum airspeed for all peregrine falcons and the mean maximum airspeed for all golden eagles is between 47.753 and 53.047 mph. Since the whole interval is positive, we can make the inference that for all such birds, peregrine falcons have a larger mean maximum airspeed than golden eagles. b) H_0: $\sigma_1=\sigma_2$, H_1: $\sigma_1 \neq \sigma_2$, $F = \frac{3.47^2}{2.98^2} = 1.3559$, critical values .2773 and 4.6658, fail to reject H_0. We did not find that golden eagles and gyrfalcons have different standard deviations in their maximum horizontal speeds. The claim that the standard deviations are the same is plausible. c) H_0: $\mu_1 = \mu_2 - 80$, H_1: $\mu_1 > \mu_2 - 80$, $t = \frac{(125.7-200.8)-(-80)}{\sqrt{\frac{2.06^2}{8}+\frac{2.77^2}{11}}} = 4.4218$, critical value 2.365, p-value < .005, reject H_0. The mean maximum diving speed of gyrfalcons less than 80 mph smaller than that of peregrine falcons. 2) We can say that the proportion of orders received by this company's customers within two business days is between .8913 and .9769. Since .95 (95%) is in the interval, the company's claim is acceptable [93.4% is $\hat{p}$, so it's irrelevant to the conclusion]. 3) H_1 would say that the average in-game life of a major league baseball is different from 6.3 pitches, p-value (.2273) > α (.05), fail to reject H_0. We did not find that the average in-game life of a major league baseball is different from 6.3 pitches. The original claim is plausible. [the critical value is not used to get the answer] 4) H_1 would say that there is a difference in the proportion of students appropriately placed when using Exam 1 versus the proportion of students appropriately placed when using Exam 2, given test statistic z = −2.635, critical values ± 1.75, p-value .0082, reject H_0. There is a difference in the proportion of students appropriately placed when using Exam 1 versus the proportion of students appropriately placed when using Exam 2. 5) H_0: $p_1=p_2=p_3=p_4=\frac{1}{4}$, H_1: partying, reading, listening to music, and exercising are not preferred equally among teenagers, critical value 7.815, .01 < p-value < .025, reject H_0. Partying, reading, listening to

activity	O	E	O−E	$(O-E)^2$	$(O-E)^2/E$
Party	20	.25(46)=11.5	20−11.5=8.5	72.25	72.25/11.5=6.2826
Read	6	.25(46)=11.5	6−11.5=−5.5	30.25	30.25/11.5=2.6304
Music	13	.25(46)=11.5	13−11.5=1.5	2.25	2.25/11.5=.1957
Exercise	7	.25(46)=11.5	7−11.5=−4.5	20.25	20.25/11.5=1.7609
total	46		test stat	$\chi^2 =$	10.8696

music, and exercising are not preferred equally among teenagers. 6) H_0: $\mu_d = 0$, H_1: $\mu_d > 0$, $t = \frac{1.1429-0}{\frac{6.1218}{\sqrt{7}}} = .4939$, critical value 3.143, p-value > .10, fail to reject H_0. We did not find that Wudmoor generally has better air quality than Statsville. 7) a) p-value (.000) < α (.005), reject H_0. Less than 80% of renters in Chef County and Twin Streams are short term renters. The claim is supported. b) p-value (.623) > α (.12), fail to reject H_0. We did not find that having children in the household is related to location. We did not find that a person is more or less likely to have children in the household based on where the person lives. c) With 95% confidence we can say that the mean amount of time that Twin Streams and Chef County residents have been at their current residence is between 4.092 and 6.433 years. Since the whole interval is less than 7.8, it would not be reasonable to conclude that Twin Streams and Chef County residents have been at their current residence for an average of 7.8 years. The average is some number of years less than 7.8. 8) a) We have to use the "`t          34   4.958     24.582`" portion of the "**Test and CI for Two Variances: job vs location**" section of output to get the needed numbers. That is giving us the size, standard deviation, and variance of the years at current job for the Twin Streams sample. H_0: $\sigma^2=40$, H_1: $\sigma^2<40$, $\chi^2=\frac{(34-1)24.582}{40} = 20.2802$, critical value 18.493, .05 < p-value < .10, fail to reject H_0. We did not find that the variance of the time Twin Streams residents have been at their current job is less than 40 years^2. b) We need to use the "**Tabulated statistics: children, location**" output to get the appropriate numbers. The proportion of Chef County residents who have children living in the household is $\frac{7}{14}$ while the sample proportion for Twin

Streams is $\frac{10}{17}$. The 14 and the 17 come from the total of the "c" and "t" columns, respectively. The "7" and the "10" come from the intersection of the "y" row with the "c" and "t" columns, respectively. $E = 1.84\sqrt{\frac{\left(\frac{7}{14}\right)\left(\frac{7}{14}\right)}{14} + \frac{\left(\frac{10}{17}\right)\left(\frac{7}{17}\right)}{17}} = .3297$; $-.4179 < p_1 - p_2 < .2415$; With 93.4% confidence we can say that the difference in the proportion of all Chef County residents who have children living in the household and the proportion of all Twin Streams residents who have children living in the household is between −.4179 and .2415. Since the interval contains zero, a logical inference that we can draw is that there is no difference in the proportion of Chef County residents who have children living in the household and the proportion of Twin Streams residents who have children living in the household. c) We need to use the "`c-residence  33  5.600  4.816`" line of the "**Paired T-Test and CI: c-residence, c-job**" output. That line is giving us the size, mean, and standard deviation for the years at current residence for the Chef County sample. $E = 2.037\frac{4.816}{\sqrt{33}} = 1.7077$; $3.8923 < \mu < 7.3077$. With 95% confidence we can say that the mean years at current residence for all Chef County residents is between 3.8923 and 7.3077 years. Since the whole interval is less than 10 years, Chef County should not be considered a stable area. 9) Interaction p-value (.330) > α (.01); no interaction. There is no difference in mean insurance premium for the different companies according to the different areas. Area main effect p-value (.000) < α (.01); there is a main effect for area. There is a difference in mean insurance premium for the different areas. Company main effect p-value (.000) < α (.01); there is a main effect for company. There is a difference in mean insurance premium for the different companies. Twin Streams (group A) customers pay a larger mean insurance premium than customers in the other areas (group B). Company I customers (group A) pay a larger mean insurance premium than Company II customers (group B). 10) a) We would use 118, 219, 157, and 67. These would be the observed frequencies for each category in the Goodness of Fit test. b) We would use 157 and 15532. These would be used to calculate $\hat{p} = \frac{157}{15532} = .01011$ as we make a confidence interval for p. 11) a) (two quantitative variables, looking for a relationship) Test for linear correlation. b) (two categorical variables, looking for a relationship) Test of independence. c) (one categorical variable with more than two categories, one quantitative variable, we'll see if there is a relationship by comparing means) One-way ANOVA. d) (two categorical variables, one quantitative variable, we'll see if there is a relationship by comparing means) Two-way ANOVA. e) (one categorical variable, testing for proportions of the categories) Goodness of Fit. f) confidence interval or hypothesis test (if we have a choice, hypothesis test is preferred when testing a claim) for μ. g) confidence interval for $\frac{\sigma_1^2}{\sigma_2^2}$ or hypothesis test where H_1 is $\sigma_1^2 > \sigma_2^2$ (or σ could be used; if we have a choice, hypothesis test is preferred when testing a claim) h) confidence interval for p. i) Since we have the population data, there is no inferential statistics procedure needed. Get the mean for each group and compare those means.

Index

Table A Standard Normal Distribution

Table provides area to the LEFT of the indicated z-score.

second decimal place of z-score

z	0	1	2	3	4	5	6	7	8	9
-3.5 and below	.0001									
-3.4	.0003	.0003	.0003	.0003	.0003	.0003	.0003	.0003	.0003	.0002
-3.3	.0005	.0005	.0005	.0004	.0004	.0004	.0004	.0004	.0004	.0003
-3.2	.0007	.0007	.0006	.0006	.0006	.0006	.0006	.0005	.0005	.0005
-3.1	.0010	.0009	.0009	.0009	.0008	.0008	.0008	.0008	.0007	.0007
-3.0	.0013	.0013	.0013	.0012	.0012	.0011	.0011	.0011	.0010	.0010
-2.9	.0019	.0018	.0018	.0017	.0016	.0016	.0015	.0015	.0014	.0014
-2.8	.0026	.0025	.0024	.0023	.0023	.0022	.0021	.0021	.0020	.0019
-2.7	.0035	.0034	.0033	.0032	.0031	.0030	.0029	.0028	.0027	.0026
-2.6	.0047	.0045	.0044	.0043	.0041	.0040	.0039	.0038	.0037	.0036
-2.5	.0062	.0060	.0059	.0057	.0055	.0054	.0052	.0051	.0049	.0048
-2.4	.0082	.0080	.0078	.0075	.0073	.0071	.0069	.0068	.0066	.0064
-2.3	.0107	.0104	.0102	.0099	.0096	.0094	.0091	.0089	.0087	.0084
-2.2	.0139	.0136	.0132	.0129	.0125	.0122	.0119	.0116	.0113	.0110
-2.1	.0179	.0174	.0170	.0166	.0162	.0158	.0154	.0150	.0146	.0143
-2.0	.0228	.0222	.0217	.0212	.0207	.0202	.0197	.0192	.0188	.0183
-1.9	.0287	.0281	.0274	.0268	.0262	.0256	.0250	.0244	.0239	.0233
-1.8	.0359	.0351	.0344	.0336	.0329	.0322	.0314	.0307	.0301	.0294
-1.7	.0446	.0436	.0427	.0418	.0409	.0401	.0392	.0384	.0375	.0367
-1.6	.0548	.0537	.0526	.0516	.0505	.0495	.0485	.0475	.0465	.0455
-1.5	.0668	.0655	.0643	.0630	.0618	.0606	.0594	.0582	.0571	.0559
-1.4	.0808	.0793	.0778	.0764	.0749	.0735	.0721	.0708	.0694	.0681
-1.3	.0968	.0951	.0934	.0918	.0901	.0885	.0869	.0853	.0838	.0823
-1.2	.1151	.1131	.1112	.1093	.1075	.1056	.1038	.1020	.1003	.0985
-1.1	.1357	.1335	.1314	.1292	.1271	.1251	.1230	.1210	.1190	.1170
-1.0	.1587	.1562	.1539	.1515	.1492	.1469	.1446	.1423	.1401	.1379
-0.9	.1841	.1814	.1788	.1762	.1736	.1711	.1685	.1660	.1635	.1611
-0.8	.2119	.2090	.2061	.2033	.2005	.1977	.1949	.1922	.1894	.1867
-0.7	.2420	.2389	.2358	.2327	.2296	.2266	.2236	.2206	.2177	.2148
-0.6	.2743	.2709	.2676	.2643	.2611	.2578	.2546	.2514	.2483	.2451
-0.5	.3085	.3050	.3015	.2981	.2946	.2912	.2877	.2843	.2810	.2776
-0.4	.3446	.3409	.3372	.3336	.3300	.3264	.3228	.3192	.3156	.3121
-0.3	.3821	.3783	.3745	.3707	.3669	.3632	.3594	.3557	.3520	.3483
-0.2	.4207	.4168	.4129	.4090	.4052	.4013	.3974	.3936	.3897	.3859
-0.1	.4602	.4562	.4522	.4483	.4443	.4404	.4364	.4325	.4286	.4247
-0.0	.5000	.4960	.4920	.4880	.4840	.4801	.4761	.4721	.4681	.4641

Table A Standard Normal Distribution

Table provides area to the LEFT of the indicated z-score.

second decimal place of z-score

z	0	1	2	3	4	5	6	7	8	9
0.0	.5000	.5040	.5080	.5120	.5160	.5199	.5239	.5279	.5319	.5359
0.1	.5398	.5438	.5478	.5517	.5557	.5596	.5636	.5675	.5714	.5753
0.2	.5793	.5832	.5871	.5910	.5948	.5987	.6026	.6064	.6103	.6141
0.3	.6179	.6217	.6255	.6293	.6331	.6368	.6406	.6443	.6480	.6517
0.4	.6554	.6591	.6628	.6664	.6700	.6736	.6772	.6808	.6844	.6879
0.5	.6915	.6950	.6985	.7019	.7054	.7088	.7123	.7157	.7190	.7224
0.6	.7257	.7291	.7324	.7357	.7389	.7422	.7454	.7486	.7517	.7549
0.7	.7580	.7611	.7642	.7673	.7704	.7734	.7764	.7794	.7823	.7852
0.8	.7881	.7910	.7939	.7967	.7995	.8023	.8051	.8078	.8106	.8133
0.9	.8159	.8186	.8212	.8238	.8264	.8289	.8315	.8340	.8365	.8389
1.0	.8413	.8438	.8461	.8485	.8508	.8531	.8554	.8577	.8599	.8621
1.1	.8643	.8665	.8686	.8708	.8729	.8749	.8770	.8790	.8810	.8830
1.2	.8849	.8869	.8888	.8907	.8925	.8944	.8962	.8980	.8997	.9015
1.3	.9032	.9049	.9066	.9082	.9099	.9115	.9131	.9147	.9162	.9177
1.4	.9192	.9207	.9222	.9236	.9251	.9265	.9279	.9292	.9306	.9319
1.5	.9332	.9345	.9357	.9370	.9382	.9394	.9406	.9418	.9429	.9441
1.6	.9452	.9463	.9474	.9484	.9495	.9505	.9515	.9525	.9535	.9545
1.7	.9554	.9564	.9573	.9582	.9591	.9599	.9608	.9616	.9625	.9633
1.8	.9641	.9649	.9656	.9664	.9671	.9678	.9686	.9693	.9699	.9706
1.9	.9713	.9719	.9726	.9732	.9738	.9744	.9750	.9756	.9761	.9767
2.0	.9772	.9778	.9783	.9788	.9793	.9798	.9803	.9808	.9812	.9817
2.1	.9821	.9826	.9830	.9834	.9838	.9842	.9846	.9850	.9854	.9857
2.2	.9861	.9864	.9868	.9871	.9875	.9878	.9881	.9884	.9887	.9890
2.3	.9893	.9896	.9898	.9901	.9904	.9906	.9909	.9911	.9913	.9916
2.4	.9918	.9920	.9922	.9925	.9927	.9929	.9931	.9932	.9934	.9936
2.5	.9938	.9940	.9941	.9943	.9945	.9946	.9948	.9949	.9951	.9952
2.6	.9953	.9955	.9956	.9957	.9959	.9960	.9961	.9962	.9963	.9964
2.7	.9965	.9966	.9967	.9968	.9969	.9970	.9971	.9972	.9973	.9974
2.8	.9974	.9975	.9976	.9977	.9977	.9978	.9979	.9979	.9980	.9981
2.9	.9981	.9982	.9982	.9983	.9984	.9984	.9985	.9985	.9986	.9986
3.0	.9987	.9987	.9987	.9988	.9988	.9989	.9989	.9989	.9990	.9990
3.1	.9990	.9991	.9991	.9991	.9992	.9992	.9992	.9992	.9993	.9993
3.2	.9993	.9993	.9994	.9994	.9994	.9994	.9994	.9995	.9995	.9995
3.3	.9995	.9995	.9995	.9996	.9996	.9996	.9996	.9996	.9996	.9997
3.4	.9997	.9997	.9997	.9997	.9997	.9997	.9997	.9997	.9997	.9998
3.50 and above	.9999									

Table B t Distribution

Table provides the positive t-score (critical value) with the indicated area to the RIGHT

Area to the RIGHT

Degrees of Freedom	.005	.01	.025	.05	.10
1	63.657	31.821	12.706	6.314	3.078
2	9.925	6.965	4.303	2.920	1.886
3	5.841	4.541	3.182	2.353	1.638
4	4.604	3.747	2.776	2.132	1.533
5	4.032	3.365	2.571	2.015	1.476
6	3.707	3.143	2.447	1.943	1.440
7	3.499	2.998	2.365	1.895	1.415
8	3.355	2.896	2.306	1.860	1.397
9	3.250	2.821	2.262	1.833	1.383
10	3.169	2.764	2.228	1.812	1.372
11	3.106	2.718	2.201	1.796	1.363
12	3.055	2.681	2.179	1.782	1.356
13	3.012	2.650	2.160	1.771	1.350
14	2.977	2.624	2.145	1.761	1.345
15	2.947	2.602	2.131	1.753	1.341
16	2.921	2.583	2.120	1.746	1.337
17	2.898	2.567	2.110	1.740	1.333
18	2.878	2.552	2.101	1.734	1.330
19	2.861	2.539	2.093	1.729	1.328
20	2.845	2.528	2.086	1.725	1.325
21	2.831	2.518	2.080	1.721	1.323
22	2.819	2.508	2.074	1.717	1.321
23	2.807	2.500	2.069	1.714	1.319
24	2.797	2.492	2.064	1.711	1.318
25	2.787	2.485	2.060	1.708	1.316
26	2.779	2.479	2.056	1.706	1.315
27	2.771	2.473	2.052	1.703	1.314
28	2.763	2.467	2.048	1.701	1.313
29	2.756	2.462	2.045	1.699	1.311
30	2.750	2.457	2.042	1.697	1.310
31	2.744	2.453	2.040	1.696	1.309
32	2.738	2.449	2.037	1.694	1.309
33	2.733	2.445	2.035	1.692	1.308
34	2.728	2.441	2.032	1.691	1.307
35	2.724	2.438	2.030	1.690	1.306
36	2.719	2.434	2.028	1.688	1.306
37	2.715	2.431	2.026	1.687	1.305
38	2.712	2.429	2.024	1.686	1.304
39	2.708	2.426	2.023	1.685	1.304
40	2.704	2.423	2.021	1.684	1.303
45	2.690	2.412	2.014	1.679	1.301
50	2.678	2.403	2.009	1.676	1.299
60	2.660	2.390	2.000	1.671	1.296
70	2.648	2.381	1.994	1.667	1.294
80	2.639	2.374	1.990	1.664	1.292
90	2.632	2.368	1.987	1.662	1.291
100	2.626	2.364	1.984	1.660	1.290
200	2.601	2.345	1.972	1.653	1.286
300	2.592	2.339	1.968	1.650	1.284
400	2.588	2.336	1.966	1.649	1.284
500	2.586	2.334	1.965	1.648	1.283

Table C χ^2 Distribution

Table provides the χ^2-score (critical value) with the indicated area to the RIGHT

Area to the RIGHT

Degrees of Freedom	.995	.99	.975	.95	.90	.10	.05	.025	.01	.005
1	-	-	.001	.004	.016	2.706	3.841	5.024	6.635	7.879
2	.010	.020	.051	.103	.211	4.605	5.991	7.378	9.210	10.597
3	.072	.115	.216	.352	.584	6.251	7.815	9.348	11.345	12.838
4	.207	.297	.484	.711	1.064	7.779	9.488	11.143	13.277	14.860
5	.412	.554	.831	1.145	1.610	9.236	11.071	12.833	15.086	16.750
6	.676	.872	1.237	1.635	2.204	10.645	12.592	14.449	16.812	18.548
7	.989	1.239	1.690	2.167	2.833	12.017	14.067	16.013	18.475	20.278
8	1.344	1.646	2.180	2.733	3.490	13.362	15.507	17.535	20.090	21.955
9	1.735	2.088	2.700	3.325	4.168	14.684	16.919	19.023	21.666	23.589
10	2.156	2.558	3.247	3.940	4.865	15.987	18.307	20.483	23.209	25.188
11	2.603	3.053	3.816	4.575	5.578	17.275	19.675	21.920	24.725	26.757
12	3.074	3.571	4.404	5.226	6.304	18.549	21.026	23.337	26.217	28.299
13	3.565	4.107	5.009	5.892	7.042	19.812	22.362	24.736	27.688	29.819
14	4.075	4.660	5.629	6.571	7.790	21.064	23.685	26.119	29.141	31.319
15	4.601	5.229	6.262	7.261	8.547	22.307	24.996	27.488	30.578	32.801
16	5.142	5.812	6.908	7.962	9.312	23.542	26.296	28.845	32.000	34.267
17	5.697	6.408	7.564	8.672	10.085	24.769	27.587	30.191	33.409	35.718
18	6.265	7.015	8.231	9.390	10.865	25.989	28.869	31.526	34.805	37.156
19	6.844	7.633	8.907	10.117	11.651	27.204	30.144	32.852	36.191	38.582
20	7.434	8.260	9.591	10.851	12.443	28.412	31.410	34.170	37.566	39.997
21	8.034	8.897	10.283	11.591	13.240	29.615	32.671	35.479	38.932	41.401
22	8.643	9.542	10.982	12.338	14.042	30.813	33.924	36.781	40.289	42.796
23	9.260	10.196	11.689	13.091	14.848	32.007	35.172	38.076	41.638	44.181
24	9.886	10.856	12.401	13.848	15.659	33.196	36.415	39.364	42.980	45.559
25	10.520	11.524	13.120	14.611	16.473	34.382	37.652	40.646	44.314	46.928
26	11.160	12.198	13.844	15.379	17.292	35.563	38.885	41.923	45.642	48.290
27	11.808	12.879	14.573	16.151	18.114	36.741	40.113	43.194	46.963	49.645
28	12.461	13.565	15.308	16.928	18.939	37.916	41.337	44.461	48.278	50.993
29	13.121	14.257	16.047	17.708	19.768	39.087	42.557	45.722	49.588	52.336
30	13.787	14.954	16.791	18.493	20.599	40.256	43.773	46.979	50.892	53.672
40	20.707	22.164	24.433	26.509	29.051	51.805	55.758	59.342	63.691	66.766
50	27.991	29.707	32.357	34.764	37.689	63.167	67.505	71.420	76.154	79.490
60	35.534	37.485	40.482	43.188	46.459	74.397	79.082	83.298	88.379	91.952
70	43.275	45.442	48.758	51.739	55.329	85.527	90.531	95.023	100.425	104.215
80	51.172	53.540	57.153	60.391	64.278	96.578	101.879	106.629	112.329	116.321
90	59.196	61.754	65.647	69.126	73.291	107.565	113.145	118.136	124.116	128.299
100	67.328	70.065	74.222	77.929	82.358	118.498	124.342	129.561	135.807	140.169

Table D F Distribution

Table provides the F-score (critical value) with area **.025** to the RIGHT

df1: degrees of freedom for the numerator
df2: degrees of freedom for the denominator

df1 → df2 ↓	1	2	3	4	5	6	7	8	9
1	647.79	799.50	864.16	899.58	921.85	937.11	948.22	956.66	963.28
2	38.506	39.000	39.165	39.248	39.298	39.331	39.335	39.373	39.387
3	17.443	16.044	15.439	15.101	14.885	14.735	14.624	14.540	14.473
4	12.218	10.649	9.9792	9.6045	9.3645	9.1973	9.0741	8.9796	8.9047
5	10.007	8.4336	7.7636	7.3879	7.1464	6.9777	6.8531	6.7572	6.6811
6	8.8131	7.2599	6.5988	6.2272	5.9876	5.8198	5.6955	5.5996	5.5234
7	8.0727	6.5415	5.8898	5.5226	5.2852	5.1186	4.9949	4.8993	4.8232
8	7.5709	6.0595	5.4160	5.0526	4.8173	4.6517	4.5286	4.4333	4.3572
9	7.2093	5.7147	5.0781	4.7181	4.4844	4.3197	4.1970	4.1020	4.0260
10	6.9367	5.4564	4.8256	4.4683	4.2361	4.0721	3.9498	3.8549	3.7790
11	6.7241	5.2559	4.6300	4.2751	4.0440	3.8807	3.7586	3.6638	3.5879
12	6.5538	5.0959	4.4742	4.1212	3.8911	3.7283	3.6065	3.5118	3.4358
13	6.4143	4.9653	4.3472	3.9959	3.7667	3.6043	3.4827	3.3880	3.3120
14	6.2979	4.8567	4.2417	3.8919	3.6634	3.5014	3.3799	3.2853	3.2093
15	6.1995	4.7650	4.1528	3.8043	3.5764	3.4147	3.2934	3.1987	3.1227
16	6.1151	4.6867	4.0768	3.7294	3.5021	3.3406	3.2194	3.1248	3.0488
17	6.0420	4.6189	4.0112	3.6648	3.4379	3.2767	3.1556	3.0610	2.9849
18	5.9781	4.5597	3.9539	3.6083	3.3820	3.2209	3.0999	3.0053	2.9291
19	5.9216	4.5075	3.9034	3.5587	3.3327	3.1718	3.0509	2.9563	2.8801
20	5.8715	4.4613	3.8587	3.5147	3.2891	3.1283	3.0074	2.9128	2.8365
21	5.8266	4.4199	3.8188	3.4754	3.2501	3.0895	2.9686	2.8740	2.7977
22	5.7863	4.3828	3.7829	3.4401	3.2151	3.0546	2.9338	2.8392	2.7628
23	5.7498	4.3492	3.7505	3.4083	3.1835	3.0232	2.9023	2.8077	2.7313
24	5.7166	4.3187	3.7211	3.3794	3.1548	2.9946	2.8738	2.7791	2.7027
25	5.6864	4.2909	3.6943	3.3530	3.1287	2.9685	2.8478	2.7531	2.6766
26	5.6586	4.2655	3.6697	3.3289	3.1048	2.9447	2.8240	2.7293	2.6528
27	5.6331	4.2421	3.6472	3.3067	3.0828	2.9228	2.8021	2.7074	2.6309
28	5.6096	4.2205	3.6264	3.2863	3.0626	2.9027	2.7820	2.6872	2.6106
29	5.5878	4.2006	3.6072	3.2674	3.0438	2.8840	2.7633	2.6686	2.5919
30	5.5675	4.1821	3.5894	3.2499	3.0265	2.8667	2.7460	2.6513	2.5746
40	5.4239	4.0510	3.4633	3.1261	2.9037	2.7444	2.6238	2.5289	2.4519
60	5.2856	3.9253	3.3425	3.0077	2.7863	2.6274	2.5068	2.4117	2.3344
120	5.1523	3.8046	3.2269	2.8943	2.6740	2.5154	2.3948	2.2994	2.2217
∞	5.0239	3.6889	3.1161	2.7858	2.5665	2.4082	2.2875	2.1918	2.1136

Table D F Distribution

Table provides the F-score (critical value) with area **.025** to the RIGHT

df1: degrees of freedom for the numerator
df2: degrees of freedom for the denominator

df1 → df2 ↓	10	12	15	20	24	30	40	60	120	∞
1	968.63	976.71	984.87	993.10	997.25	1001.4	1005.6	1009.8	1014.0	1018.3
2	39.398	39.415	39.431	39.448	39.456	39.465	39.473	39.481	39.490	39.498
3	14.419	14.337	14.253	14.167	14.124	14.081	14.037	13.992	13.947	13.902
4	8.8439	8.7512	8.6565	8.5599	8.5109	8.4613	8.4111	8.3604	8.3092	8.2573
5	6.6192	6.5245	6.4277	6.3286	6.2780	6.2269	6.1750	6.1225	6.0693	6.0153
6	5.4613	5.3662	5.2687	5.1684	5.1172	5.0652	5.0125	4.9589	4.9044	4.8491
7	4.7611	4.6658	4.5678	4.4667	4.4150	4.3624	4.3089	4.2544	4.1989	4.1423
8	4.2951	4.1997	4.1012	3.9995	3.9472	3.8940	3.8398	3.7844	3.7279	3.6702
9	3.9639	3.8682	3.7694	3.6669	3.6142	3.5604	3.5055	3.4493	3.3918	3.3329
10	3.7168	3.6209	3.5217	3.4185	3.3654	3.3110	3.2554	3.1984	3.1399	3.0798
11	3.5257	3.4296	3.3299	3.2261	3.1725	3.1176	3.0613	3.0035	2.9441	2.8828
12	3.3736	3.2773	3.1772	3.0728	3.0187	2.9633	2.9063	2.8478	2.7874	2.7249
13	3.2497	3.1532	3.0527	2.9477	2.8932	2.8372	2.7797	2.7204	2.6590	2.5955
14	3.1469	3.0502	2.9493	2.8437	2.7888	2.7324	2.6742	2.6142	2.5519	2.4872
15	3.0602	2.9633	2.8621	2.7559	2.7006	2.6437	2.5850	2.5242	2.4611	2.3953
16	2.9862	2.8890	2.7875	2.6808	2.6252	2.5678	2.5085	2.4471	2.3831	2.3163
17	2.9222	2.8249	2.7230	2.6158	2.5598	2.5020	2.4422	2.3801	2.3153	2.2474
18	2.8664	2.7689	2.6667	2.5590	2.5027	2.4445	2.3842	2.3214	2.2558	2.1869
19	2.8172	2.7196	2.6171	2.5089	2.4523	2.3937	2.3329	2.2696	2.2032	2.1333
20	2.7737	2.6758	2.5731	2.4645	2.4076	2.3486	2.2873	2.2234	2.1562	2.0853
21	2.7348	2.6368	2.5338	2.4247	2.3675	2.3082	2.2465	2.1819	2.1141	2.0422
22	2.6998	2.6017	2.4984	2.3890	2.3315	2.2718	2.2097	2.1446	2.0760	2.0032
23	2.6682	2.5699	2.4665	2.3567	2.2989	2.2389	2.1763	2.1107	2.0415	1.9677
24	2.6396	2.5411	2.4374	2.3273	2.2693	2.2090	2.1460	2.0799	2.0099	1.9353
25	2.6135	2.5149	2.4110	2.3005	2.2422	2.1816	2.1183	2.0516	1.9811	1.9055
26	2.5896	2.4908	2.3867	2.2759	2.2174	2.1565	2.0928	2.0257	1.9545	1.8781
27	2.5676	2.4688	2.3644	2.2533	2.1946	2.1334	2.0693	2.0018	1.9299	1.8527
28	2.5473	2.4484	2.3438	2.2324	2.1735	2.1121	2.0477	1.9797	1.9072	1.8291
29	2.5286	2.4295	2.3248	2.2131	2.1540	2.0923	2.0276	1.9591	1.8861	1.8072
30	2.5112	2.4120	2.3072	2.1952	2.1359	2.0739	2.0089	1.9400	1.8664	1.7867
40	2.3882	2.2882	2.1819	2.0677	2.0069	1.9429	1.8752	1.8028	1.7242	1.6371
60	2.2702	2.1692	2.0613	1.9445	1.8817	1.8152	1.7440	1.6668	1.5810	1.4821
120	2.1570	2.0548	1.9450	1.8249	1.7597	1.6899	1.6141	1.5299	1.4327	1.3104
∞	2.0483	1.9447	1.8326	1.7085	1.6402	1.5660	1.4835	1.3883	1.2684	1.0000

Table D F Distribution

Table provides the F-score (critical value) with area **.05** to the RIGHT

df1: degrees of freedom for the numerator
df2: degrees of freedom for the denominator

df1 → df2 ↓	1	2	3	4	5	6	7	8	9
1	161.45	199.50	215.71	224.58	230.16	233.99	236.77	238.88	240.54
2	18.513	19.000	19.164	19.247	19.296	19.330	19.353	19.371	19.385
3	10.128	9.5521	9.2766	9.1172	9.0135	8.9406	8.8867	8.8452	8.8123
4	7.7086	6.9443	6.5914	6.3882	6.2561	6.1631	6.0942	6.0410	6.9988
5	6.6079	5.7861	5.4095	5.1922	5.0503	4.9503	4.8759	4.8183	4.7725
6	5.9874	5.1433	4.7571	4.5337	4.3874	4.2839	4.2067	4.1468	4.0990
7	5.5914	4.7374	4.3468	4.1203	3.9715	3.8660	3.7870	3.7257	3.6767
8	5.3177	4.4590	4.0662	3.8379	3.6875	3.5806	3.5005	3.4381	3.3881
9	5.1174	4.2565	3.8625	3.6331	3.4817	3.3738	3.2927	3.2296	3.1789
10	4.9646	4.1028	3.7083	3.4780	3.3258	3.2172	3.1355	3.0717	3.0204
11	4.8443	3.9823	3.5874	3.3567	3.2039	3.0946	3.0123	2.9480	2.8962
12	4.7472	3.8853	3.4903	3.2592	3.1059	2.9961	2.9134	2.8486	2.7964
13	4.6672	3.8056	3.4105	3.1791	3.0254	2.9153	2.8321	2.7669	2.7144
14	4.6001	3.7389	3.3439	3.1122	2.9582	2.8477	2.7642	2.6987	2.6458
15	4.5431	3.6823	3.2874	3.0556	2.9013	2.7905	2.7066	2.6408	2.5876
16	4.4940	3.6337	3.2389	3.0069	2.8524	2.7413	2.6572	2.5911	2.5377
17	4.4513	3.5915	3.1968	2.9647	2.8100	2.6987	2.6143	2.5480	2.4943
18	4.4139	3.5546	3.1599	2.9277	2.7729	2.6613	2.5767	2.5102	2.4563
19	4.3807	3.5219	3.1274	2.8951	2.7401	2.6283	2.5435	2.4768	2.4227
20	4.3512	3.4928	3.0984	2.8661	2.7109	2.5990	2.5140	2.4471	2.3928
21	4.3248	3.4668	3.0725	2.8401	2.6848	2.5727	2.4876	2.4205	2.3660
22	4.3009	3.4434	3.0491	2.8167	2.6613	2.5491	2.4638	2.3965	2.3419
23	4.2793	3.4221	3.0280	2.7955	2.6400	2.5277	2.4422	2.3748	2.3201
24	4.2597	3.4028	3.0088	2.7763	2.6207	2.5082	2.4226	2.3551	2.3002
25	4.2417	3.3852	2.9912	2.7587	2.6030	2.4904	2.4047	2.3371	2.2821
26	4.2252	3.3690	2.9752	2.7426	2.5868	2.4741	2.3883	2.3205	2.2655
27	4.2100	3.3541	2.9604	2.7278	2.5719	2.4591	2.3732	2.3053	2.2501
28	4.1960	3.3404	2.9467	2.7141	2.5581	2.4453	2.3593	2.2913	2.2360
29	4.1830	3.3277	2.9340	2.7014	2.5454	2.4324	2.3463	2.2783	2.2229
30	4.1709	3.3158	2.9223	2.6896	2.5336	2.4205	2.3343	2.2662	2.2107
40	4.0847	3.2317	2.8387	2.6060	2.4495	2.3359	2.2490	2.1802	2.1240
60	4.0012	3.1504	2.7581	2.5252	2.3683	2.2541	2.1665	2.0970	2.0401
120	3.9201	3.0718	2.6802	2.4472	2.2899	2.1750	2.0868	2.0164	1.9588
∞	3.8415	2.9957	2.6049	2.3719	2.2141	2.0986	2.0096	1.9384	1.8799

Table D F Distribution

Table provides the F-score (critical value) with area **.05** to the RIGHT

df1: degrees of freedom for the numerator
df2: degrees of freedom for the denominator

df1 → df2 ↓	10	12	15	20	24	30	40	60	120	∞
1	241.88	243.91	245.95	248.01	249.05	250.10	251.14	252.20	253.25	254.31
2	19.396	19.413	19.429	19.446	19.454	19.462	19.471	19.479	19.487	19.496
3	8.7855	8.7446	8.7029	8.6602	8.6385	8.6166	8.5944	8.5720	8.5494	8.5264
4	5.9644	5.9117	5.8578	5.8025	5.7744	5.7459	5.7170	5.6877	5.6581	5.6281
5	4.7351	4.6777	4.6188	4.5581	4.5272	4.4957	4.4638	4.4314	4.3985	4.3650
6	4.0600	3.9999	3.9381	3.8742	3.8415	3.8082	3.7743	3.7398	3.7047	3.6689
7	3.6365	3.5747	3.5107	3.4445	3.4105	3.3758	3.3404	3.3043	3.2674	3.2298
8	3.3472	3.2839	3.2184	3.1503	3.1152	3.0794	3.0428	3.0053	2.9669	2.9276
9	3.1373	3.0729	3.0061	2.9365	2.9005	2.8637	2.8259	2.7872	2.7475	2.7067
10	2.9782	2.9130	2.8450	2.7740	2.7372	2.6996	2.6609	2.6211	2.5801	2.5379
11	2.8536	2.7876	2.7186	2.6464	2.6090	2.5705	2.5309	2.4901	2.4480	2.4045
12	2.7534	2.6866	2.6169	2.5436	2.5055	2.4663	2.4259	2.3842	2.3410	2.2962
13	2.6710	2.6037	2.5331	2.4589	2.4202	2.3803	2.3392	2.2966	2.2524	2.2064
14	2.6022	2.5342	2.4630	2.3879	2.3487	2.3082	2.2664	2.2229	2.1778	2.1307
15	2.5437	2.4753	2.4034	2.3275	2.2878	2.2468	2.2043	2.1601	2.1141	2.0658
16	2.4935	2.4247	2.3522	2.2756	2.2354	2.1938	2.1507	2.1058	2.0589	2.0096
17	2.4499	2.3807	2.3077	2.2304	2.1898	2.1477	2.1040	2.0584	2.0107	1.9604
18	2.4117	2.3421	2.2686	2.1906	2.1497	2.1071	2.0629	2.0166	1.9681	1.9168
19	2.3779	2.3080	2.2341	2.1555	2.1141	2.0712	2.0264	1.9795	1.9302	1.8780
20	2.3479	2.2776	2.2033	2.1242	2.0825	2.0391	1.9938	1.9464	1.8963	1.8432
21	2.3210	2.2504	2.1757	2.0960	2.0540	2.0102	1.9645	1.9165	1.8657	1.8117
22	2.2967	2.2258	2.1508	2.0707	2.0283	1.9842	1.9380	1.8894	1.8380	1.7831
23	2.2747	2.2036	2.1282	2.0476	2.0050	1.9605	1.9139	1.8648	1.8128	1.7570
24	2.2547	2.1834	2.1077	2.0267	1.9838	1.9390	1.8920	1.8424	1.7896	1.7330
25	2.2365	2.1649	2.0889	2.0075	1.9643	1.9192	1.8718	1.8217	1.7684	1.7110
26	2.2197	2.1479	2.0716	1.9898	1.9464	1.9010	1.8533	1.8027	1.7488	1.6906
27	2.2043	2.1323	2.0558	1.9736	1.9299	1.8842	1.8361	1.7851	1.7306	1.6717
28	2.1900	2.1179	2.0411	1.9586	1.9147	1.8687	1.8203	1.7689	1.7138	1.6541
29	2.1768	2.1045	2.0275	1.9446	1.9005	1.8543	1.8055	1.7537	1.6981	1.6376
30	2.1646	2.0921	2.0148	1.9317	1.8874	1.8409	1.7918	1.7396	1.6835	1.6223
40	2.0772	2.0035	1.9245	1.8389	1.7929	1.7444	1.6928	1.6373	1.5766	1.5089
60	1.9926	1.9174	1.8364	1.7480	1.7001	1.6491	1.5943	1.5343	1.4673	1.3893
120	1.9105	1.8337	1.7505	1.6587	1.6084	1.5543	1.4952	1.4290	1.3519	1.2539
∞	1.8307	1.7522	1.6664	1.5705	1.5173	1.4591	1.3940	1.3180	1.2214	1.0000

Table E Pearson Correlation Coefficient r

Table provides the positive critical value for the indicated significance level (The other critical value is the opposite of the one shown.)

n	α = .05	α = .01
4	.950	.990
5	.878	.959
6	.811	.917
7	.754	.875
8	.707	.834
9	.666	.798
10	.632	.765
11	.602	.735
12	.576	.708
13	.553	.684
14	.532	.661
15	.514	.641
16	.497	.623
17	.482	.606
18	.468	.590
19	.456	.575
20	.444	.561
25	.396	.505
30	.361	.463
35	.335	.430
40	.312	.402
45	.294	.378
50	.279	.361
60	.254	.330
70	.236	.305
80	.220	.286
90	.207	.269
100	.196	.256

CPSIA information can be obtained
at www.ICGtesting.com
Printed in the USA
LVHW10s0022190918
590620LV00006B/14/P

9 781974 008209